Proceedings of the 2020 DigitalFUTURES

Philip F. Yuan · Jiawei Yao ·
Chao Yan · Xiang Wang ·
Neil Leach
Editors

Proceedings of the 2020 DigitalFUTURES

The 2nd International Conference on Computational Design and Robotic Fabrication (CDRF 2020)

Editors
Philip F. Yuan
College of Architecture and Urban Planning
Tongji University
Shanghai, China

Jiawei Yao
College of Architecture and Urban Planning
Tongji University
Shanghai, China

Chao Yan
College of Architecture and Urban Planning
Tongji University
Shanghai, China

Xiang Wang
College of Architecture and Urban Planning
Tongji University
Shanghai, China

Neil Leach
College of Architecture and Urban Planning
Tongji University
Shanghai, China

Funded by National Key R&D Program of China (Grant No.2016YFC0702104, 51908410), Shanghai Science and Technology Committee (Grant No.17dz1203405) and College of Architecture and Urban Planning (CAUP), Tongji University, China.

ISBN 978-981-33-4399-3 ISBN 978-981-33-4400-6 (eBook)
https://doi.org/10.1007/978-981-33-4400-6

This Springer imprint is published by the registered company Springer Nature Singapore Pte Ltd.
The registered company address is: 152 Beach Road, #21-01/04 Gateway East, Singapore 189721, Singapore

Preface

DigitalFUTURES is an annual series of academic events, consisting of a conference, workshops and exhibition, hosted by the College of Architecture and Urban Planning, Tongji University, Shanghai. In 2020, DigitalFUTURES celebrates its 10th anniversary. The aim of DigitalFUTURES is to encourage international collaboration and interaction and to promote theoretical and scientific research into computational design, robotic fabrication and other areas of architectural intelligence.

In the article 'Computing Machinery and Intelligence' (1950), Alan Turing predicted that machines would soon be able to learn and develop intelligence. 70 years later, it is clear that Turing was absolutely correct. Machine learning now plays a central role within artificial intelligence and has become an important new area of research for architects.

The theme of DigitalFUTURES 2020 is 'Machine Intelligence'. This refers most directly to machine learning and machines involved in robotic fabrication. But it also refers to all other forms of architectural intelligence, such as material intelligence, swarm intelligence and artificial intelligence, along with intelligent design and fabrication, and intelligent techniques for visualizing, monitoring and controlling the built environment.

The discipline of architecture is being transformed by machine intelligence. So too, the architect is also being transformed. No longer limited by human intelligence, the architect is becoming a cyborg-like creature, drawing upon machine intelligence as a prosthetic extension to the architectural imagination. Likewise, the built environment is also being transformed. From smart thermostats through to intelligent buildings and entire smart cities, the built environment is becoming an intelligent environment.

The intention behind this year's events is to promote a series of scholarly dialogues on these technological transformations in architecture, in an attempt to redefine the notion of intelligence and its role in architectural production from both a theoretical and technological standpoint.

DigitalFUTURES Executive Committee

Committees

Honorary Advisors

Jiaping Liu	Xi'an University of Architecture and Technology, China
Zhiqiang Wu	Tongji University, China
Weiping Shao (Executive Chief Architect)	Beijing Institute of Architectural Design, China
Guoqiang Li	Tongji University, China
Bernard Stigler (Director)	Centre Georges-Pompidou, France
Philippe Block	ETH Zurich, Switzerland
Achim Menges	University of Stuttgart, Germany
Antoine Picon	GSD, USA
Patrik Schumacher	Zaha Hadid Architects (ZHA), UK
Yi Min (Mike) Xie	RMIT University, Australia

Organization Committees

Philip F. Yuan (Workshop Coordinator)	Tongji University, China
Neil Leach (Conference Coordinator)	Tongji University, China
Yi Min (Mike) Xie (Paper Selection Coordinator)	RMIT University, Australia
Guohua Ji (Award Coordinator)	Nanjing University, China

Scientific Committees (list by surname)

Contents

Machine Seeing

Machine Learning

Machine Thinking

Machinic Phylum and Architecture

Andrej Radman(✉)

Faculty of Architecture and the Built Environment, Delft University of Technology, Julianalaan 134, 2628, BL Delft, The Netherlands
a.radman@tudelft.nl

Abstract. The chapter draws on the anti-substantivist and anti-hylomorphic legacy of two significant Deleuze and Guattari's interlocutors: Raymond Ruyer and Gilbert Simondon. Ruyer vehemently opposed the logic of mechanicism without regressing to (active) vitalism. His masterpiece *Neofinalism*, yet to be fully appreciated in architectural circles, is an ode to multiplicity or 'absolute form'. The title is to be read as a challenge to the hegemony of the step-by-step causation and partes-extra-partes mereology. According to Ruyer, non-locality is the key, not only to the question of subjectivity, but to the problem of life itself. Simondon too shies away from the metaphysics of presence. For him, the process of individuation cannot be grasped on the basis of the fully formed individual. In other words, the knowledge of individuation *is* the individuation of knowledge. Simondon's highest ambition in *On the Mode of Existence of Technical Objects* was to integrate culture and technics (*tekhne*). The conviction that culture need not be antagonistic to technology is particularly pertinent to the ecologies of architecture. In the second half of the chapter, the affordance theory meets contemporary neurosciences.

Keywords: Schizoanalytic cartography · Machinic desire · Ecologies of architecture · Ethico-aesthetics

Once it is no longer the goal of the architect to be the artist of built forms but to offer his services in revealing the virtual desires of spaces, places, trajectories and territories, he will have to undertake the analysis of the relations of individual and collective corporeality by constantly singularizing his approach. Moreover, he will have to become an intercessor between these desires, brought to light, and the interests that they thwart. In other words, he will have to become an artist and an artisan of sensible and relational lived experience (Guattari 1989) [1].

'Culture' is everything we don't have to do. We have to eat, but we don't have to have 'cuisines' […]. We have to cover ourselves against the weather, but we don't have to be so concerned as we are about whether we put on Levi's or Yves Saint-Laurent. We have to move […], but we don't have to dance. […] I call the 'have-to' activities functional and the 'don't have to's stylistic. […] The first thing to note is that the whole bundle of stylistic activities is exactly what we would describe as 'a culture' […] (Eno 1996) [2].

P. F. Yuan et al. (Eds.): CDRF 2020, *Proceedings of the 2020 DigitalFUTURES*, pp. 3–16, 2021.
https://doi.org/10.1007/978-981-33-4400-6_1

1 Nips and Bites

The chapter draws on the anti-substantivist and anti-hylomorphic legacy of two significant Deleuze and Guattari's interlocutors: Raymond Ruyer and Gilbert Simondon. Ruyer vehemently opposed the logic of mechanicism without regressing to (active) vitalism. He concurred with Alfred North Whitehead who famously dismissed the concept of 'simple location' as a bias in favour of the tangible and self-presence [3]. Ruyer's masterpiece *Neofinalism*, yet to be fully appreciated in architectural circles, is an ode to multiplicity or 'absolute form' [4]. The title is to be read as a challenge to the hegemony of the step-by-step causation and partes-extra-partes mereology. According to Ruyer, non-locality is the key, not only to the question of subjectivity, but to the problem of life itself [5]. Simondon too shies away from the metaphysics of presence. For him, the process of individuation cannot be grasped on the basis of the fully formed individual. In other words, the knowledge of individuation *is* the individuation of knowledge [6]. Simondon's highest ambition in *On the Mode of Existence of Technical Objects* was to integrate culture and technics (*tekhne*). The conviction that culture need not be antagonistic to technology is particularly pertinent to the ecologies of architecture. To paraphrase Marshall McLuhan, ecology starts where nature ends [7]. Simondon opposed structuralism with the theory of operations that he named *allagmatics* [8]. The transition from operation to structure is machinic rather than structural insofar as it is system making rather than systematic. The 'machinic' conception of consistency is thus determined neither by the naïve 'organic' autonomy of the vitalist whole, nor by the crude reductionist expression of the whole in the sum of its mechanical parts. While structures are by definition balanced, the thought must venture beyond the given – far from the equilibrium. The term 'plane of consistency' is in itself a sufficient clue to what is primarily at stake in the thought, namely the reality of abstraction. Tessellation (*planification*) of the Planomenon is an abstraction without being an achievement of reason. Consequently, (machine) intelligence may be defined by the (unconscious and impersonal) capacity to insert an interval between the cause and effect – a margin of indetermination related to the non-entailment of open systems.

Let us draw an ethological diagram consisting of two diverging lines (resembling an image of a rail track in central linear perspective) (Fig. 1). The top part **S-R** (close to the vanishing point) draws the stimulus (**S**) and response (**R**) close together as in the deterministic, i.e. mechanical mode of operation. The 'conceptual persona' dwelling in this range is a simple organism that cannot afford to break away from linear causality, such as a tick [10]. The further apart the two lines the more severed the causal chain. Before we reach the bottom of the diagram where the stimulus transforms into perception (**P**) and the response into action (**A**), the gap is sufficiently wide to be occupied by a more complex organism capable of play, like a cat. As Gregory Bateson rightly insists, a cat's nip is very different from its bite [11]. It does not conform to the (functional) if-then logic: *if* a tick smells a warm-blooded animal *then* it latches onto it. Rather, the nip is pretense or acting *as-if*, i.e. doing what it doesn't have to do. According to the second epigraph, play may qualify as (proto)culture, a style. Finally, at the base of the diagram (**P—A**), a more complex non-mechanical (recurrent) causality pushes perception and action further apart. Its 'telos' is not subject merely to the material-energetic constraints but also to the informational or epistemic semiosis. In other words, ends and means may

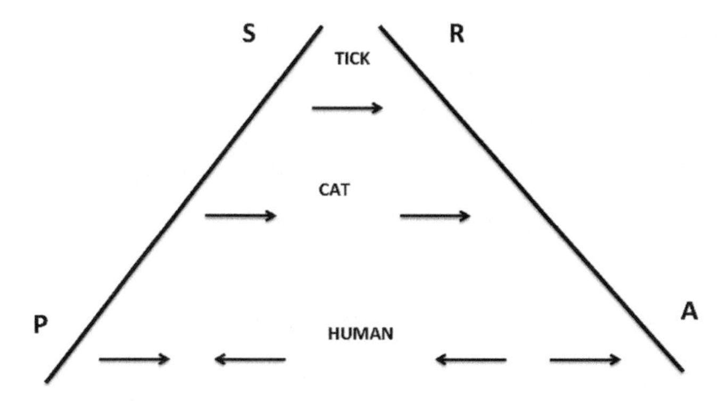

Fig. 1. Inserting the interval between stimulus (**S**) and response (**R**). The degree of mnemonic detachability is measured by the width between the two poles and the 'direction' of causality. The recursive causality designates the cause (**P**) coming into being with the effect (**A**). In the words of Simondon, this is "a [neofinalist] conditioning of the present by the future, or by what up to now does not exist [9]."

come to be reversed. Take Hannah Arendt's reference to the profoundly paradoxical Christian concept of 'turning the other cheek', which radically disrupts the cause-and-effect inevitability. In doing so, it steps out of simple determinism towards Ruyerian *neofinalism* by way of Simondonian *technicity* defined as a force of psychosocial invention and cultural transformation [12]. It may be argued that the diagram runs from the Spinozian *natura naturata* at the top towards *natura naturans* at its ever-widening bottom [13]. It brings to mind the apex-base relation from the famous Bergsonian cone of (pure) memory [14]. The divergence of lines effectively measures the (degree of) detachability of virtual wholes from the actual parts, memory from matter (time from space). Yes, there is isomorphism between the two, but without resemblance. This means that we can happily leave behind the skyhook category of the 'imaginary'. Contrary to our deepest prejudice, the visible is no more real than the invisible and memory is not a property of bodies. For Ruyer, bodies may be said to be properties of memory:

> *The main difference between physical beings and the most complex organisms does not probably derive from the instantaneity or the absence of memory in the former but from a lack of detachment of this memory, which in physical beings is always inherent to the rhythm of activity, which is only ever 'the form in time' and does not constitute a transspatial 'reserve' clearly detached from the actual* [15].

The co-determination of the actual and the virtual has been a life-long occupation of Guattari's. His neologism *ethico-aesthetics* aptly dramatises the entanglement of action (**A**) and perception (**P**). Putting experience first relegates the sciences to the second order of expression. The collective architectural enunciation (wrongly attributed to the will of the architect) renders the full coincidence of the body and its territory (as a simple location) impossible. Guattari went on to develop a schizoanalytic cartography where heterogeneous ontological domains – actuality, virtuality, possibility and reality – had to be thought together [16] (Fig. 2). Metamodelling was his strategy to prevent things from becoming systemic and thus stratified (closed system). The four 'unconsciousnesses' are: existential territory (**T**), universes of value (**U**), energetic and semiotic flows (**F**), and the machinic phylum (**P**). The purposeless purpose of P is to draw the endo-referential and endo-consistent body ever further away from itself in the direction of exo-referentiality and exo-consistency. The fourfold offered a way out of the deadlock between the ostensible immediacy of the subject (**T**), and the constitutive distance of the system (**P**).

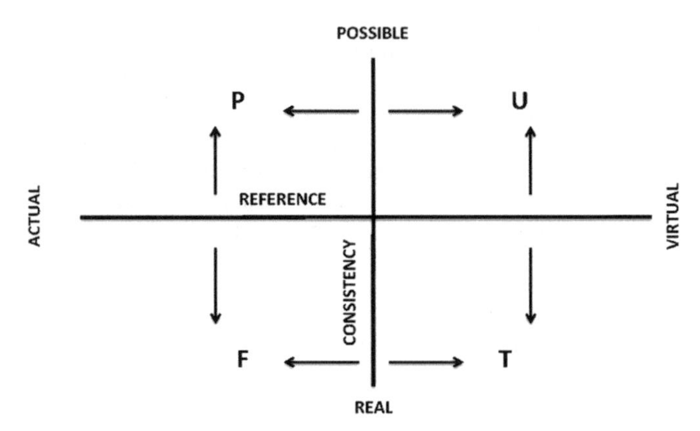

Fig. 2. Any architectural collective enunciation worthy of its ecological attribute can be said to consist of quadruple ontological domains: efficient Territory (**T**) and final Universes of Value (**U**) as non-discursive, and material energetic and semiotic Flows (**F**) and formal machinic Phylum (**P**) as discursive. These are four quasi-causes of the assemblages that are always articulated together.

In contrast to the evolutionary mechanism of passive adaptation, the quasi-Lamarckian machinism is 'accelerationist' [17]. It is as cultural as it is natural given the ideality and materiality of its flows that reach far beyond the anthropic. We may have too easily dismissed an early naturalist who anticipated modern epigenetics and whom Darwinists have long disparaged. Jean-Baptiste Lamarck (1744–1829) argued that evolution could occur within a generation or two. According to Philip Steadman, the theory of Darwin is an 'elective' theory of evolution, where the environment chooses appropriate changes in organism from the range offered by variation. By contrast, Lamarckism is an 'instructive' theory where the environment is imagined to be able to exercise a direct effect on organisms and 'teach' them to change themselves in appropriate ways [18]. This revelation is paramount for the 'niche constructionists' or those in the business of associating milieus: architects and urbanists.

T is an ethological concept that designates vital familiar space, the ground, an individual or collective body. **U** are nascent quasi-subjective ideas before they are objectified or expressed. **T** and **U** belong to the *virtual* (giving) half of the fourfold diagram. **T-U** may be said to be quasi-subjective and *pathic* in comparison to their ontic counterpart of **F-P**. The former *non-discursive* and the latter *representational*. From the point of view of psychopathologies, neurosis is associated with the actual and psychosis with the virtual pole of the horizontal axis of reference [19]. The vertical axis of consistency stretches from the *real* (**F** and **T**) to the *possible* (**P** and **U**). Guattari's urge to substitute schizoanalysis for psychoanalysis originates from the necessity to expand the operation beyond the real to the realm of the possible. It is important to underscore that Guattari's 'possible' is not to be mistaken for the retroactive hypostatisation of the real. It simply designates that which is further from the equilibria (the real) where genuine modulation of territorialisation occurs. Ethological plasticity would not be possible without the *ritornello*. Paradoxically, while **U** provides for the rhythm (repetition and difference), **F** is segmented. As already stated, the ever-proliferating rhizome **P** (quasi-objective ideas) opens up the possibility of resingularisation of desire and values. Qua Deleuze's ventriloquism, Michel Foucault offers a helpful architectural example: the (machine) prison, as an endo-referential and exo-consistent form of *content* (**U**), is inconceivable without the prisoner as its substance (**T**). On the side of *expression*, the exo-referential and endo-consistent concept of 'delinquency' is its substance (**F**) and penal law its form (**P**) [20]. According to Foucault, environments enunciate, just as enunciations determine environments, but they remain heterogeneous with no direct causality, no common totalising form. "The diagram is no longer an [...] archive but a map, a cartography that is coexstensive with the whole social field. It is an abstract machine [21]". Deleuze continues:

> *[E]very diagram is intersocial and constantly evolving. It never functions in order to represent a persisting world but produces a new kind of reality, a new model of truth. It is neither the subject of history, nor does it survey history. It makes history by unmaking preceding realities and significations, constituting hundreds of points of emergence or creativity, unexpected conjunctions or improbable continuums. It doubles history with a sense of continual evolution* [22].

The focus on singularities in Guattari's *Schizoanalitic Cartographies* should not come as a surprise given their inbuilt resistance to calculation or instrumental use of representation. The shortcoming of binary systems like linguistic semiology is that, like capitalism, they render everything translatable according to the standard of general equivalence [23]. If the asignifying process of decoding **F>P** and deterritorialisation **T>U** were not possible, the diagram would be reducible to discrete calculable quantities that could be assigned a place in a pre-ordered transcendent structure. Thanks to the non-programmable immanent movement of de-re-stratification, the fourfold remains sufficiently unstable and open to the multiple (multiplicity as a critique of structuralism). The diagram is emancipatory for as long as it sustains the 'rhythm', but it might as well become a map of discipline and control if the movement is arrested and its domains petrified [24]. By the same token, and in conjunction with the first epigraph, there is a way to circumvent the ready made Oedipal structure and instead engage in the cartography of subjectification:

> *I consider that it is the architect who finds he is in the position of having to analyse certain specific functions of subjectification himself. In this way and in the company of numerous other social and cultural operators, he could constitute an essential relay at the heart of multiple-headed Assemblages of enunciation*, able to take analytic and pragmatic responsibility for contemporary productions

> of subjectivity. *As a consequence, one really is a long way here from only seeing the architect in the simple position of critical observer!* [25]

This is an account that grants ontological priority to the machinic desire and is of utmost political, social, and existential importance [26]. In the present condition of the digital turn, it has become necessary to resist the self-fulfilling prophecy of reducing the world to the (socially constructed) code. The Simondonian material-discursive concept of technicity taught us that nature did not exist prior to the machine. Evoking the latest discoveries in evolutionary biology – it is better to biologise than to structuralise – Guattari referred to the worlding technicity as the 'machinic phylum'. Crucially, machines speak to machines before they speak to humans [27]. In other words, they are social before they are technical [28].

2 Ducks and Rabbits

We will now turn from the production of production to the production of recording and, finally, production of consummation (larval subject) [29]. The second half of the chapter, where the affordance theory meets contemporary neurosciences, starts from the brain that becomes a subject in the 'absolute survey' [30]. Its near synonym – 'self-enjoyment' – does not designate pleasure but an immediacy without immediate objectification.

> *[It] was a very important discovery that the brain wasn't entirely determined. Some anatomic structures of the brain are, of course, genetically programmed, but a significant part of the neural organization is open to outside influences and develops itself consequently to these influences or interactions. It means an important part in the structure of your brain depends on the way you're living and on your experience. History is inscribed within the biological. That is what 'plastic' means when applied to the brain* [31].

According to the biologist and Nobel Prize laureate Gerald Edelman, the brain is first and foremost a selectionist system [32]. The importance of selectivity as the defining characteristic of knowing cannot be overemphasised [33]. Perception is context-dependent and adaptive. It is not a Turing process, Edelman insists, because the world is a non-labelled place. *Data does not equal information.* The ecological approach to perception knows no such thing as 'sense data'. Ecological, it must be qualified, stands for reciprocity between the life form and its environment. Their mutual relation is not one of computing but of resonance or affective attunement. The reality is not 'chunked' [34]. This premise should fundamentally reconfigure the debate on nature and nurture, and on the (im)possibility of 'carving nature at the joints' [35]. Our categories are retroactively imposed as a result of analytic reflection. Most importantly, our cognition depends utterly on motion, that is, sensori-motor interaction. "Begin in the middle! [...] Don't assume to know in advance how the chunking will resolve! [36]".

The famous Hebb rule stipulates that the neurons that *fire* together – *wire* together. As a result, synaptic connections either get strengthened or weakened. Their excitement and inhibition are not 'decided' by the genes but at the epi-genetic level. By this we mean that the whole virtual experience is responsive to the significance of the actual stimulus. When a new pattern is selected the attractor landscape is rearranged and new basins of attraction are added. There is no ready-made memory storage, no pre-established compartments or clear-cut boundaries. Experience is relational, non-local and perpetually updated. In a word, encephalisation is *machinic*. This is the gist of Edelman's critique of representation. He is not alone in tapping into the resources of topological field theory [37]. Yet the habit to *overcode* is difficult to shake off. In the words of Erin Manning:

> *What we perceive is always first a relational field. [...] Still, given the quickness of the morphing from the relational field into the objects and subjects of our perceptions, many of us neurotypicals feel as though the world is 'pre-chunked' into species, into bodies and individuals. This is the shortcoming, as autistics might say, of neurotypical perception* [38].

Not only are the neurotypicals too quick to chunk compared with the autistics, they are also incapable of self-tickling [39]. The barrier to self-tickling is akin to the barrier to telling oneself a joke. Unlike schizophrenics, neurotypicals deprive themselves of the ability to self-stimulate in a sufficiently unpredictable fashion by dampening their own sensory responses to the ongoing stimulation. From this perspective it is perhaps true that to see is indeed to forget the name of the thing one sees [40].

Building upon the work of the neuroscientist Walter Freeman, his disciple Michael Spivey studies cognition as a self-organising process (auto-affection) that involves phase transitions, criticality and autocatalysis. In this light, *affordances* appear not as the mapping of external features but as a creative form of enacting significance on the basis of the organism's embodied history [41]. They retain ontogenetic independence from the cognitive schema. Consider Spivey's example of the Necker cube [42]. (Fig. 3) One cannot instantaneously perceive both implicit depictions that the 'axonometric wireframe' of a cube offers – a box from above *and* from below. The same applies to the rabbit/duck illusion: it is either one or the other. In other words, the ecological view maintains that there exists, in any such (two-dimensional) figure, information about a number of (three-dimensional) shapes. The perceiver merely *selects* one; the perceiver's attention is directed to that information. Spivey's explanation is that the transition between perceptual states (two in the cases of the Necker cube and rabbit/duck) is in fact a phase transition (singularity) [43].

Experimental evidence suggests that it takes time for a trajectory across a 'high dimensional phase space' to settle in one or the other attractor, depending on the vicinity to the 'event horizon' – defined as 'the point of no return' – where the actual threshold for overt response is located. The attractor is the box viewed from above *or* from below (rabbit *or* duck). It is important to stress that potentiality is never a fully accrued value. As Francisco Varela explains: "Given the myriad of contending subprocesses in every cognitive act, how are we to understand the moment of negotiation and emergence when one of them takes the lead and constitutes a definitive behavior? [44]" In the field of visual perception, a fraction of a second is a substantial amount of time to spend between two possible perceptual states (as in the case of the Necker cube) afforded by a stimulus:

> *These transitions are not instantaneous, but take at least a couple hundred milliseconds. What this reveals is that on the way toward achieving a stable percept, the*

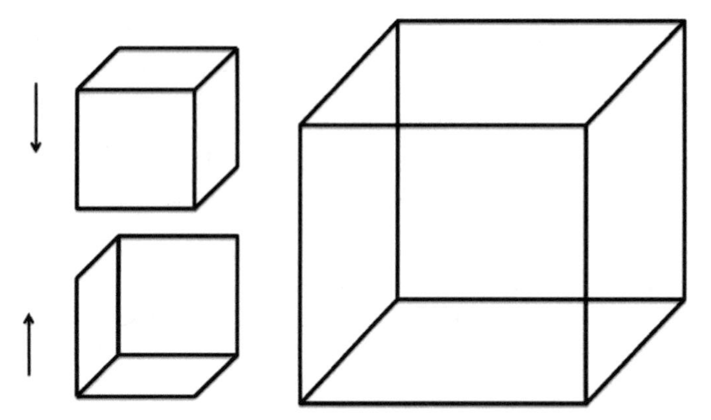

Fig. 3. The Necker cube is not an illusion but a kinematically-motivated perception. Because the image is one-sided (i.e., no tinkering is possible), the optical information about occlusion (i.e., which overlapping lines are nearer and which are farther) is unavoidably ambiguous.

> *brain spends a significant amount of time in regions of phase space that* do not neatly correspond to any of the labelled categories that language, or the experimenter, or society itself, has laid before it [45].

This proves that sharp transitions in behaviour need not be attributed to formally discrete logical processes, but can emerge instead from nonlinear dynamics in continuous modulations of a machinic assemblage. Such a 'fibrous' approach offers a welcome update to the Gibsonian information theory [46]: picking up the invariances to 'select' the most advantageous course of action out of the transspatial 'virtual phase space'. To paraphrase Massumi, which came first – the picker or the picked? Which is the chicken and which is the egg? [47]. The answer is neither. They both come last. To start with an affordance is to start from the middle by endorsing a theoretical model of decision-making and attention-control at the pre-reflective machinic level [48]. To speak of affordance is to break with the stifling notions of culture as representation or as reflection. It is to break with properties for capacities and, finally, to break with signification for the speculative-pragmatist significance. Dare we say, it is to break with the Two Cultures, micro- and macro-reductionism, in favour of an ethics of transversality and experimentation. In the words of Kwinter: "It is a fundamentally bourgeois idea to live the 'critical' life, to assess the value of objects and practices when the processes of production are themselves wild and alive and doing their business semi-independently elsewhere [49]". It amounts to megalomania.

The selectionist approach is fully compatible with evolutionary biology (evo) and developmental systems theory (devo), insofar as the emphasis is on plasticity and adaptation (evo-devo), rather than an already given essence or striving towards some proper form [50]. The Gibsonian theory gives credence to an alternative account of the phenomena of retention and expectation without recourse to memory. Recall how experience 'consults' itself when, for example, anticipating the taste of an expected flavour one is surprised to taste an unexpected one. There is neither logical mediation nor interpretation involved in this foreshadowing. Retention leads into and feeds anticipation.

Anticipation, in turn, rests and draws upon retention. It is not implausible that the emergence of an immune system owes to the incorporated expectation of injury or risk of potential harm. As stated above, the embodied, enactive cognition, may be best described *not* as a sequence of logical computational states, but as a continuous trajectory through virtual state space – absolute or non-dimensional survey – flirting with 'meaningful' attractors but rarely settling into them. "What exist are processes of change, [emergent] constraints exhibited by those processes, and the statistical smoothing and the attractors (dynamical regularities that form due to self-organizing processes) that embody the options left by these constraints [51]".

Constraints channel broad possibilities into narrow probabilities. Consider the following example. When stringing letters together to form a word (a – ar – arch – architecture), we start from an undifferentiated (flat) attractor landscape where a single letter can lead to anything. Yet, as information builds up, as in 'arch', the phase space gets ever more differentiated (constrained) until we end up with a single basin of attraction, that of 'architecture'. Hide and seek works the same way. If an object is always hidden in one specific place instead of several, the attractor landscape gets rearranged to bear a single basin [52]. It is arguably for the same reason that typefaces are recognisable despite there being a great variety of them [53]. The same applies to the invariant facial features in spite of the continuous transformation through the aging process [54].

There is an enormous plasticity in the nervous system, or else it would never be able to handle the complexity and novelty of the ever-changing environment, be it non-organic, artificial or technological. In any case, *activity is dominated more by experience than by stimuli.* It is for this reason that meta-stable affordances are sought out and detected so as to help coordinate behaviour. This is achieved through the operationally specific variability based on the capacity to vary the means to achieve the ends, i.e. flexibility, prospectivity and retrospectivity [55]. Nevertheless, it would be a fatal mistake to break up the task of action-coordination into purely internal neural circuitry. The ethico-aesthetic affordance theory recognises that organisms use *both* internal and external means of coordinating behaviour:

> *Moving from place to place is supposed to be 'physical' whereas perceiving is supposed to be 'mental', but this dichotomy is misleading. Locomotion is guided by visual perception. Not only does it depend on perception but perception depends on locomotion inasmuch as a moving point of observation is necessary for any adequate acquaintance with the environment. So we must perceive in order to move, but we must also move in order to perceive* [56].

We tend to think of the visual content of an image as a representation of the object's form or, beyond this naïve approach, as an acquired cultural code enabling us to recognise percepts as referencing objective forms. However, neither of these approaches to image-content works in terms of (built) environment. According to Massumi, it is precisely movement and not message that is the actual content of architecture [57]. Gibson is explicit:

> *The visual world is a kind of experience that* does not correspond to anything, *not any possible picture, not any motion picture, and not even any 'panoramic' motion picture. The visual world is not a* projection *of the ecological world. How could it be? The visual world is the outcome of the picking up of invariant information in an ambient optic array by an exploring visual system, and the awareness of the observer's own body in the world is a part of the experience* [58].

Having sensations does not simply amount to perceiving. The useful dimensions of sensitivity are those that specify the environment and the observer's relation to the environment (*umwelt*) [59]. An infant does not have to learn to convert sensations into lawful perception, both extero- and proprio-ception.

The fault lies, according to Tim Ingold, with understanding cultural production as a number of discrete, finite processes, each with a beginning and an ending: "production, and the meaning of production, must therefore be understood *intransitively*, not as a transitive relation of image to object [60]". This is to say that life cannot be understood mechanistically. According to Ruyer, it has to be understood axiologically. The 'axiological subject' values (affordances) rather than knows (objects). The lure of the virtual, towards which all our acts are directed, is the world of values. Yet, tending to the future, which is fibrously connected to the past, always comes with the dynamic potential for divergence from the present.

We have yet to shake off the 'bad habit' of representationalism in order to rightfully embrace a unity in multiplicity. A beginner's guide to metamodelling worthy of its machinic reputation rests on the following injunctions: 1) Insert an interval between **A** and **P** (**S** and **R**); 2) Sustain the movement between **T**, **U**, **P** and **F**; 3) Start from the middle! The irreducible triad may be parsed in the three syntheses from *Anti-Oedipus*: the connective – partial objects and flows, the disjunctive – singularities and chains, and the conjunctive – intensities and becomings [61]. It is by activating the transversal operations, each time anew, that we may hope to see the parochial culture of hylomorphism (covert idealism) give way to the life-affirming creative environmental, social and psychic teleodynamics [62].

References

1. Guattari, F.: Schizoanalytic Cartographies, p. 232. Bloomsbury, London (2013)
2. Eno, B.: Culture. In A Year with Swollen Appendices, pp. 317–21. Faber and Faber, London (1996). (317)
3. The term 'simple location' was coined by Alfred North Whitehead. It designates the fallacy of the attempt to locate concrete particulars in definite portions of space and time. See: Whitehead, A.N.: Science and the Modern World. Pelican Mentor Books, New York (1948)
4. Neo- or quasi-finalism is akin to Spinoza's 'conatus' and Nietzsche's 'will to power'. See: Ruyer, R.: Neofinalism. Minnesota University Press, Minneapolis (2016)

5. Ruyer, R.: Neofinalism, p. 94. Minnesota University Press, Minneapolis (2016). The self-contained, sovereign subject is but a 'zombie concept'. Such concepts "carry a presuppositional force of such staying power that they tend to return no matter how many times you slay them." See: Massumi, B.: Immediation unlimited. In. Manning, E., Munster, A., Stavning Thomsen, B.M. (eds.) Immediation II, pp. 501–43 (502). Open Humanities Press, London (2019)
6. Simondon, G.: Genesis of the Individual. In: Incorporations, pp. 297–319. Zone Books, New York (1992)
7. "'Ecological' thinking became inevitable as soon as the planet moved up into the status of a work of art." See: McLuhan, M.: At the moment of Sputnik the planet became a global theater in which there are no spectators but only actors. J. Commun., 48–58 (49) (1974)
8. Simondon defines an operation as a conversion of a structure in another structure. See: Adkins, T.: A Short List of Gilbert Simondon's Vocabulary. Fractal Ontology Blog (2007). https://fractalontology.wordpress.com/2007/11/28/a-short-list-of-gilbert-simondons-vocabulary/. Accessed 13 May 2020
9. Simondon, G.: On the Mode of Existence of Technical Objects, p. 62. Univocal Publishing, Minneapolis (2017)
10. von Uexküll, J.: A Stroll through the worlds of animals and men: a picture book of invisible worlds (1957). In Instinctive Behavior: The Development of a Modern Concept. New York: International Universities Press, Inc., pp. 5–80. Cf. Bateson, G. (1977). Afterword. In About Bateson. New York: E.P. Dutton, pp. 233–47 (241)
11. Bateson, G.: A theory of play and fantasy. In: Steps to an Ecology of Mind; Collected Essays in Anthropology, Psychiatry, Evolution, and Epistemology, pp. 138–48 (141–46). Ballantine, New York (1972)
12. Arendt, H.: The Human Condition, pp. 74–75. The University of Chicago Press, Chicago (1998)
13. For Spinoza, natura naturans refers to the self-causing activity of nature, while natura naturata, meaning 'nature natured', refers to nature considered as a passive product of an infinite causal chain
14. Bergson's 'pure memory' (rhythms and frequencies of duration) is opposed to the most relaxed level of duration, that is, space or matter in the most condensed contraction of the whole (of time) into the present of understanding. The leap into a virtual or pure past (not psychological) is an ontological and not a chronological move. See: Bergson, H.: Matter and Memory, p. 197. Dover Publications, New York (2004)
15. Ruyer, Neofinalism, p. 149
16. Guattari, Schizoanalytic Cartographies
17. Radman, A.: Involutionary architecture: unyoking coherence from congruence. In: Braidotti, R., Bignall, S. (eds.) Posthuman Ecologies: Complexity and Process after Deleuze, pp. 61–86. Rowman & Littlefield International, London (2019)
18. Steadman, P.: The consequences of the biological fallacy; functional determinism. In: The Evolution of Designs: Biological Analogy in Architecture and the Applied Arts, pp. 179–200. Routledge, London (2008)
19. While language is the neurotic's preferred medium of expression, most psychotics express themselves best using non-linguistic semiotic material. Deleuze stages an encounter between the surface of Lewis Carroll and the depth of Antonin Artaud as a paradigm for the logic of sense, where sense only makes sense in relation to 'non-sense'. See: Deleuze, G.: The Logic of Sense, p. 93. Columbia University Press, New York (1990)
20. Form can have two meanings: it either organises matter (content), or forms functions (expression). See: Deleuze, G.: Foucault, pp. 23–44. Minnesota University Press, Minneapolis (1988)

21. See: Deleuze, G.: Foucault, p. 34. Minnesota University Press, Minneapolis (1988). [emphasis in the original].
22. See: Deleuze, G.: Foucault, p. 35. Minnesota University Press, Minneapolis (1988)
23. Hauptmann, D., Radman, A. (eds.) Asignifying Semiotics: or How to Paint Pink on Pink, Footprint, 8/1(14). Architecture Theory Chair in partnership with Stichting Footprint and Techne Press, Delft (2014). https://doi.org/10.7480/footprint.8.1. Accessed 13 May 2020
24. There are two basic diagrams [...]: that of regulation by negative feedback which suppresses difference and seeks equilibrium, or that of guidance by positive feedback which reinforces difference and escapes equilibrium. See: Land, N.: Machinic desire. Textual Pract. **7**(3), 471–82 (475) (1993)
25. Guattari, Schizoanalytic Cartographies. [emphasis added]
26. As used by Franz Brentano and then Husserl, 'intentionality' means that mental states like perceiving are always about something, that is, directed towards something. By contrast, for Deleuze intentionality does exist but it is always multiple. In other words, there is never a single originator of the intention. Desire itself is a multiplicity of competing drives
27. Guattari, F.: Machinic heterogenesis. In: Conley, V. (ed.) Rethinking Technologies, pp. 13–27 (22). Minnesota University Press, Minneapolis (1993)
28. Deleuze, Foucault., p. 39
29. The connective synthesis of production, the disjunctive synthesis of recording, and the conjunctive synthesis of consummation, i.e. nothing is given, everything is produced. The larval subject is a residuum or spare part that sits alongside the desiring-machine. See: Deleuze, G., Guattari, F.: Anti-Oedipus, p. 338. Penguin, New York (2008)
30. Bains, P.: Subjectless subjectivities. In: Massumi, B. (ed.) A Shock to Thought: Expression after Deleuze and Guattari, pp. 101–116. Routledge, London (2002)
31. Vahanian, N.: A conversation with catherine malabou. JCRT **9**(1), 1–13 (2008)
32. Edelman, G.M.: Second Nature: Brain Science and Human Nature. Yale University Press, New Haven and London (2006)
33. Heft, H.: Ecological Psychology in Context: James Gibson, Roger Barker, and the Legacy of William James's Radical Empiricism, p. 28. L. Erlbaum, Mahwah (2001)
34. Manning, E.: Always More Than One: Individuation's Dance. Duke University Press, Durham and London (2013)
35. Plato employed the carving metaphor as an analogy for the reality of Forms (Phaedrus 265e): like an animal, the world comes to us pre-divided. Ideally, our best theories will be those which "carve nature at its joints"
36. Manning, Always More Than One, p. 220
37. Smith, B.: Topological foundations of cognitive science. In: Eschenbach, C., et al. (eds.) Topological Foundations of Cognitive Science. Graduiertenkolleg Kognitionswissenschaft, Hamburg (1994)
38. Manning, Always More Than One, p. 219
39. Clark, A.: Surfing Uncertainty: Prediction, Action, and the Embodied Mind, pp. 112–115, 213. Oxford University Press, Oxford (2016)
40. Weschler, L.: Seeing is Forgetting the Name of the Thing One Sees: A Life of Contempporary Artist Robert Irwin. University of California Press, Berkley and Los Angeles (1982)
41. Varela, F.J., Thompson, E., Rosch, E.: The Embodied Mind: Cognitive Science and Human Experience, p. 175. The MIT Press, Cambridge (1991)
42. Spivey, M.J., Anderson, S.E., Dale, R.: The phase transition in human cognition. New Math. Nat. Comput. **5**(1), 197–220 (2009)
43. Phase transition in a broad sense, is a transition of a substance from one phase to another (e.g. solid – liquid – gas) upon a change in external conditions, such as temperature, pressure, etc.; in a narrow sense applied here, it is an abrupt change in perceptual states

44. Varela, F.J.: The reenchantment of the concrete. In: Steels, L., Brooks, R. (eds.) The Artificial Life Route to Artificial Intelligence: Building Embodied, Situated Agents, pp. 11–20. Lawrence Erlbaum Assoc., New Haven (1995)
45. Spivey, Anderson and Dale, The Phase Transition in Human Cognition., p. 205. [emphasis added] See also: Spivey, M.J.: The Continuity of Mind. Oxford University Press, New York (2007)
46. Following the lines of continuity is consistent with the Ruyerian fibrous conception of the universe. See: Ruyer, Neofinalism, pp. 140–53
47. Massumi, B.: The political economy of belonging and the logic of relation. In Parables for the Virtual; Movement, Affect, Sensation, pp. 68–88 (68). Duke University Press, Durham (2002)
48. Radman, A.: Deep Architecture: an ecology of hetero-affection. In: Jobst, M., Frichot, H. (eds.) Architectural Affects after Deleuze and Guattari, pp. 63–80. Routledge, London (2021)
49. Kwinter, S.: There is no such thing as 'post-critical' (only good and bad criticism). Prax. Des. Crime Forum **5**, 17, 19, 21 (21) (2003)
50. Oyama, S.: Sustainable Development: Living with Systems. In: Clarke, B. (ed.) Earth, Life, and System: Evolution and Ecology on a Gaian Planet. Fordham University Press, New York (2015)
51. Deacon, T.: Incomplete Nature: How mind emerged from matter, p. 197. W.W. Norton & Company, New York and London (2012)
52. Thelen, E., Smith, L.B.: Dynamic Systems Theories. In: Lerner, R.M. (ed.) Handbook of Child Psychology, pp. 258–312. Wiley, New Jersey (2006)
53. Kwinter, S.: A discourse on method (for the proper conduct of reason and the search for efficacity in design). In: Geiser, R. (ed.) Explorations in Architecture; Teaching, Design, Research, pp. 34–47. Birkhäuser, Basel (2008)
54. Kugler, P.N., Shaw, R.: Symmetry and symmetry-breaking in thermodynamic and epistemic engines: a coupling of first and second laws. In: Synergetics of Cognition, pp. 296–331. Springer, Heidelberg (1990)
55. In retrospective control, adjustments are made in respect to what has occurred; in prospective control, in respect to what will occur. See: Turvey, M.T.: Lectures on Perception: An Ecological Perspective, pp. 305, 376–377. Routledge, New York and London (2019)
56. Gibson, J.J.: The Ecological Approach to Visual Perception, p. 223. Lawrence Erlbaum Associates, New Jersey (1986)
57. Massumi, B.: Building Experience; The Architecture of Perception. In: Benjamin, A., Spuybroek, L. (eds.) NOX Machining Architecture, pp. 322–331. Thames & Hudson, London (2004)
58. Gibson, The Ecological Approach to Visual Perception., p. 207. [emphasis added]
59. Uexküll: A Stroll Through the Worlds of Animals and Men

60. Ingold, T.: The architect and the bee: reflections on the work of animals and men. Man New Ser. **18**(1), 1–20 (15) (1983)
61. Deleuze and Guattari, Anti-Oedipus, p. 338
62. Guattari, F.: The Three Ecologies. Continuum, London (2008)

Pipes of AI – Machine Learning Assisted 3D Modeling Design

Chuan Liu[1], Jiaqi Shen[2], Yue Ren[3], and Hao Zheng[2](✉)

[1] Zhejiang University of Technology, Hangzhou, China
[2] University of Pennsylvania, Philadelphia, USA
zhhao@design.upenn.edu
[3] University College London, London, UK

Abstract. Style transfer is a design technique that is based on Artificial Intelligence and Machine Learning, which is an innovative way to generate new images with the intervention of style images. The output image will carry the characteristic of style image and maintain the content of the input image. However, the design technique is employed in generating 2D images, which has a limited range in practical use. Thus, the goal of the project is to utilize style transfer as a toolset for architectural design and find out the possibility for a 3D modeling design. To implement style transfer into the research, floor plans of different heights are selected from a given design boundary and set as the content images, while a framework of a truss structure is set as the style image. Transferred images are obtained after processing the style transfer neural network, then the geometric images are translated into floor plans for new structure design. After the selection of the tilt angle and the degree of density, vertical components that connecting two adjacent layers are generated to be the pillars of the structure. At this stage, 2D style transferred images are successfully transformed into 3D geometries, which can be applied to the architectural design processes. Generally speaking, style transfer is an intelligent design tool that provides architects with a variety of choices of idea-generating. It has the potential to inspire architects at an early stage of design with not only 2D but also 3D format.

Keywords: AI in design · Neural Networks · Style transfer

1 Principle of CNN

Convolutional Neural Networks (CNN) is a class of deep neural networks that are most commonly used to analyze visual images. Figure 1 shows the image that analyzes how an input image is processed by CNN. Firstly, the computer reads the image as pixels and represents it as a matrix, which will then be processed by the convolutional layer. This layer uses a set of learnable filters that convolve across the width and height of the input file and compute the dot product to give the activation map. Different filters that detect different features are convolved in the input file and output a set of activation maps that are passed to the next layer in the CNN. The pooling layer between the convolutional layers can be found in the CNN architecture. This layer substantially reduces the number of

P. F. Yuan et al. (Eds.): CDRF 2020, *Proceedings of the 2020 DigitalFUTURES*, pp. 17–26, 2021.
https://doi.org/10.1007/978-981-33-4400-6_2

parameters and calculations in the network and controls overfitting by gradually reducing the size of the network [7].

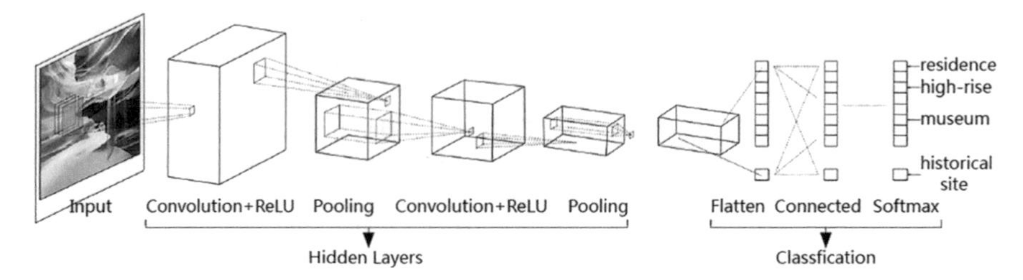

Fig. 1. Introduction of CNN structure.

The next layer is the fully connected layer, in which the neurons are fully connected to all activation of the previous layer. Therefore, their activation can be calculated by matrix multiplication followed by offset. This is the final stage of the CNN network [2, 3]. In general, CNN uses relatively less preprocessing than other image classification algorithms. This means that the network learns manually designed filters in traditional algorithms. This independence from prior knowledge and human effort in feature design is a major advantage.

1.1 Principle and Applications of Style Transfer

CNN-based Style Transfer is a type of algorithm for processing digital images or video, using the look or visual style of another image. In the paper *A Neural Algorithm of Artistic Style,* Gatys introduces a Style Transfer Network to create high-quality artistic images [5]. Moreover, it has been successfully applied in key areas such as object and face generation. The style transfer process assumes an input image and a sample style image. The input image is fed through the CNN and the network activation is sampled at each convolutional layer of the VGG-19 architecture [6]. The content image is then obtained as a resulting output sample. The style image is then fed through the same CNN and the network activation is sampled. These activations are encoded as a matrix representation to represent the "style" of a given style image (Fig. 2). The goal of Style Transfer is to synthesize an output image that shows the content of a content image to which a style image style is applied.

To fully understand Style Transfer, Ostagram, an online style transfer tool, is used to generate images. A pair of images is selected as the content image and style image for the experiment, then switch the role of the image to another for another set of experiments. It's evident that the output image presents the geometry of its content image and the style of the style image (Fig. 3).

1.2 Project Goal

Style Transfer allows us to create new images with high perceived quality, combining the content of any photo with the look of many well-known artworks. Gatys [4] and

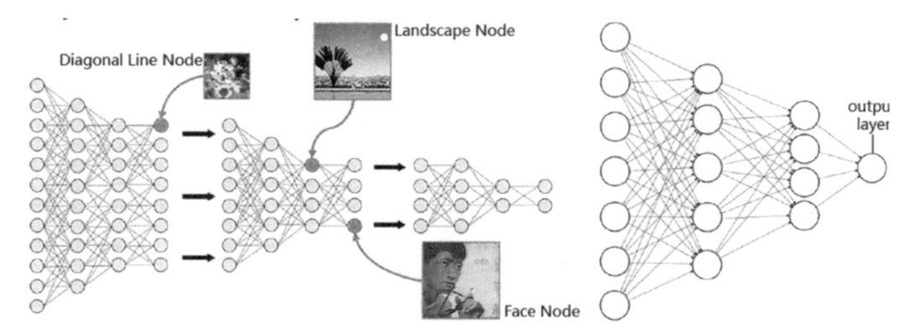

Fig. 2. Style deconstruction of the Style Transfer Neural Network.

Fig. 3. Example of style transfer.

Mordatch [8] mention the versatility of the Neural Network. For example, a map is generated in light of the features of the target image, such as image density, to constrain the texture synthesis process. Besides, image analogy is applied to transfer textures from already stylized images to target images.

Despite the remarkable results of Style Transfer, design techniques are only used to design images to be generated at the 2D level, such as graphic design [1]. In order to broaden its scope of use, it is necessary to design 3D levels in order to apply the generated results to architectural design. In order to achieve the project goal, 3D geometry is required as the result of the Style Transfer. With the completed 3D geometry, further designs like 3D structures or buildings can be used.

In order to put this idea into practice, it is necessary to answer questions about the nature of creativity and the criteria for evaluating creativity. Can style transfer create a novel sensibility, can we as humans perceive and understand it? In order to answer these questions, the project objectives presented here need to solve the problem not only from the aesthetic aspect - the idea that style transfer can produce fascinating 2D images - and from the point of view of emphasizing practical value: considering about the practicality of image-generated content and the usability of converting it to a 3D format.

2 2D Image Representation of 3D Volume

Since Style Transfer processes images on a 2D scale, the 3D model is needed to be converted into 2D images at first. An existing project is selected as the carrier of the design boundary. The project is a theater design, which is diverse in interior spaces, has a variety of floor plans of different heights (Fig. 4). By adjusting the position of section plan in Sketchup, a series of floor plans are gained according to the height increment of the building. Then, the ten-floor plans are converted into images that only carry the information of the outline, which is the boundary between the interior and exterior space of every floor plan (Fig. 5). Black and white colors are used to distinguish the interior and exterior space. At the stage, two sets of color-filling methods are applied as one set black to interior space and white to exterior space while one set the opposite. The methods will bring different results for output images since the color is an influential factor of Style Transfer. Thus, the most suitable result will be selected and be employed in further design.

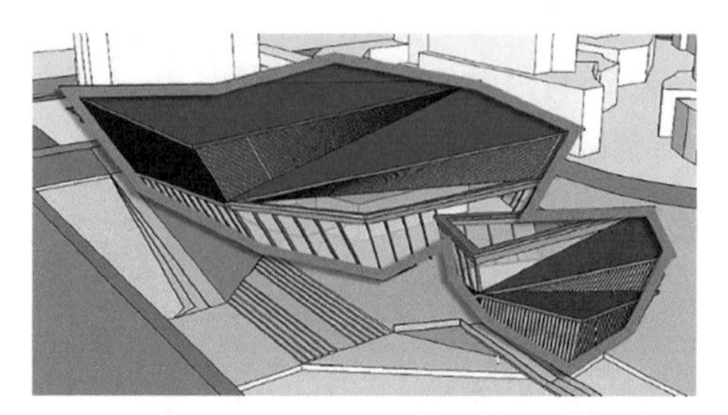

Fig. 4. Origin model of the content images.

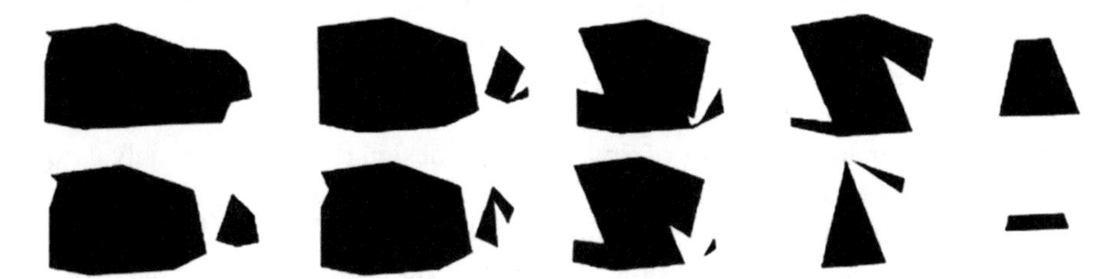

Fig. 5. The geometry of floor plans.

In general, by translating 3D volume into a 2D image, content for the input image is obtained as the basic element of the Style Transfer Neural Network. Later on, the style image will be imported into the network with the content images to start the generating process.

2.1 The Effect of Style Weight in Style Transfer

After the content images are obtained, a style image should be imported so that the generating phase can be activated. For style image, facade images and landscape images are firstly employed to the Neural Style Transfer Neural Network. The result of the output images turns out not available for further design because of its indistinguishable geometry. To eliminate the chance of an unavailable output image, style images that contain a clearer outline and distinctive color contrast are applied. Thus, a framework of a truss structure is set as the style image (Fig. 6).

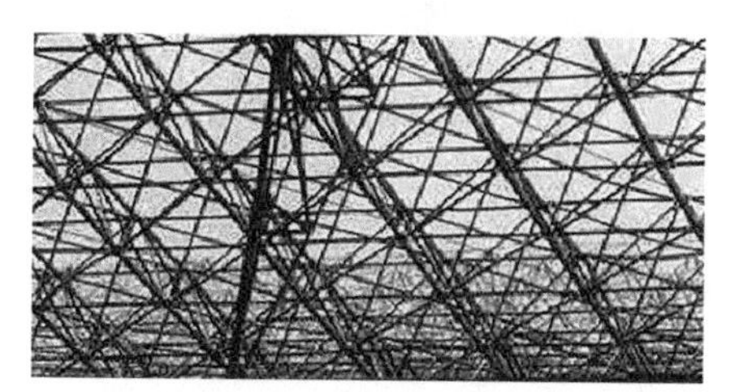

Fig. 6. Style image.

Before importing style image and content image into the network, several input parameters are needed to be confirmed. These parameters including “Style Weight”, “Content Weight” and “tv Weight”. There will be different effects presented by the output images by adjustments of every input parameter separately. With the awareness, a content image is imported into the neural network while adjusting a single parameter regularly to receive a series of output images. From the three sets of output images, it is easier to tell the difference between every set of output images with the influence of every parameter. At the same time, the regular pattern is easier to be observed of every set so that we can decide which input parameter is more suitable for further design.

After comparing each set of output images, we noticed that “Style Weight” is the input parameter that brings the most ideal generating result. Figure 7 shows the images that exported as the output images while adjusting the value of only style weight. From the images, the difference is obvious between every single image and the surprising found is the regular pattern of the set of images. With the increase of the value on style weight, the part of generated texture in the black area, which represents interior space, is more evident as well as taking up more percentage of the black area. Those phenomena are not shown in the other two sets of experiments with the adjustment of “tv Weight” and “Content Weight”. Thus, “Style Weight” is selected as the input parameter for design.

Fig. 7. Output images generated by adjusting style weight.

With the conclusion above, firstly, “tv Weight” and “Content Weight” is given a specific value in the design. For “Style Weigh”, a range from 500 to 5000 is given to the content images since the range shows the most ideal case for the contents of new generating images. For instance, the “Style Weight” value of the 1st-floor plan is 500, and that of the 2nd-floor plan is 1000. With the set of rules, values are distributed for every content image so that the contents of output images are well-regulated and persuasive as well (Fig. 8).

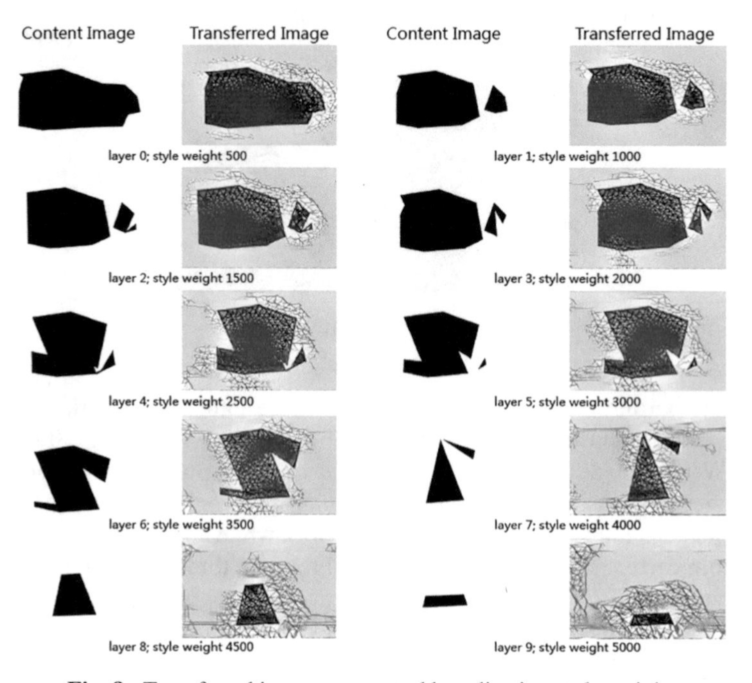

Fig. 8. Transferred images generated by adjusting style weight.

2.2 Transformation of Image to Geometry

After the processing phase of Style Transfer, output images are obtained for further design. At the stage, the major concern is how to utilize the images in architectural design, since the 2D image is unavailable for geometry-oriented modeling software. Thus, 2D images need to be converted into geometry to fit it in architectural language. Rhinoceros, which is a geometry-based modeling software, is applied to accomplish the conversion.

To translate the image into geometry, a PDF version of the output images is imported to Rhino. After the import phase is completed, the geometry like polyline is used for geometric generation. There are several rules at the editing phase, the first is to delete the area that is useless for the coming design. The area in white color, which represents the exterior part of the floor plan, is defined as the nugatory space of the design.

After eliminating the white area in every image, there is another set of rules for further edition. Since the generated texture in the black area presents as intersected line segments from the top perspective, the polyline making tool is utilized to translate the texture into geometry. By connecting all the line segments together with polylines, a geometry version of the image is completed (Fig. 9). The rest of the output images will be edited according to the same set of rules until all the geometric layers are obtained (Fig. 10).

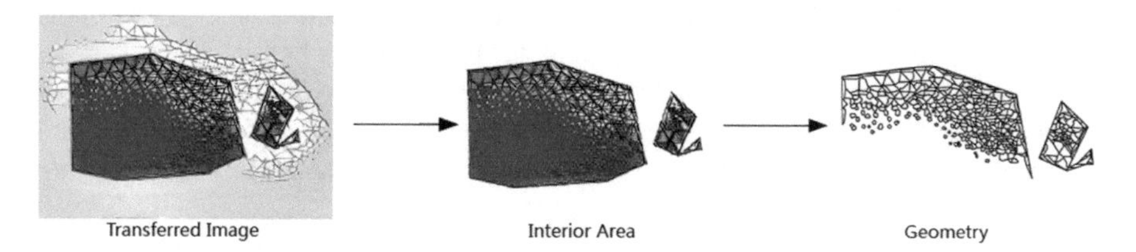

Fig. 9. Transferred image translation.

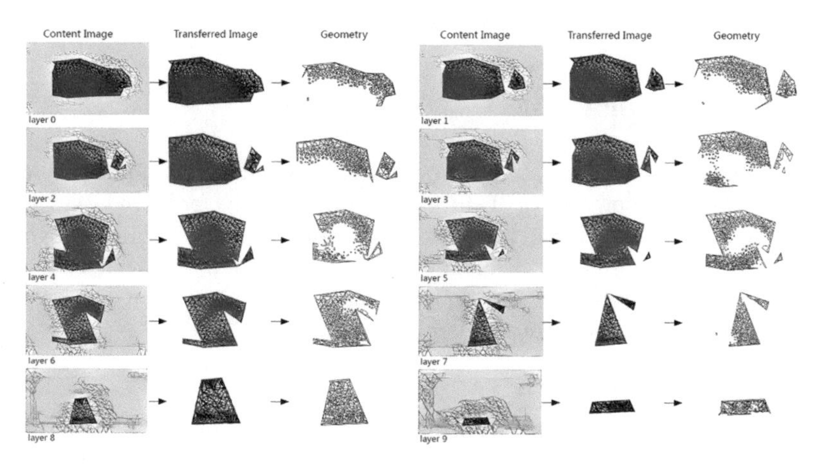

Fig. 10. Translation of all the geometric layers.

2.3 Algorithm Analysis of Geometry Generation Between Adjacent Layers

The 3D model will be accomplished after all the geometric layers are stacked together according to the original sequence of the floor plans. To make the model applicable for architectural design, pillars should be added between the layers so that the load condition of the structure is reasonable.

There are lots of possibilities for the geometry of the pillars, since the only rule a line need to obey is started from an intersection point and end on another one at the upper layer. However, if it is the only rule for the pillar's geometry, the design is likely unavailable for forming an architecture model. Realizing the problem of connecting uncertainty, Grasshopper, a plug-in based on Rhinoceros, is applied to explore the best way of connection from an algorithmic perspective.

After the script is completed, the information of two adjacent layers is imported into Grasshopper. Then, there are two input parameters, which decide the number and tilt angle of lines, needed to be modified for generating the vertical geometry. The input parameter which decides the line's number is being modified at first since a determined value of the parameter is a prerequisite for further adjustment like a tilt angle. A number slider, which domain is from 0–1, is employed to adjust the number of lines between the layers. In the experiment, the domain is divided into five groups of subdomains, which are 0–0.2, 0.2–0.4, 0.4–0.6, 0.6–0.8, and 0.8–1. Then the median is selected from every subdomain to be the value of the input parameter for each set of experiments.

After the number of geometry is fixed, a panel is set for the domain of the tilt angle. To determine which domain is the most suitable, another five sets of experiments are conducted according to the same rule of the former experiment. A domain from 0–10 is divided into five subdomains, which are 0–2, 2–4, 4–6, 6–8, and 8–10. Then the subdomains are processed by the same kind of rule. In general, the vertical geometry is determined based on the generated lines, which are baked according to the selected values of two input parameters. With the vertical and horizontal geometry, the structure will be further designed in a more aesthetic way.

3 Result of Section Plans

To translate the geometry into architectural language, piping, a design tool in Rhinoceros, is used to convert geometry into a 3D format. By adjusting the radius of both horizontal and vertical geometry, the optimal format is generated. Figure 11 shows the image of two section plans, one is the view from the right and the other one is from the front. The views are selected according to the structural complexity, which presents the distribution and relationship of vertical geometry between layers.

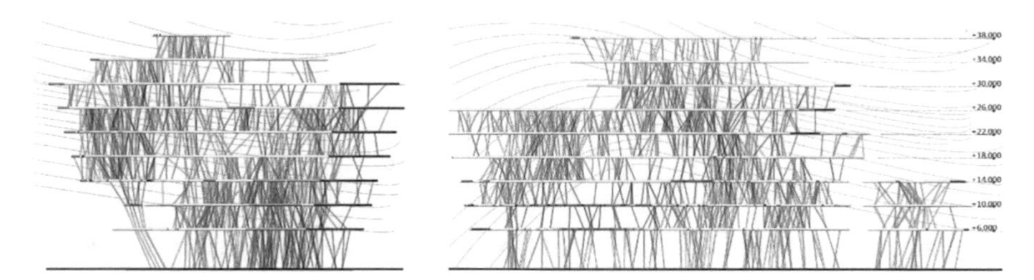

Fig. 11. Front facade (left) and right facade (right).

Moreover, the human scale is considered when designing the height between the adjacent layers so as to meet the requirement of use. The height of the 1st floor is 8 meters, which is two times higher than the rest of the floor heights which are 4 meters. Public spaces can be positioned at the area where has a more open vertical space with fewer pillars while private spaces the opposite.

3.1 Result of Perspective View

The vertical and horizontal pipes are presented in different textures so as to provide a more comfortable space experience for visitors. Vertical pipes are designed as pillars, thus grey is used with a hint of transparency to mimic the format of pillars. Horizontal pipes are designed as floors since there is no given requirement for the texture, white is used to distinguish themselves with vertical pipes.

Figure 12 shows the image that depicts the relationship between the building and its surroundings. The original building is evolved into several variants, which are achieved by stretch, compression, and superimposition of the original format. This step helps to expedite the design process, which translates the original building into a series of new formats so as to fit in different circumstances of a city.

Fig. 12. Panorama.

4 Conclusion

Generally speaking, 2D geometry is successfully converted into a 3D format that is available for architectural design. Since 2D images are the basis for 2D geometry, Style Transfer, a neural network to generate 2D images, is of great significance for the whole design. From the project, Style Transfer not only successfully generates new images that inspire architects for further 3D model design but also provides a series of choices by adjusting input parameters so that architects can select out the one based on sensibility and aesthetics.

All of this brings to an envisagement that if Style Transfer can be employed as a post-human approach to architecture. The neural network used in the project successfully blends the style into the content image. Therefore, it is possible to generate a new style in light of two input images that are of distinctive design styles of two architects. It would be more interesting if the new style can be applied in architectural language so that Style Transfer will proceed into a creating level.

Future application in architectural design with Style Transfer involves in the creation of architectural images as well as vectorized data, for example the iterative optimization of building data with similar loss functions. Apart from application in architecture, it is in the prospect of being utilized in fields like graphic design and animation design. Thus, it is highly possible that Style Transfer Neural Network will be an indispensable tool for designers at the creating stage in the coming future.

References

1. del Campo, M., Manninger, S., Sanche, M., Wang, L.: The church of ai-an examination of architecture in a posthuman design ecology. In: CAADRIA 2019 (2019)
2. Fukushima, K.: Neocognitron: a self-organizing neural network model for a mechanism of pattern recognition unaffected by shift in position. Biol. Cybern. **36**(4),193–202 (1980)
3. Fukushima, K.: Neocognitron. Scholarpedia **2**(1), 1717 (2007)
4. Gatys, L.A., Ecker, A.S., Bethge, M.: A neural algorithm of artistic style.arXiv preprint arXiv: 1508.06576 (2015)
5. Gatys, L.A., Ecker, A.S., Bethge, M.: Image style transfer using convolutional neural networks. In: Proceedings of the IEEE Conference on Computer Vision and Pattern Recognition, pp. 2414–2423 (2016)
6. Kümmerer, M., Wallis, T.S.A., Bethge, M.: Deepgaze ii: reading fixations from deep features trained on object recognition. arXiv preprint arXiv:1610.01563 (2016)
7. LeCun, Y., Bengio, Y., Hinton, G.: Deep learning. Nature **521**(7553), 436–444 (2015)
8. Mordatch, I., Abbeel, P.: Emergence of grounded compositional language in multi-agent populations. In: Thirty-Second AAAI Conference on Artificial Intelligence (2018)

Developing a Digital Interactive Fabrication Process in Co-existing Environment

Chun-Yen Chen, Teng-Wen Chang(✉), and Chi-Fu Hsiao

National Yunlin University of Science and Technology, Douliou, Taiwan
tengwen@softlab.tw, chifu.research@gemail.yuntech.edu.tw

Abstract. In the stage of prototype practice, the maker mainly works by himself, but it needs to test and adapt to find correct fabrication method when maker didn't have clearly fabrication description. Therefore, rapid prototyping is very important in the prototype practice of the maker. "Design- Fabrication-Assembly" (DFA)- an integration prototyping process which helps designers in creating kinetic skin by following a holistic process. However, DFA lacks a medium for communication between design, fabrication and assembly status. This paper proposes a solution called co-existing Fabrication System (CoFabs) by combining multiple sensory components and visualization feedbacks. We combine mixed reality (MR) and the concept of digital twin (DT)–a device that uses a virtual interface to control a physical mechanism for fabrication and assembly. By integrating virtual and physical, CoFab allows designers using different methods of observation to prototype more rigorously and interactively correct design decisions in real-time.

Keywords: Digital fabrication · Digital twin · Mixed reality · Interactive design

1 Introduction

In recent years, the maker movement has arisen a trend of handmade implementation around the world. As makers work by their own hands, practical problems or design errors will not appear until they really trying to combine pieces together in the real world. In order to lower the manufacturing threshold and obtain more rapid prototyping method, makers often manufacture in digital ways. But It requires a professional division of labor in the construction process. Therefore, the prototyping process needs a lot of time to consultation, waiting, and manufacturing. Design-Fabrication-Assembly (DFA)- an integrated prototyping process proposed by [5] which enables users to take advantage of fabrication as part of a collaborative process. But there's a need for the communication media in DFA frame [5]. Therefore, we via combine mixed reality and the fabrication machine–a device that uses a virtual interface to control a physical mechanism for fabrication and assembly. We use a "seeing-moving-seeing" design thinking model, which allows designers to refine their ideas [10]. Designers can analyze and present these ideas with the use of design media making it easier to explain their thinking patterns A co-existing space representation [7, 8] is used for modeling such interplay transitions from virtual to physical spaces [13]. The possibilities of multiple operation by designing,

P. F. Yuan et al. (Eds.): CDRF 2020, *Proceedings of the 2020 DigitalFUTURES*, pp. 27–35, 2021.
https://doi.org/10.1007/978-981-33-4400-6_3

manufacturing and integrating tools through digital means, such as robotic-arm were expanded [2, 4].

Three aspects of the interactive fabrication process based on "seeing-moving-seeing" model are human behavior, digital stream, and physical entity. Additional three interplays comprise of control, communication and computation and are the interface between human behavior, digital sensing, physical entities. By Interacting with the digital stream and the physical entity via combine digital twin (DT) with mixed reality technology, a designer can embed himself/herself into a network of a manufacturing system [11]. We implement a Coexisting Fabrication System (CoFabs) to support an interactive process for designers that allows for real-time modification and manufacturing in the design sequence. The process integrates both real-world and virtual environments based on the "Digital Twin" (DT) concept, which proposes a system where physical entities and virtual information are referenced to each other in a recursive way through a series of physical changes, information analysis, and generative fabrication suggestions. In this way, the workflow is optimized [3]. We also analyze each thread that results and create a system model to simplify the workflow of controlling physical machines and lower its technical threshold.

2 Related Work

Digital manufacturing is changing the way people design, produce, and interact with objects and devices. The diversity of current manufacturing processes includes laser cutting, 3D printing, CNC milling and printed circuit board (PCB) manufacturing and they may produce parts in a variety of forms and materials. Because of the rise of digital fabrication technology, it redefines and integrates industrial manufacturing logic. Designers must be capable of abstract design thinking, describing and keeping design results under control. With the advent of a series of rapid prototyping (RP) technologies, designers are able to see design results more quickly.

2.1 Fabrication Process of Maker

The prototyping process of interactive surfaces is divided into three stages: Design, Fabrication and Assembly, DFA framework (as shown in Fig. 1). Two main issues are: (1) Lack of unified communication media across D-F-A stages. While design is changed, the fabrication will adapt according to the geometry and methods. The assembly methods and sequence should also adapt. Most digital tools are for synchronizing the design and fabrication, not assembly methods. (2) As mentioned above, participants during assembly stage are often unskilled and need to be trained in order to adapt to the changes from fabrication stage. Such condition is often not possible. DFA is a collective prototyping process that will provide a framework to allow many participants to join the process, experiment with the concept and respond to the adaptation of interactive surfaces. Three stages of DFA are: (1) design stage: design installation prototype; (2) fabrication stage: convert design into components and carriers, and find the suitable fabrication methods; (3) assembly stage: assemble fabricated components and carriers for final installation [6]. There are three retrospective-modification process and two communication tools required for completion of DFA process [5].

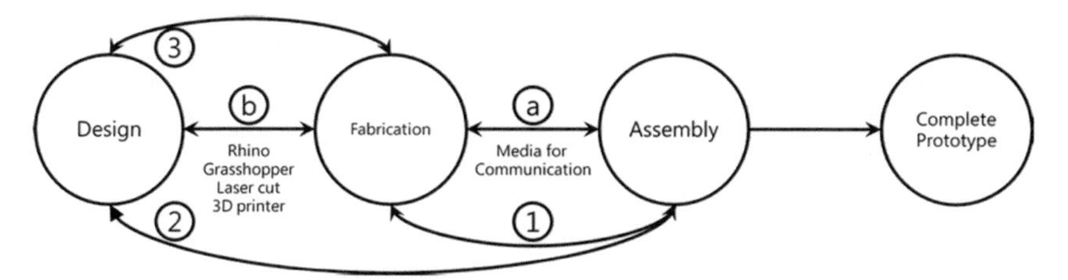

Fig. 1. DFA process

2.2 Towards Co-existing Environment

Mixed reality is the integration of both real and virtual worlds to create new environments and visualizations of physical and digital objects coexisting and interacting in real time. Mixed reality occurs not only in the physical or virtual world, but also in the combination of reality and virtual reality, including augmented reality and enhanced virtual immersive technology. Currently, the range of mixed reality technology and applications has expanded to include entertainment and interactive arts as well as engineering and medical applications.

In 2014, Weichel et al. established a mixed reality environment that lowered the barrier for users to engage in personal fabrication (Fig. 2). Users design objects in an immersive augmented reality environment, interact with virtual objects in a direct gestural manner and can introduce existing physical objects effortlessly into their designs [12].

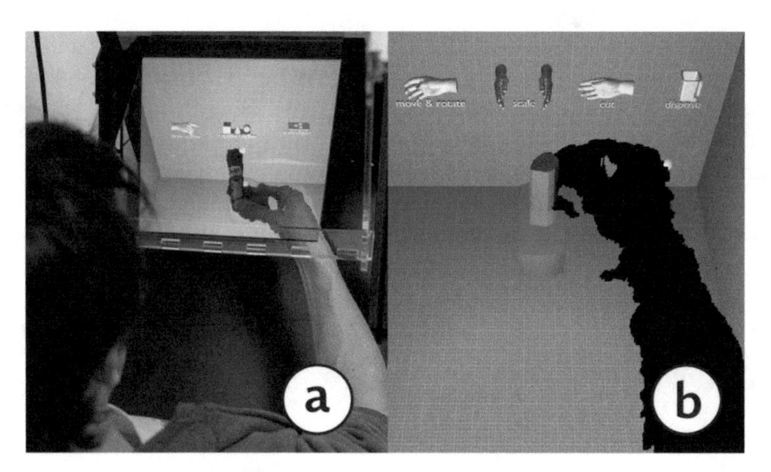

Fig. 2. MixFab's user interface

2.3 Automation Digital Fabrication Tools

In the course of explicit bricks in 2010, the robot arm was first time combine with hot-wire cutting to fabricate. The architectural potential of the developed system was tested through applying it on the design and fabrication in a 1:1 scale prototypical building structure. The participants were challenged to analyze and test different interlocking systems by making simple prototypes in order to apply the outcome to an overall system. The prototypes were fabricated using a robot in combination with a hot wire cutting machine. Existing constraints set by the fabrication process hat do be considered as well as to check stability and flow of forces (Fig. 3). Gaining stability due to friction and interlocking system were seen as the potential of the fabrication process and had to be exploited as far as possible [1].

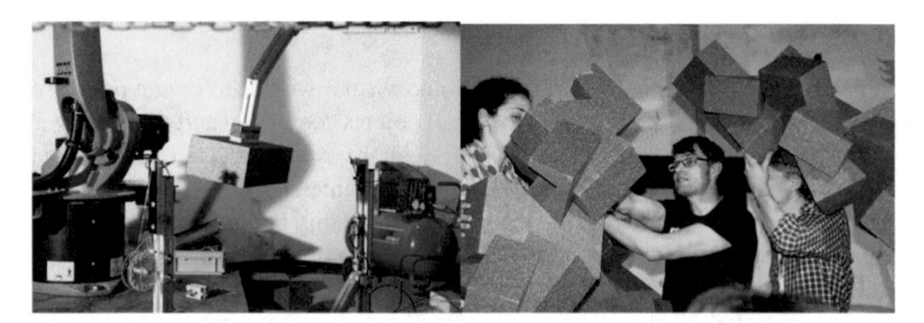

Fig. 3. Robot arm hot-wire cutting process (left) and result (right).

2.4 Summary

Through the analysis of above literature, the concept of the DFA process was introduced into the digital fabrication process. Under the operating conditions of rapid prototyping and virtual reality blending, it allows designers to join the process, experiment with the concept and respond to the adaptation of interactive surfaces. Generally, the devices in a mixed reality are limited to their own resolution and interactivity in the display of the interface. The HoloLens [9] developed by Microsoft not only provides a wireless head-mounted display, but also gave the ability to assemble AR commands and the opportunity for users to face the appearance of such applications. Therefore, HoloLens was used in this research as a mixed reality interface device for the system.

3 Methodology

Participatory behavioral observations were made and explored in this paper. Hence, makers with cross domains were the main targets and various manufacturing methods capable of rapid prototyping were analyzed. It included metal bending, metal printing, hot wire cutting, incremental sheet forming, CNC engraving and panel striking. In order to allow CoFabs extend rapid prototyping to the other fabrication process. We divide the fabrication process into 4 steps: (1) Interface setting, (2) Component setting, (3) fabrication and (4) finish, for verifying DFA process (seen in Fig. 4). But maker needs technical knowledge in every stage, such as tools assembly and parametric adjustment. Therefore, we analyzed the different tasks for each stage and systematic via computation design in this paper, and control the physical machine by combining the co-existing technology and digital twin to reduce the fabrication technical threshold. We analyzed the work area setting, initial model and tool setting in the interface setting stage. The stage of component setting is combined co-existing technology and fabrication system to control machines and perform the task. Moving path and angle were analyzed by the user's drag gesture in the fabrication stage. In the final finish stage, we used color to display the assemble position and check for any part that didn't match.

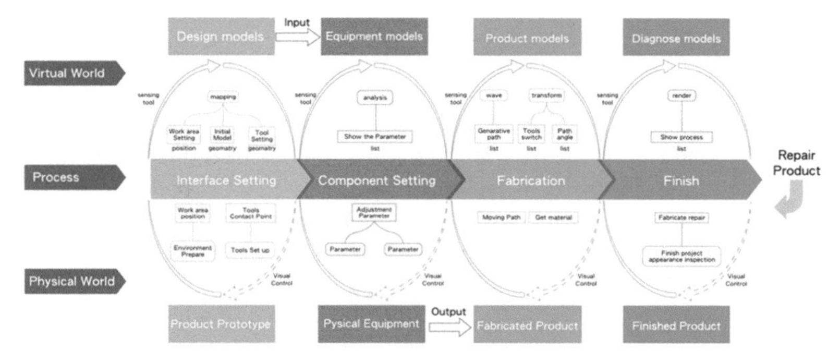

Fig. 4. CoFabs process

4 The Experiment

In this paper, we use HoloLens to develop a co-existing fabrication system (CoFabs). Currently, we focus on research between HoloLens GUI and interactive behavior. The virtual graphic computation system can make users see the combination of digital and physical model environments more efficiently. But it needs detection material to achieve an acceptable simulation. We have to optimize the digital model when we are deforming virtual objects in this co-existing fabrication process.

We used Fab Car to perform a small test on a unit component, utilizing robot arm hot-wire cutting fabrication. Fab Car is a collaborative project aimed towards the implementation of a modular car. It gives designers the ability to modify, customize and adapt the vehicle to their specific needs at any given time and allows designers to iterate on the project in real-time.

First, we set up three different Fab Car templates model at the left of the interface (seen in Fig. 5-a), to allow for fabrication customization. System will compute the tire position and revise template when the user selects one of the templates. and the user can also revise the length and width to decrease space at the right of interface (seen in Fig. 5-b), to adjust Fab Car architecture (Fig. 5-c).

We provide the user with an internal structure component to select and assemble individual components, such as tire, engine, transmission shaft etc. We divide work area into 4 workplaces: (1) Frame setting, (2) structure setting, (3) Car shell design, (4) Assembly component, to make it easier for the user to select and place component and

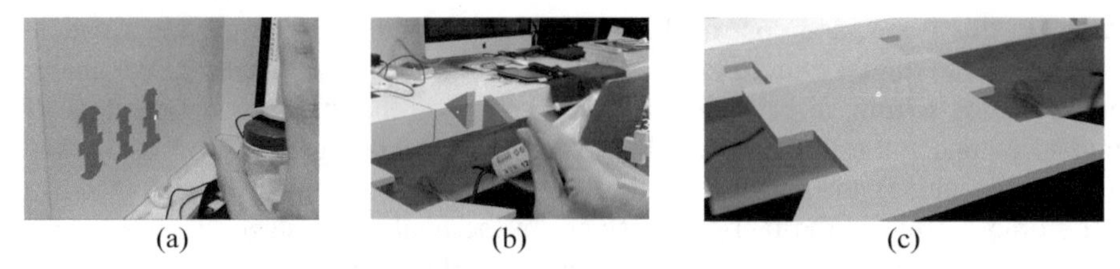

(a) (b) (c)

Fig. 5. (a) Fab Car template, (b) template parameter and (c) Fab Car architecture.

allow step-by-step fabrication process. we connect component of every workplace to allow for iterations.

First, we reset the three different architectures and tire templates. The template model will import to second workplace by selection, and system will compute the space to place by selected size of the tire (seen in Fig. 6-a). We set up the component of engine, seat, transmission shaft. User can click and drag to pick up the component, and customized Fab Car internal structure by placing the components in the desired position. (seen in Fig. 6-b)

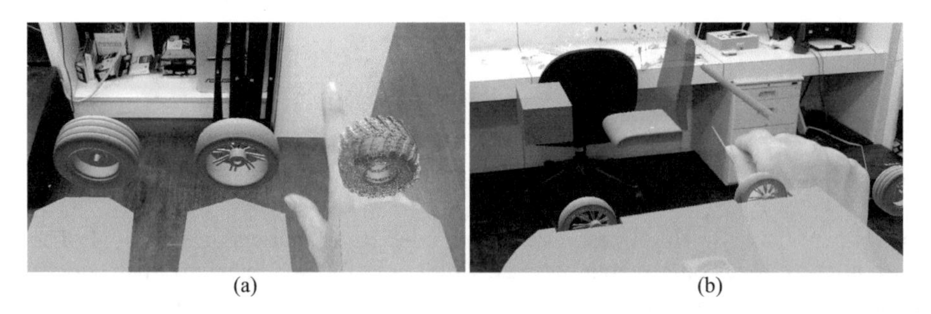

(a) (b)

Fig. 6. (a) first workspace: select template, (b) second workspace: pick and place component

In the third workspace, we extruded the model of the base upward into a cube to provide user with the initial model of cutting Styrofoam, and computation with Boolean delete on the reserved space in the previous step to limit the cutting scope. Then, through track two hand positions and generate the line from the middle as cutting tool in virtual. Users can generate the cutting path through click and moving two hands to reduce path unsmooth due to muscle fatigue when hand-floating. Each click once to generate a line as forming initial cutting surface and path (Fig. 7-a). In order to allow the user can modify the model and precisely cutting, we set up a few buttons to provide user modify, edit and divide modifiable, modifying and confirm status. In the process, we through the initial interface for distinguishing the use of gestures as follow:

- UndoCrv: Provide user to undo and regenerate curve.
- Confirm: Transform multi-curve to the surface after generating curves and proceed to the next curve and surface.
- UndoSrf: Provide user to undo and rebuild the surface after generating surfaces.

- Edit: Click to generate the edit point on the surface, and modify the surface via click and drag the point.
- Clear: Clear all models on screen for initial setting.

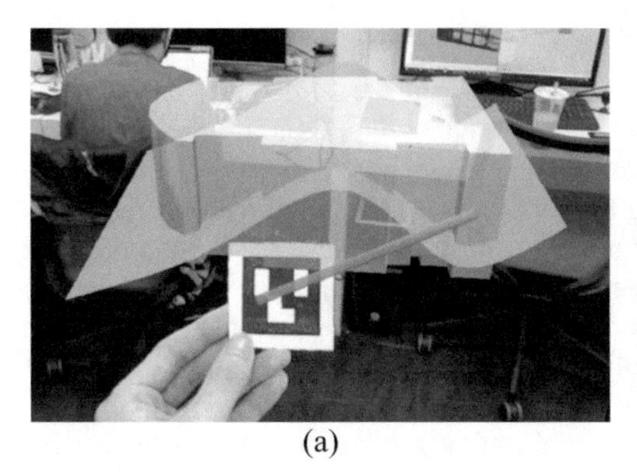

(a)

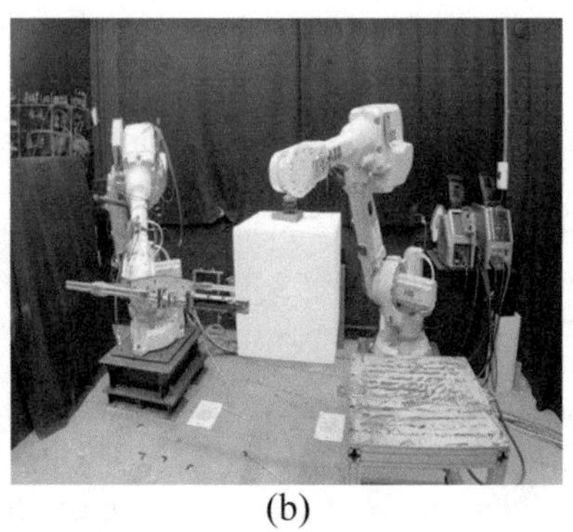

(b)

Fig. 7. (a): Click gesture to generate cutting surface, (b): Robot arm according moving path to hot-wire cutting.

(a)

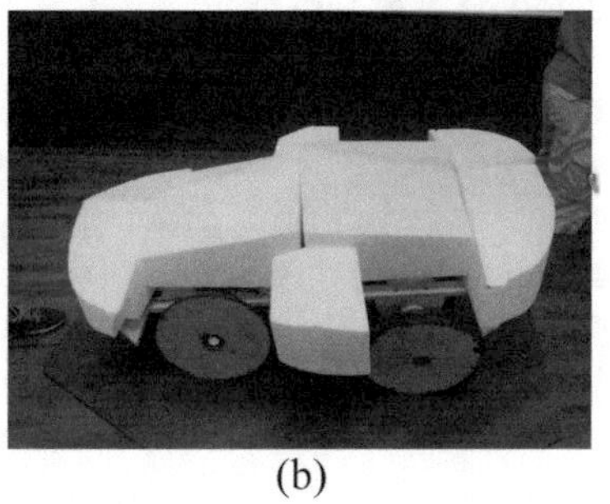

(b)

Fig. 8. (a): Used red to distinguish collision position, (b): Fab Car finish project

We use the editor to transfer the script and parameter of Fab Car to the MR device when the user finished set cutting surface and starts to perform the hot-wire cut. Through preview every piece of Styrofoam that needs to be cut, this allows the user design car shell component step-by-step. Finally, user can select the component which to be cut, and generate cutting path according to the set cutting surface via the editor to generate RAPID code to control the robot arm perform hot-wire cut (Fig. 7-b).

In the fourth stage, we through MR device simulate the component, assemble sequence and position of the final project in physical environment and via click and drag gesture to preview and assemble the component. It provided the user clear assembly process and position of each component. In addition, we used color to distinguish car shell and internal structure collision position. It allows the user to check and make sure all the component is the match, and modify the component via the display of collision position (Fig. 8-a). Finally, we assembled a Fab Car step-by-step after modifying the components (Fig. 8-b). We observed the user don't need to wait for the final project to make sure it's available when they operated easy interface to get stronger custom tools.

5 Conclusions

This paper via "seeing- moving- seeing" design model with robot arm combine to MR technology for digital fabrication. Under the operating conditions of rapid prototyping and virtual reality blending, and using the robot arm which is a three-axis fabrication tool, and integrating mixed reality for remote control to achieve the DFA prototyping process, users can utilize different "seeing methods" to observe the finished product. To reduce complications of robot arm operation, allowing the designer to operate the robotic arm more simply. Thus, the CoFabs process immerses users into an interactive co-existence environment more suitable for user immersion.

Not only did the user witness the model building progress, but also determined the manufacturing detail of fabrication with prompted feedback information. Designers can also modify several specific parameters to change the shape from simulated model to real materials. Through this human-machine collaboration task, we can make design decisions and fabricate designs more simply, intuitively, as well as easily communicating design intent in real-time. In the result of experiment for the development of Cofabs prototype, we learned:

Distributed Fabrication
In the situation which don't need oral instructions and diagram, user can share fabrication process and working on different projects together. In this process, we'll face the interface and interactivity problem. It can via the user experience research method to solve the problem. Then, in order to allow the user easier to operate, design and fabricate, we will automatic the fabrication process via digital computation in future work. It makes user control the design development process.

Co-existing Fabrication Accuracy
In the co-existing fabrication, imprecise virtual and physical information delays interactive feedback. We use HoloLens to solve this problem but it still needs fault tolerance.

Information Synchronizes
Virtual and physical information cannot be complete synchronize through the transmission of network signals. In the future, it can via 5G network to increase the wireless transmission speed to solve this problem.

References

1. Bärtschi, R., et al.: Wiggled brick bond. Adv. Arch. Geom. **2010**, 137–148 (2010)

2. Chen, C.-Y., et al.: Developing an interactive fabrication process of maker based on "seeing-moving-seeing" model. In: Proceedings of the 2019 DigitalFUTURES. CDRF 2019, Shanghai, China. Springer (2019)
3. Grieves, M., Vickers, J.: Digital twin: mitigating unpredictable, undesirable emergent behavior in complex systems. In: Transdisciplinary Perspectives on Complex Systems, pp. 85–113. Springer (2017)
4. Hsiao, C.-F., et al.: A co-existing interactive approach to digital fabrication workflow. In: 25th International Conference of the Association for Computer-Aided Architectural Design Research in Asia (CAADRIA), Bangkok, Thailand (2020)
5. Hsieh, T.-L., Chang, T.-W.: ViDA: a visual system of DFA process for interactive surface. In: 23 International Conference Information Visualisation. Flinders University, Adelaide, Australia (2019)
6. Hsieh, T.-L., Chang, T.-W.: How to collective design-and-fabricating a weaving structure interaction design—six experiments using a design-fabrication-assembly (DFA) approach. Presented at the 4th RSU National and International Research Conference on Science and Technology, Social Sciences, and Humanities 2019. (RSUSSH 2019), Rangsit University, Thailand (2019)
7. Lai, I.-C., Chang, T.-W.: Companying physical space with virtual space-a co-existence approach. In: The 8th Annual Conference of Computer Aided Architectural Design Research in Asia (CAADRIA), Bangkok, Thailand (2003)
8. Lu, K.-T., Chang, T.-W.: Experience montage in virtual space. In: Proceedings of the 10th International Conference on Computer Aided Architectural Design Research in Asia 2005. CAADRIA, New Delhi, India (2005)
9. Microsoft. HoloLens: a new way to see your world. [website] (2019). https://www.microsoft.com/microsoft-hololens/en-us/hardware. Accessed 10 Oct 2019
10. Schon, D.A., Wiggins, G.: Kinds of seeing and their functions in designing. Des. Stud. **13**(2), 135–156 (1992)
11. Teng-Wen, C., et al.: A fabricating behavior sensor computing approach for a co-existing design environment. Sens. Mater. (2020, to be published)
12. Weichel, C., et al.: MixFab: a mixed-reality environment for personal fabrication. In: Proceedings of the SIGCHI Conference on Human Factors in Computing Systems. ACM (2014)
13. Wesugi, S., et al.: Interactive spatial copy wall for embodied interaction in a virtual co-existing space. In: RO-MAN 2004 13th IEEE International Workshop on Robot and Human Interactive Communication (IEEE Catalog No. 04TH8759). IEEE (2004)

Real-Time Defect Recognition and Optimized Decision Making for Structural Timber Jointing

Dan Luo[1(✉)], Joseph M. Gattas[2], and Poah Shiun Shawn Tan[3]

[1] School of Architecture, University of Queensland, Brisbane, Australia
d.luo@uq.edu.au
[2] School of Civil Engineering, University of Queensland, Brisbane, Australia
[3] School of Engineering and Applied Science, University of Pennsylvania, Philadelphia, USA

Abstract. Non-structural or out-of-grade timber framing material contains a large proportion of visual and natural defects. A common strategy to recover usable material from these timbers is the marking and removing of defects, with the generated intermediate lengths of clear wood then joined into a single piece of full-length structural timber. This paper presents a novel workflow that uses machine learning based image recognition and a computational decision-making algorithm to enhance the automation and efficiency of current defect identification and re-joining processes. The proposed workflow allows the knowledge of worker to be translated into a classifier that automatically recognizes and removes areas of defects based on image capture. In addition, a real-time optimization algorithm in decision making is developed to assign a joining sequence of fragmented timber from a dynamic inventory, creating a single piece of targeted length with a significant reduction in material waste. In addition to an industrial application, this workflow also allows for future inventory-constrained customizable fabrication, for example in production of non-standard architectural components or adaptive reuse or defect-avoidance in out-of-grade timber construction.

Keywords: Out-of-grade timber · Machine learning · Decision tree · Optimization · Simulation · Manufacture

1 Introduction

Timber is a renewable construction material with a low carbon footprint. It holds significance in being both a sustainable modern-day building material, as well as a traditional material that intertwines with cultures all over the world. Though there are a range of positive environmental and engineering properties associated with timber, as natural material, it contains a high level of variation in material properties as the result of uneven natural growth and non-uniform environmental conditions. Defects such as knots, checks, splits, and wane are common [1], but their presence can only usually be identified and assessed after the fabrication of the sawn board. In Australia, sawmill reports by Harding have found up to 57.5% of sawn board can fail structural grading

P. F. Yuan et al. (Eds.): CDRF 2020, *Proceedings of the 2020 DigitalFUTURES*, pp. 36–45, 2021.
https://doi.org/10.1007/978-981-33-4400-6_4

requirements [2]. As compared to structural timber, out-of-grade timber is far less desirable for most construction or manufacturing operations, and consequently it has a low market demand and is often woodchipped and/or sold at a loss [3].

A common method in the timber industry for value-recovery in out-of-grade timbers is through the cutting out of defective parts, and the subsequent joining of the non-defective (clear wood) segments into a full-length structural board (Fig. 1). Joining is normally achieved with finger joints, as they have adequate strength for typical structural timber applications, such as wood trusses and laminated beams [4]. Though the pipeline is mostly automated, it still often requires humans to visually identify the defect and mark them out for removal. This removal process is largely dependent upon the experience of the worker and is also the time-limiting operation in the material-recovery process. The segment jointing method itself is also is a source of leftover clear wood material, as pieces are jointed to a length based on sequential segments, and trimmed down to a pre-determined final length.

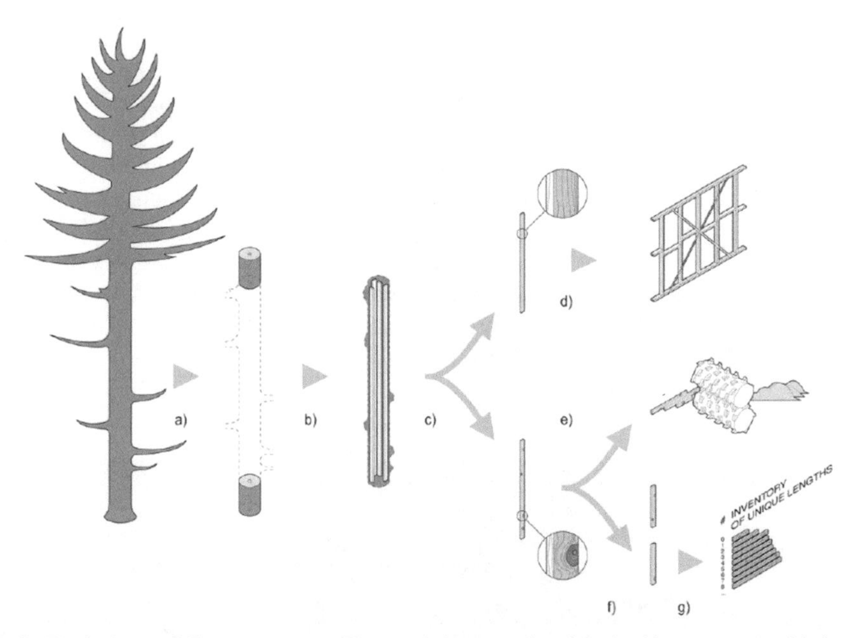

Fig. 1. Typical sawmilling process: a) The tree is harvested and docked into transportable lengths (approx. 6 m) with varying branch distribution according to: silvicultural practice, climate, soil, planting density etc.; b) each log is sawn into usable framing members and is subject to the naturally occurring structural defects; c) sawn members are sorted into those with and without defects; d) members free from defects are certified for use in structural applications; e) members with defects are used in non-structural application or woodchipped; f) defects are removed, resulting in short lengths, which provide: g) an inventory of short, unique lengths [3].

This research investigates the possibility of enhancing this process with machine learning based image recognition and decision-making algorithms. Image recognition is widely adapted within different industries, such as the classification of packages for delivery [5]. The advantage of using image recognition and machine vision is that it can be achieved with minimum hardware investment cost, often simply by adding a camera to an existing adaptive industrial pipeline [5]. The images of timber collected from a machine vision manufacturing process can also feed back into the image recognition and defect detection system, allowing its accuracy improve over time with expanded training datasets, and allowing the system to self-improve and or evolve with changing timber stock. Additionally, introducing a machine vision system allows a live data record to be collected for each clear wood segment generated. Such information provides a potential to introduce a decision-making algorithm that can optimize the joining process to reduce left over material, or for *customization* of components with different target lengths.

2 Defect Recognition and Removal

2.1 Pre-process the Image for Segmentation

The images of timber used in this research study are collected from an inline Machine Stress Rating (MSR) system model 720 HCLT. For any testing timber board, it collects a series of images and also directly measures the board apparent Modulus of Elasticity (E), with the two datasets then used to assign a final board grade. For the visualization requirements of the present project, the collected image required prepossessing by decreasing the noise, removing distortion, and removing the background:

- Image de-noise: Spikes of light noise at the border of the timber is widely seen in the scan images, owing to the rapid-rolling scanner used in the industrial setting. This noise can be significantly reduced by reading the horizontal proximity of each pixel to evaluate if it's a 'spike region', which is characterized by a small area of foreground (board) pixel, bordered on both sides by background pixel (Fig. 2a–b).

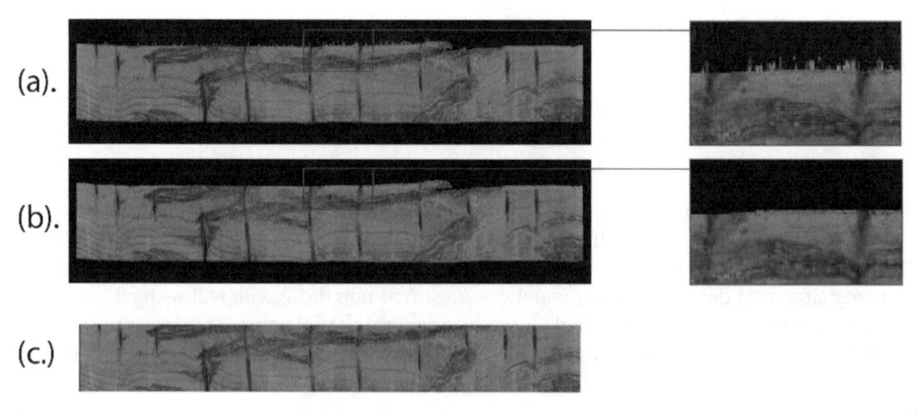

Fig. 2. (a). Raw image collected from MSR; (b). Image after denoise; (c). Image after un-distortion and background removal

- Un-distortion and background removal: the central axis of the scanned timber member is often slightly un-even or non-linear. A pixel column line-scan process aligns the foreground pixels in the denoised image to a linear axis, and removes any background noise. The final image is a consistent and formatted visualization of board material and defects (Fig. 2c).

Currently, this pre-processing workflow is developed based on image collected from scans from the machine grading system. When compared to images collected from the 7200 HCLT, images collected from typical camera hardware would contain far more noise, an uneven background, and uneven lighting quality. When connecting with an industrial manufacture environment, though following similar principles, the system can be further calibrated based on the image collected by the camera feed, accommodating the local lighting and environmental conditions.

2.2 Preparation of the Classifier

Based on the collected image, a classifier is developed to identify the location of defects. This classifier is trained to assign a discrete categorized label for each pixel in the timber scan image, to identify if the pixel belongs to the following three classes: defective timber, non-defective timber, or background (non-timber). In addition, the classifier also outputs a probability map that describes the probability of each of the three classes at each pixel.

The collected image is interactively labelled by defining the region of "Background", "Defective Timber", and "Non-Defective Timber" in the Fuji distribution of Weka. The training of the classifier took place using the API of the Trainable Weka Segmentation [6]. The classifier is trained based on a FastRandomForest model, which is an enhanced multi-tread version of the ensemble learning method for classification, regression, etc., developed by Fran Supek [7]. The model has a batch size of 100, 8 thread, 200 trees. The training feature includes Gaussian Blur, Sobel Filter, Hessian, Difference of Gaussians, and Membrane Projections. The features are extracted, converted and formatted to a set of vectors for the Weka classifier. The features are calculated with 8 threads after the labelling of training data in Weka.

2.3 Preparation of the Classifier

After the training of the classifier in the intuitive Trainable Weka Segmentation GUI, the classifier is applied in the backend of a simulation interface. It takes in an image of the timber piece to generate classification and probability maps for the image. The classification map returns the classification of each individual pixel in one of three classes: "Background", "Defective Timber", and "Non-Defective Timber" (Fig. 3c–e). The classification map set includes a probability map for each of the classes, where each pixel is assigned a 0–1 value that describes the probability of the selected pixel being in given class (Fig. 3a–b). In this research, the raw probability value is used to determine the defectiveness of each section of the timber. The classifiers can further be trained to increase accuracy, by collecting additional labelled images in the process.

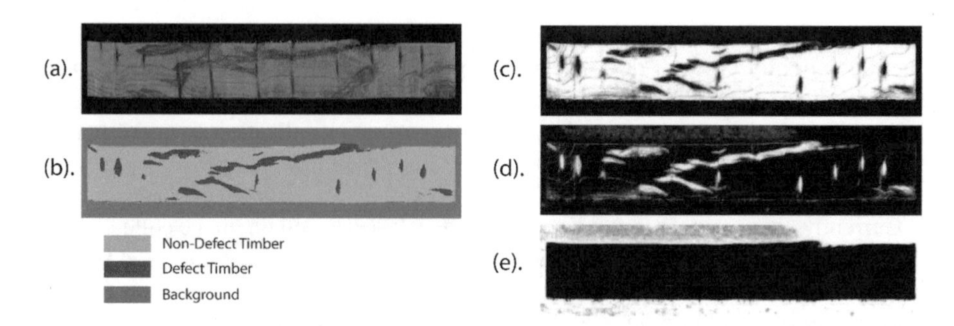

Fig. 3. (a) Raw image collected from MSR; (b) Binary classification map; (c) Probability map for non-defective timber; (d) Probability map for defective timber; (e) Probability map for background.

With the probability map for defective timber, the image is further processed to simulate the removal of timber defects. There are two parameters used to define the threshold of identifying a timber defect, giving a user the option to calibrate the range and sensitivity threshold for defect removal. The first parameter defines the threshold of probability for pixels to be considered as defective, which is currently set at 75%; the second parameter is the percentage of defect pixel in a 1 pixel-wide column slice, which is currently set at 5%. Exceeding which, the corresponding slice of timber will be considered as defective area.

The defect removal process will scan through the processed timber image and remove any defect areas composed of more than 5 consecutive defective slices, which translates into a defect with horizontal dimension of 3 cm or greater. For the remaining material, graded timber segments that are 20 cm or longer are stored in a dynamic stock of clear wood timber segments (Fig. 4).

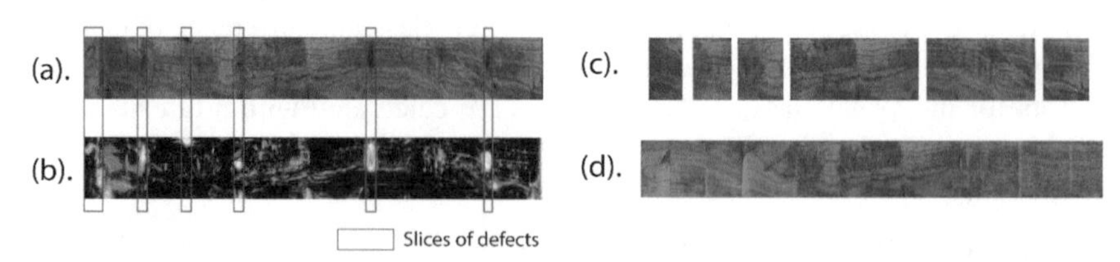

Fig. 4. (a) Raw image collected from MSR; (b) Probability map for defective timber; (c) Non-defect timber segments; (d). Jointed piece of standard length.

3 Decision Making for Joining Timber Segments

Each iteration of inputting a new piece of timber will add an indefinite number of segments of non-uniform lengths into a 'dynamic stockpile' of segments, ready for jointing. Owing to manufacturing standardisation, there is a uniform set target length for a final jointed timber piece, taken as the common industry length of 4800 mm for present study. However, into jointing up to the target length, a new challenge is encountered in the selection of timber segments from the dynamic stock pile: how to select segments to exceed the target length, but with a minimum of overrun so as to minimise the waste generated from final trimming.

A new system is proposed to resolve this challenge. The system criteria for deciding which pieces of timber segments should be joined together follows premises as summarised below:

- Generate different length combinations from a selection of a set of timber pieces that provide the user's target length (4800 mm) and a small overrun allowance (target total length +200 mm).
- Of all combinations, those with the closest proximity to the target length are prioritized for joining.
- On receiving a new input of segments, the new segments will join the left over segments int the dynamic stock pile waiting to be jointed. The dynamic stock is re-scanned to update the possible combinations and joining priority that meets the previous two criteria.
- Monitor the amount of wastage that comes from the difference between the user's target length and the actual total length of the selected combination. This wastage may be the result of the total length being longer than the user's target length.

The most extensive algorithm to explore the subspace of potential combination of segments is the Exhaustive recursion method, where all possible solutions are explored and compared. However, the computational cost for such method explodes exponentially as the pieces in the stockpile increase, causing an infeasibly-long processing time.

Given the restriction in computational power and processing time, this study instead implements a backtracking logic to select the appropriate combination of timber segments, which is far less computationally demanding, but similarly effective to the exhausted recursion method. Figure 5 provides a breakdown of the jointed outcome, from receiving the first 10 randomly selected out-of-grade timber piece as input to the system sequentially.

The pseudocode below represents the backtracking selection logic behind the scene:

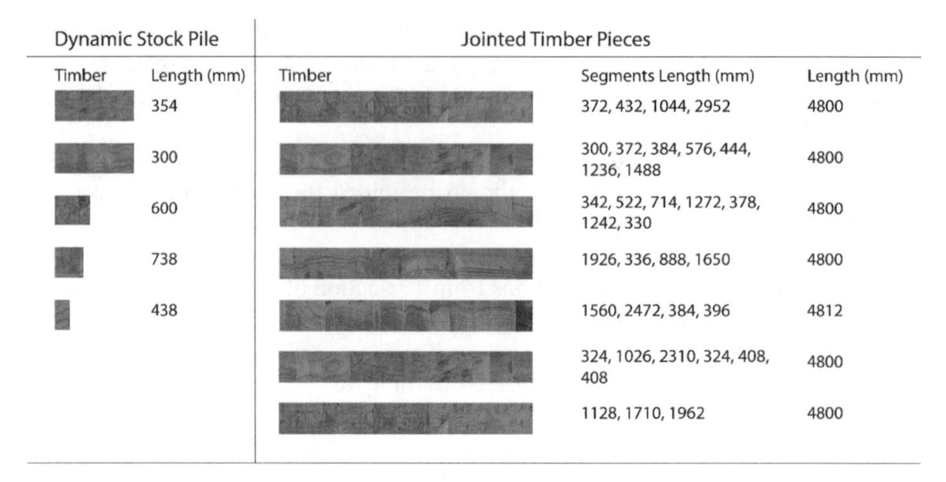

Dynamic Stock Pile		Jointed Timber Pieces		
Timber	Length (mm)	Timber	Segments Length (mm)	Length (mm)
	354		372, 432, 1044, 2952	4800
	300		300, 372, 384, 576, 444, 1236, 1488	4800
	600		342, 522, 714, 1272, 378, 1242, 330	4800
	738		1926, 336, 888, 1650	4800
	438		1560, 2472, 384, 396	4812
			324, 1026, 2310, 324, 408, 408	4800
			1128, 1710, 1962	4800

Fig. 5. Stock condition with 10 under graded timber pieces as input sequentially.

```
ArrayList<Timber> connectedTimber
ArrayList<Timber>  segmentStock
Boolean combineTimberPiece (Timber timberSegment){
        If (no more segments available) { //base case
                Return connectedTimber;
        }

        For (all timber in segmentStock){
                Try one choice of timberSegment c:
                //try to see if timber segment c could contribute to
                        a optimized combination.
                If (combineTimberPiece (c)) {
                        add segmetnt to connectedTimber;
                    return true;
                }
                Unmake choice c
        }
        return false;
        //tried all timber segments, none of the combination meets the criteria.
}
```

After processing a randomly selected 20 pieces of defective timber, the system manufactures 14 pieces of 4,800 mm long full length structurally finger joined timber without defect, with a total length of 67,200 mm. At same time, this system generates only 58 mm of total waste leaves 6 pieces in the dynamic stock pile, 4,308 mm in total length. This proves the efficiency of the proposed system in salvaging non-defective timber, with a utilisation rate of over 99.96%. The track of inventory change is as below in Fig. 6:

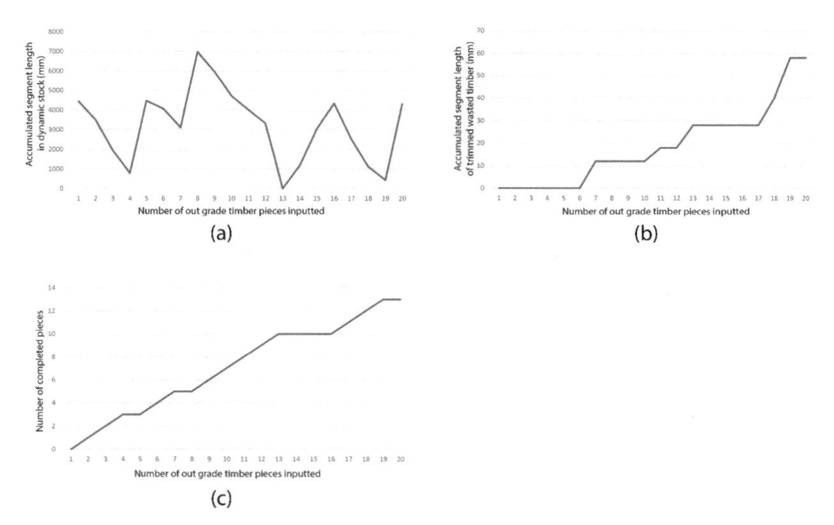

Fig. 6. (a). Change in dynamic stoke pile; b). Changes in the accumulated waste; c) Changes in the number of completed pieces.

4 User Interface

In this project, a platform is developed to simulate the backend process of image recognition and decision-making to evaluate the feasibility of the proposal and its efficiency. By calling to get next piece of timber, the system will pop in the scan image of the next timber from the database to simulate the collected data from camera in a timber mill. This image is been processed with 4 steps to recognize and remove defective area. Each of the 4 steps including image denoise, background removal with un-distortion, probability calculation, and defect removal is visualized sequentially in the left column for the user to understand and evaluate the system.

The segments without defect are stored in a dynamic stock. The next column provides the visualization of the longest 5 segments in the dynamic stock, waiting to be jointed. At each input of new segments, the decision-making algorithm evaluates the dynamic stock of segments, and picks the appropriate segments to joint into a full piece with target length. The latest 5 completed pieces are displayed at the right column, along with access to the folder keeping the entire stock of completed jointed timber pieces (Fig. 7).

In the interface backend, the program tracks the percentage of wasted timber, salvaged timber, and the distribution of segments for the completed timber. Thus, efficiency of the system is tracked throughout the process. Also, the target length can be changed at any time to test the robustness of the system.

The pre-training of the image classifier is performed within the interface of Fuji distribution of ImageJ. It enables the interactive tracing for different classes of elements that can be performed by worker without any computational background, thus offering an intuitive interface to translate the experience and judgement of worker into the classifier that provides a similar process as a visual grading process.

Fig. 7. Graphic interface of the system

5 Discussion and Future Development

The system developed in this research provided the simulation of the manufacturing of structural finger jointed. This simulation system is a reflection on an established industrial manufacture system, and a further investigation the potential of applying machine learning and decision-making algorithm to improve its efficiency in terms of increased processing speed and reduced material waste. Such development can be potentially achieved with very low-cost hardware investment, by simply adding an additional camera into an processing line, for capture of material properties in real-time.

The effectiveness of the system arises from visually-identifying the variation in material property and making intelligent, algorithm-driven decisions for material processing based on the property of individual material. The current training data is based on a manual labelling process similar to the industry. Reflecting on the workflow, there is potential to further improve the accuracy and efficiency of the system by using the MOE data currently collected from machine stress grading processes.

Compared to the traditional manual process of labelling and categorizing, the proposed system is able to systematize the decision-making process, quantify its output, and provide a consistent evaluable structure for the workflow. By embracing a digitized process for machine learning based defect detection, accuracy could improve overtime with the increasing real time collection of datasets, contributing to more accurate removal of defects. With future development, this feature of real-time updating would pave the way for a self-evolving dynamic classifier. A defect detection classifier can adaptively transfer between different scenarios, while the manual labelling process relies on the non-transferrable experience of individual. With the digital data for each piece collected during the manufacture process, a live inventory-based optimization for the dynamic stock becomes possible. This increases the theoretical material efficiency to as much as 99.9%, and provides possible avenues for future adaption of the system to provide customizable parts of different targeted lengths.

6 Conclusion

This research discussed the possibility of using machine vision and decision-making algorithm to optimize a standard industrial workflow in salvaging defective timbers. Such system work with the real-time processing of material data in a pipeline, instead of a static database. It allows for a versatile adaption into the existing industrial workflow with minimum cost.

By introducing this digital and intelligent workflow, the system gains additional potential to increase in material efficient and environmental friendliness by significantly reducing the material wasted during the manufacture process. Also, embracing the intelligence in the system, it pave the way for future upgrade of the production system for optimized mass customization of non-uniform members.

References

1. Cherry, R., Manalo, A., Karunasena, W., Stringer, G.: Out-of-grade sawn pine: a state-of-the-art review on challenges and new opportunities in cross laminated timber (CLT). Constr. Build. Mater. **211**, 858–868 (2019)
2. Harding, K.: Review of wood quality studies in Queensland and northern New South Wales exotic pine plantations (2008)
3. Baber, K.R., Burry, J.R., Chen, C., Gattas, J.M., Bukauskas, A.: Inventory constrained funicular modelling. In: Proceedings of IASS Annual Symposia, vol. 2019, no. 5, pp. 1–10 (2019)
4. Özçifçi, A., Yapıcı, F.: Structural performance of the finger-jointed strength of some wood species with different joint configurations. Constr. Build. Mater. **22**(7), 1543–1550 (2008)
5. Sheth, S., Kher, D.R., Shah, R., Dudhat, P., Jani: Automatic sorting system using machine vision (2010)
6. Kaynig, V., Schindelin, J., Arganda-Carreras, I.: Trainable Weka Segmentation (2019). https://imagej.net/Trainable_Weka_Segmentation
7. Levatić, J., Džeroski, S., Supek, F., Smuc, T.: Semi-supervised learning for quantitative structure-activity modeling. Informatica **37**, 173–179 (2013)

On-Site BIM-Enabled Augmented Reality for Construction

Adam Chernick[1(✉)], Christopher Morse[3(✉)], Steve London[1], Tim Li[1], David Ménard[2], John Cerone[1], and Gregg Pasquarelli[1]

[1] SHoP Architects, New York, NY, USA
ajc@shoparc.com
[2] Unity Technologies, San Francisco, CA, USA
[3] Assembly OSM, New York, NY, USA
cwm@assemblyosm.com

Abstract. We describe a prototype system for communicating building information and models directly to on-site general contractors and subcontractors. The system, developed by SHoP Architects, consists of a workflow of pre-processing information within Revit, post-processing information outside of Revit, combining data flows inside of a custom application built on top of Unity Reflect, and delivering the information through a mobile application on site with an intuitive user interface. This system incorporates augmented reality in combination with a dashboard of documentation views categorized by building element.

Keywords: Augmented reality (AR) · BIM · On-site construction

1 Introduction

1.1 Motivation

The process of constructing a building is complex. A typical deliverable for the architect is a set of construction documents that outline design intent, which can easily consist of upwards of 2,000 pages.

Flipping pages and cross-referencing drawings in order to gain understanding of design intent can be inefficient and confusing, especially within the ever-changing context of a construction site. In 2D drawings, it is inherently difficult to understand three-dimensional depth and how new construction connects to the as-built or context. Navigating these documents on site is unintuitive and can be spatially misleading, leading to errors and costly change orders. "It is becoming increasingly necessary to develop new ways to leverage our project data to better manage the complexity of our projects and allow the many stakeholders to make better more informed decisions" [1].

At the same time, new technologies are emerging that create opportunities to change the way this information is delivered and consumed on site. Mobile devices such as smartphones and tablets are increasingly pervasive and provide more computational power than ever before. Pre-existing cloud computing and data streaming infrastructure

P. F. Yuan et al. (Eds.): CDRF 2020, *Proceedings of the 2020 DigitalFUTURES*, pp. 46–56, 2021.
https://doi.org/10.1007/978-981-33-4400-6_5

is making data connections to these devices faster and more seamless. Camera improvements and image processing algorithms allow for mobile devices to continually scan their physical locations and analyze the results with increasing precision. The combination of these advances creates the ability to use augmented reality (AR) on site and in real time as construction proceeds. While still in their infancy, studies of the uses of new technology methods within construction have already shown productivity gains of 14 to 15% and cost reductions of 4 to 6% [2].

1.2 Related Work

There are many products on the market that look to solve related issues. Procore and Fieldwire are similar solutions in that they focus heavily on the management aspect of the construction process. Their solutions have robust tools for handling submittals, RFIs, financials, and schedules of a project. Bluebeam is a session-based PDF markup tool largely used for construction documents, with some functionality built around navigating a drawing set efficiently. Bluebeam has found a way to link sheet number callouts within a drawing as a sort of hyperlink to the next PDF drawing, reducing the time it takes to navigate a drawing set. Their collaboration tools also let teams jump into a live "Google doc"-like session that can be edited simultaneously. InsiteVR provides an immersive multi-user VR solution for construction and project design coordination. Insite's tools let users jump into 3D models before they are built to overlay multiple trade models to find conflicts and track issues. These solutions address similar information gaps as our solution, but in fundamentally different ways and within different scopes.

Other related work within the field of AR has also advanced in recent years. In their Rocky Vault Pavilion, Sun and Zheng describe a hybrid fabrication paradigm for onsite free-form construction using Unity3D and immersive AR (specifically, the Hololens) [3]. Fologram has also been used with the Hololens to assist in complex assemblies, and it's on-site capabilities have been demonstrated in the bending of steel and wood [4, 5]. Unlike these solutions we are not focusing on construction management as a whole, nor documentation viewing, but on a specific process for efficiently obtaining project information and spatial understanding.

1.3 Our Solution

At SHoP, we have developed a prototype application for mobile devices that addresses these issues. By combining interactive AR on site with a digital set of linked drawings and models, we aim to enhance the ability of construction workers and contractors to more quickly and comprehensively understand design intent within the direct physical environment of the existing construction. They are able to easily access all the information they need within the actual building context. The BIM model can be used as a central repository of information to which external tools for analysis, interaction, and representation can be linked [6].

In this paper we discuss two primary features. First is the ability to access relevant two-dimensional drawings from the traditional drawing set. Rather than jumping from sheet to sheet and view to view based on reference numbers, the user instead is able to access all drawings that relate to a specific building element of interest by selecting that element.

The second feature is the ability to overlay models digitally onto the physical space of a construction site using AR. This feature requires alignment of digital and physical spaces, interaction with the digital models to display the appropriate geometry of interest at any given time, and well-optimized digital models that maintain a high level of visual fidelity and information accuracy while also being computationally efficient to display on low-powered mobile devices.

In order to implement this application, we use the Unity 3D real-time engine for development and Unity Reflect for data import from Revit into our application. For the first round of development, we use a static model, but we anticipate taking further advantage of Unity Reflect in order to establish a live connection as part of our future work. Additionally, we implement custom data preparation workflows in order to create additional metadata relating to the building elements as well as process the existing drawings from Revit into formats usable within our application.

2 AR Application

Our solution enables an on-site user to obtain associated drawings and details about a specific building component in one click. AR accomplishes this in an immersive way, in a 1:1 scale that gives the user spatial understanding within context. Additionally, we note the importance of providing a simple user interface and intuitive user experience, as well as offline capabilities to account for challenges inherent at construction sites. The fundamental differences between other solutions and ours can be broken into two key pieces: contextual model overlay using AR, and model interaction as a query system.

2.1 Model Overlay Using Augmented Reality

By using AR to overlay our BIM model on site, the user can see conflicts with the contextual, or as-built, environment that would not have been possible to see using typical methods. The one-to-one scale enables instant spatial understanding that is difficult and time consuming to distill from the typical drawing set itself.

3D model localization is the process of "pinning" a virtual 3D model at a specific position, scale, and rotation within the physical world. Our application uses image recognition for localization; however, a single image recognition solution does not work within the context of a construction site (Fig. 1). As construction sites can be very large and complex, a single image target will allow the 3D model overlay to "drift" as the user moves around the site, causing inaccuracy between the digital and physical environments. We therefore create a solution that re-localizes at different positions around the construction site. Multiple unique image targets are set at specific intervals around the site which work to negate drift between the digital model overlay and physical environment, as well as reset the model's origin to the new target position.

Fig. 1. Image recognition system for localization of digital model overlay

2.2 Model Interaction as Query System

The established connection between model and drawing data enables the 3D model to become an interactive query system to obtain relevant information. This is a fundamental shift from flipping physical pages or searching through a PDF. By embedding metadata into the model, we are able to list and give access to associated information related to a specific selected element. This is in contrast to navigating a traditional construction document set, which is typically a sequence-based approach for locating and referencing applicable drawings and details.

In a typical construction document system, to understand and build a small section of a simple partition wall, the navigation sequence can take many steps, requiring a subcontractor to jump back and forth between many different sheets, each of which contains only some of the information required and requires references to other sheets for a full understanding. Our system is different: The user selects the building component in question from a 3D view and instantly gets a list of all associated drawings and information (Fig. 2).

Fig. 2. A screenshot showing the information and associated views of a selected model element

2.3 Abstraction of Drawings

"Sheets" within a construction document set are based entirely on their physical namesake: a printable sheet of paper. Typical sheet sizes include ARCH D 24" × 36" and ARCH E 36" × 48". PDFs are still based on what can fit in those predefined areas. We propose to maintain the idea of "sheets" as a collection of sub-drawings, but also to expand it beyond its traditional page boundaries. Our system does not include sheets but will in future development feature a view dashboard that performs the same function of presenting multiple unique yet related drawing views.

In the current version of the application, these unique drawing views are presented independently and individually to the user. These unique views have associated metadata which contains their view name, view number, and associated sheet name and sheet number, to be used in the future development of our view dashboard system. This dashboard will lean on its roots in current typical sheet systems to allow easy adoption.

2.4 Additional Features

Additional features that have been developed to a proof-of-concept level include augmented versions of typical drawing elements. These drawing elements include augmented section cut marks and callouts. The iconography of these augmented elements mimic their 2D drawing counterparts to facilitate adoption and comprehension for end users (Fig. 3). This allows users who are accustomed to a certain way of working to easily transition to this new system. The section cut marker bears the same view and sheet number as the section cut drawing within the construction document set; in this sense, it is a literal extension of the drawing set (Fig. 4).

Fig. 3. Left, a screenshot of an interactive section cut callout in AR. Right, the counterpart call-out on within the traditional drawing set.

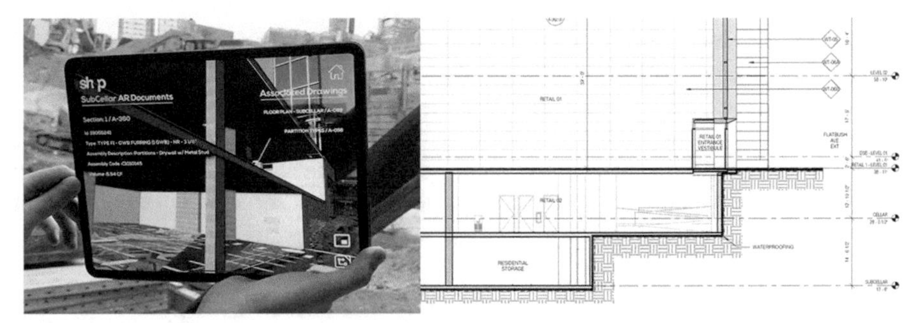

Fig. 4. Left, a screenshot of the section cut interactive BIM model in AR. Right, the counterpart section cut drawing within the traditional drawing set

3 Data Pipeline

In order to provide direct and intuitive access to the appropriate drawings, we developed a workflow to extract necessary information from the Revit model and bring it into our application (Fig. 5).

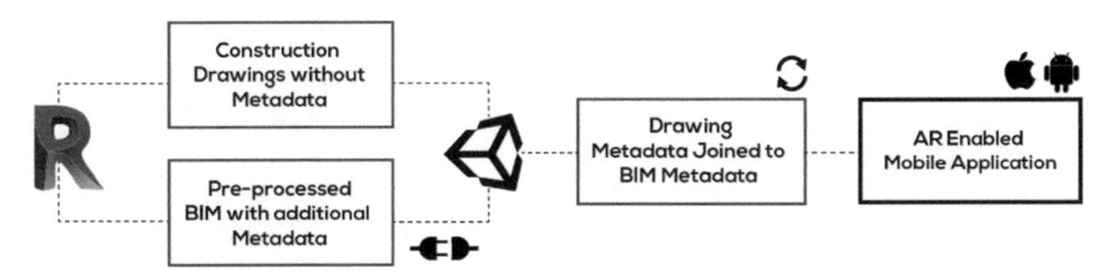

Fig. 5. Data flow diagram from the Revit model to the final mobile application.

3.1 BIM Pre-processing, Custom Parameter Creation and Population

The AR application requires a relatively simple data set. For each element selected by the user, the application requires a list of every Revit View that the element is visible in; This information is not readily populated in Revit. Although each View object contains an array of each element within it, the elements themselves are blind to the Views in which they occur.

We perform a brute force operation as a pre-processing step to generate an association of elements and views and save the results for later use in the application. We iterate through every View, then iterate again through each View's elements, ultimately using the Revit API to generate a custom 'Views Associated' parameter populated on each element containing the list of views that element is referenced in. For large models with many Views, this method can be extremely time intensive and not viable for a live application.

In order to respect Revit as an authorship tool and the origin of our data, rather than using an external file or data format, the list of Views is populated directly into a custom

Family Parameter embedded in each element. This embedded parameter also allows us to access the information after importing the model into our application through Unity Reflect. One downside to this strategy is that the resulting parameter cannot contain a list of information, so the view IDs are concentrated into a single string, then re-parsed once in Unity, requiring additional processing time. Although Revit doesn't currently maintain this data set itself, we think it should, and this approach can serve as a placeholder until it does.

3.2 Construction Document Export and Metadata Post-processing

3.2.1 Construction Document Export

Without the physical limitations of construction documents (CDs), the concept of paper sheets can be expanded to accommodate faster and more intuitive navigation. Our proposed solution entails a dashboard of views dynamically displayed as the user selects an element in the overlaid 3D model. The dynamic grouping of views replaces the role of static CD sheets, and as a result, the individual views defined in Revit become the basic unit of a set of construction documents rather than full sheets.

The collection of views is exported from Revit as high-resolution PNG images and are named according to Revit's default naming convention of the family and type of the view and the view name. In order to provide access to these views, the application associates the View ID generated in the metadata processing step with the file names generated here. Revit offers a limited ability to customize the prefix and suffix to the file name and does not include other useful identifiers such as ViewID.

To provide the information needed to connect elements to the resulting views, as well as to maintain the option for conventional sheet number identification, we use the view schedule functionality of Revit and pull together useful attributes of the views, including view ID, view name, associated sheet number, sheet name, and family and type into a single lookup table. This table is exported as a CSV file, which can then be parsed into our application, and the appropriate fields can be recombined to generate the corresponding file name for each View ID.

3.2.2 Metadata Post-processing

To reduce runtime loading and processing of data, the CSV file is pre-processed in the Unity editor using an Editor Script to generate a hierarchy representing sheets and views with associated metadata components. We found this structure to be useful for organization while developing the application, as well as for maintaining relationships between views that can be utilized in addition to the element-based view sorting described above. Additionally, we populate a dictionary using the View ID as a key in order to efficiently search for the relevant information after selecting an element.

The large number of high-resolution images presents performance challenges for the use of Unity with low-powered mobile devices. A system is developed to load and unload images at runtime. In order to reduce development time for a proof of concept application, we first attempted to use the Resources system built into Unity to allow for runtime loading of data, but found that this system generated uncompressed image storage which was then too large for the application to handle. As a workaround, we currently utilize Unity's Sprite Atlas feature instead. We have found this system to work well to organize and reference images, without running into file size issues when building the application.

Alternative systems were investigated that would allow files to be stored remotely in cloud storage and downloaded to the application when needed. Such systems would allow the overall file size to remain at a minimum, while also allowing for individual runtime loading of images. This could potentially be done with built-in systems such as Asset Bundles or Addressables, or developed independently to allow for direct and automated delivery of images generated from Revit without requiring pre-processing in Unity. Such methods, however, would require near-constant internet connectivity on the job site. For both this reason and to reduce development time, remote storage of views is not currently implemented. This is an area of ongoing refinement and development, and we expect that a combination of strategies will ultimately be required for optimal performance.

4 Unity Reflect

Our solution leverages Unity Reflect in order to translate Revit data into a real-time context. Since Revit is a parametric tool, its data needs to be tessellated into geometry for a real-time engine to consume. This process would typically be done by exporting the model to a geometry format like FBX or OBJ, but there are three main issues with this workflow. First, the exported data loses all the BIM information associated with it in the process. Second, when going from parametric models to geometry, the resulting geometry is massive and not adapted for real-time rendering. Third, every change in the source model would need a fresh export.

Reflect allows for all the data to be dynamically generated, while maintaining the metadata attached to the geometry. In Revit's case, this metadata is the BIM information added to the model. The connection with the source data is also maintained so that any upstream changes to the Revit model are transferred to the game engine. Reflect also handles the complexity of the generated geometry by simplifying the resulting meshes and merging them together when appropriate.

Our solution leverages this connection to import data into the Unity Editor. The incoming metadata is then accessible through the Unity engine, both for the metadata post-processing step as mentioned above, and for displaying it at runtime to inform the end user. Reflect also handles the complexity of the data coming into a real-time engine so that no additional work is needed after the import process.

5 Results

5.1 Case Study/User Testing

We have tested this on site at the supertall tower designed by SHoP currently under construction at 9 Dekalb Avenue in Brooklyn, NY. Our internal construction administration team, as well as the general contractor and owner, have tested the application within a limited scope and have expressed that it could positively impact their work efficiency.

The project director for the general contractor and owner JDS Development, Michael Jones, outlined the difficulties of communication on a construction site. He and his team noted that they believe this application could prevent expensive mistakes on site. Michael and his team mentioned a few improvements that could increase the usability of the application, including the option to drop the opacity of the 3D model overlay to see the underlying context site, as well as the ability to add comments at specific positions within the digital overlay.

5.2 Future Development

The system described in this paper is an initial proof of concept application. There are many areas of future development planned based on initial development roadmaps, as well as user feedback from testing the existing application on site.

We have identified several potential bottlenecks that delay the import of information from Revit into our system. This includes pre-processing information to establish relevant references, as well as producing and processing the 2D drawings. We are looking into implementing workflows to automate these processes.

We note that the generation of lists connecting Revit Views with elements is not a particularly novel problem to solve. Having elements that contain references to their View contexts is something that we believe could be implemented as a native feature of Revit. One reason why it may not already exist is that Revit was ultimately designed to produce drawing sets. The idea that individual model elements may be interactive on site is a relatively new concept, so this contextual awareness was unnecessary. It is worth noting that during our development, we discovered that the Autodesk Forge API does contain a "__viewable_in__" property associated with each element which, as the name suggests, lists Views where each element occurs [7]. A future effort may take advantage of this API by uploading our model to BIM360, letting it do the processing, and then accessing this parameter via the Forge API.

Another area of future development is connecting the Unity Reflect system to a live Revit model to allow automatic updates for the end user. This would allow the transference of dynamic views in context, while keeping the complexity of the model abstracted away, only delivering to the end user what is needed in their current context. While the Reflect system already handles the information transfer well, the future work in this area is more focused on legal questions around how to validate and approve changes within this non-traditional method of information delivery.

One of the largest areas of improvement is the user interface and user experience. While we continue to work on the dynamically generated view dashboard, we will maintain the ability of our dynamic system to tailor the grouping and sorting of this set of information to specific users. There is the potential of integrating sorting algorithms such as Brin and Page's PageRank Citation Ranking based on the number of elements referencing each view, or the number of times all users view them [8]. This dashboard system will break the rigid drawing set structure into something that can be expanded upon, re-organized, and integrated with other information. Instead of pages collected by type and level of detail, we can present drawing information deliberately grouped together to answer questions as they are asked.

The goal of this project is to facilitate information delivery on-site to the people who need it. Without an intuitive user interface and seamless user experience, we run the risk of losing the advantages of real time communication because of a hard-to-use application. This level of development requires rigorous user testing to establish how people expect to use such a system, and uncover what features are most useful.

5.3 Connections

In addition to the specific application described in this paper, we see larger-scale potential to change the ways building designs are delivered, communicated, and built. The prevalence of devices and software that can visualize three-dimensional geometry in increasingly immersive and interactive environments suggests a fundamental shift away from static two-dimensional drawings sets and towards connected digital models augmented by metadata. Such a shift allows for more direct communication between designers and contractors and has the potential to cut out the middle stage of converting models into two-dimensional projections that lose information.

The amount of data that can be generated and delivered will continue to increase, and the filtering and selective display of this data will thus become increasingly important. Methods of data visualization and user interaction will help communicate this data in effective ways to produce better outcomes.

With this in mind, we expect that some of the lessons learned from this effort involving data interoperability and user experience testing will be increasingly useful for other bespoke applications for future projects and future scenarios.

References

1. Aparicia, G., Kontovourkis, O.: Sustainable computational workflows. In: 6th eCAADe Regional International Workshop Proceedings (2018)
2. Koeleman, J., Ribeirinho, M., Rockhill, D., Sjödin, E., Strube, G.: Decoding Digital Transformation in Construction. 20 August 2019. [Online]. Available https://www.mckinsey.com/industries/capital-projects-and-infrastructure/our-insights/decoding-digital-transformation-in-construction. (Accessed 05 2020)
3. Sun, C., Zheng, Z.: Rocky vault pavilion: a free-form building process with high onsite flexibility and acceptable accumulative error. In: Yuan, P.F., Xie, Y.M.M., Yao, J., Yan, C. (eds.) CDRF 2019, pp. 27–36. Springer, Singapore (2020). https://doi.org/10.1007/978-981-13-8153-9_3
4. Jahn, G., Newnham, C., Beanland, M.: Making in mixed reality. holographic design, fabrication, assembly and analysis of woven steel structures. In: Proceedings of the 38th Annual Conference of the Association for Computer Aided Design in Architecture (ACADIA), pp. 88–97 (2018)
5. Jahn, G., Wit, A.J., Pazzi, J.: [BENT]. In: Proceedings of the 39th Annual Conference of the Association for Computer Aided Design in Architecture (ACADIA), pp. 438–447 (2019)
6. Abdelmohsen, S.: Genres of communication interfaces in bim-enabled architectural practice. In: 6th International Conference Proceedings of the Arab Society for Computer Aided Architectural Design (ASCAAD), pp. 81–91 (2012)
7. Goncalves, G.: Navigating Between 2D Views. 21 June 2017. [Online]. Available https://forge.autodesk.com/blog/navigating-between-2d-views. (Accessed May 202)
8. Page, L., Brin, S.: The PageRank Citation Ranking: Bringing Order to the Web. (1998). [Online]

Recycling Construction Waste Material with the Use of AR

Caitlyn Parry[1](✉) and Sean Guy[2]

[1] RMIT University, Melbourne, Australia
Caitlyn.parry@rmit.edu.au
[2] Fologram, Melbourne, Australia
Sean@fologram.com

Abstract. This paper aims to present a methodology for reusing and recycling scrap timber from building sites using augmented reality and flexible digital models. The project we present describes a process that enables existing material to be reused and repurposed such that the designed model is updated by the digital inventory of digitised offcuts/waste elements.

Keywords: Material reuse · Material recycling · Augmented reality fabrication · Digitization of physical parts

1 Introduction

The construction industry is one of the greatest contributors to global pollution via both the carbon emissions from the production of construction material and its direct contribution to landfill. From 1900 to 2010, the amount of materials accumulated in buildings and infrastructure across the world increased 23-fold (Krausmann et al. 2017). Annually, the construction and demolition industry produce over 20 million tonnes of waste. The waste material from these industries is mainly constituted from timber, metal, concrete, bricks, rock and soil (Shooshtarian 2019).

The development of the digital twin pertaining to smart cities has enabled the capacity to model, simulate and observe over time, the complex interactions of the city and technology (Mohammadi et al. 2017) however it has yet to harness any future potential inherent in the embodied material resources of the city. The Smart City is an advanced piece of infrastructure, and until every built form is digitised specifically as an inventory of materials, down to the lengths of timber, there remains an opportunity to work with the collection of site offcuts and landfill bound materials.

2 Aims

This paper seeks to present a methodology for using mixed reality in the documentation, fabrication and construction using waste building material. There is a gap in the current construction industry which, until each building has shifted to also exist as it's digital twin

P. F. Yuan et al. (Eds.): CDRF 2020, *Proceedings of the 2020 DigitalFUTURES*, pp. 57–67, 2021.
https://doi.org/10.1007/978-981-33-4400-6_6

means that buildings are demolished with crude attempts to recycle and subsequently no real scalable or economically viable pathways for reuse. By using mixed reality fabrication methodologies for the digitization and assembly of mass-customized scrap timber pieces, we aim to demonstrate how scrap material can be effectively re-used in construction without the need for advanced material manipulation and wastage.

This paper presents two approaches for the re-use of waste material.

The first project categorizes timber scrap material pieces of unique lengths but with similar cross section profiles that are used as a best fit for a designed timber aggregation. This research is explored in a workshop for the AA Visiting School Melbourne 2019 & Design Studio at RMIT University 2019 with the intention to reduce the need to machine odd shaped offcut materials through the construction of a timber assembly of more than 600 uniquely shaped pieces.

The second project, initially set up as a workshop for the 2020 Melbourne Design Week, extends the logic further to develop a workflow that eliminates any material wastage. It also expands the usability of differentially sized timber from the similar elements used in project 1. A dynamic digital model is updated in response to the input of each timber part's dimensions. This digital model updates and adapts to the kit of parts fed into it. In addition to developing a method for a developable digital model, a process for real like geo located inventory of the scanned directs assembly workers to the location of each part to be retrieved. 'Digital twin' models are typically developed as a digitised model first, or computed from large data sets. The aim of this paper is to present a simpler method for digitisation and digital model making.

3 Method

Both of the following projects and methodologies discussed here make use of the Fologram plugin for Grasshopper (Jahn et al. 2020) which enables the augmented visualisation of geometry via either smart phones or AR headsets.

3.1 Method 1 | Mass Customisation and Working to a Fixed Digital Model

3.1.1 Recycled Offcut Timber Scrap Material

This project developed a methodology of working with offcuts of scrap timber that were machined to fit to a relatively fixed digital model. As discussed by Jahn et al., the digital model acts as a template to guide material location. The timber offcuts were sourced from a local furniture company, with no repeatability in their length, width or height from piece to piece. A variety of breeds of timber were used in the project, ranging from Tasmanian Oak, Ash and Beech. The timber pieces were categorized into groups based on self-similar cross section profiles, but with significantly varying lengths measuring as short as 50 mm and as long as 2000 mm. Profiles with cross sections between 20–30 mm in both dimensions were selected for mark-up in augmented reality. These cross section sizes were chosen for a visual density of the timber in the design, as well as for self-similar surface areas for face on face joints (Fig. 1).

Fig. 1. A variety of timber offcuts categorized by their cross section profiles

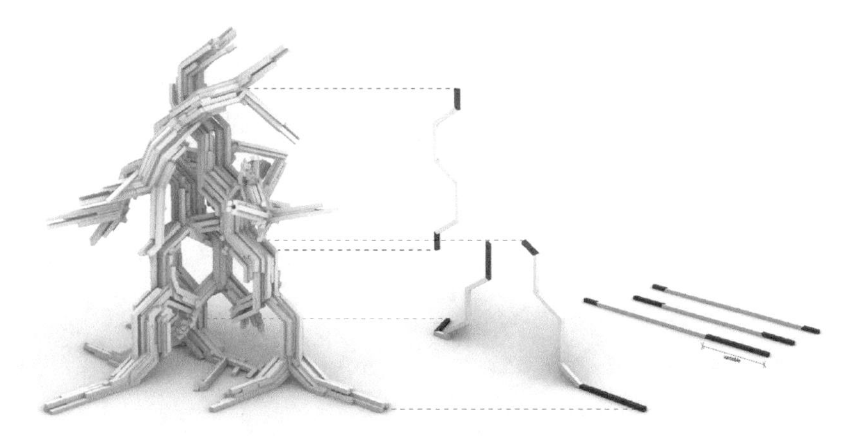

Fig. 2. L-System Field aggregation with dynamically nested assemblies for custom timber lengths

3.2 Mass Customized Aggregation Geometry

A simple parametric model was created from a rule based L-System aggregation. A customized field system was simulated using the L-System as a guide to create variation in the parts of the aggregation. These field lines were broken down into smaller discrete elements to approximate the locations of the offcut scrap timber. These design decisions were made to transform a typically standardized design language of an L-System into an aggregation of mass customized adaptable parts. Every angle in the aggregation was standardized to 30° to reduce material wastage in the cutting process. A standardized angle meant that each piece could be dynamically nested to one another, and then fit to the odd shaped timber sizes. Because these tessellations would never fit the timber lengths perfectly, a customizable 'End piece' (shown in red in Fig. 2) was designed within the aggregation to be variable in length to eliminate timber offcut wastage in the marking of the timber lengths. These end pieces affected the final outcome, so rather than the final construction representing a physical twin of the digital, the process led to an adaptable outcome that changed during the fabrication process.

3.3 Holographic Part Nesting

Rather than writing a complex algorithm to nest the timber pieces in each odd length, assistants to the project were able to use the Microsoft HoloLens to project holographic cutting templates and dynamically fit pieces inside each offcut timber part. Assistants were able to live stream each part from the aggregation as an overlay on each timber offcut through Fologram. The holographic overlay provided a wireframe of each part, the part number, information about the location of each cutting angle, the length of each edge on the part, and a sanity check for each angle to ensure the digital model adhered to a standardized angle of 30°. Assistants to the project were able to accurately measure and mark the timber scrap lengths against this holographic guide of each part in the digital model without the use of any drawings (see Fig. 3). Once a part had been marked, the next part would be aligned at the cutting location of the previous mark to tessellate and nest the pieces together. Participants worked from the inside (middle) of each offcut scrap and outward, so the end pieces could be variable in length. This process allowed for a dynamic nesting of parts, so rather than relying on a complex nesting algorithm, assistants were able to use their own intuition based on the holographic guides in front of them to decide which pieces would best fit along each timber length.

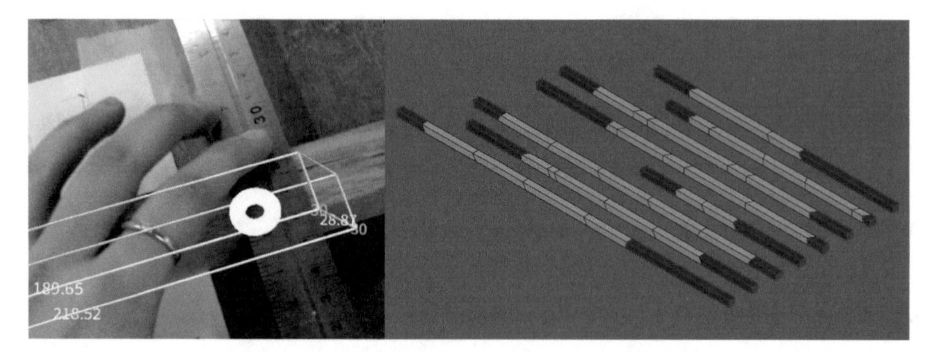

Fig. 3. Parts were dynamically nested within offcut scraps using a holographic guide and user intuition

3.4 Mixed Reality Interface

An interactive mixed reality interface was designed by assistants to change marking instructions for each timber offcut piece directly inside the Microsoft HoloLens (see Fig. 4). This meant changes to the holograms being displayed could be adjusted without the need for direct interaction with the Rhino & Grasshopper interface. The interface contained a series of buttons that allowed assistants to cycle through each piece in the aggregation, nest the hologram to their timber length and rotate each holographic guide to align with previous pieces to mark each part end for cutting. The interface also included different shading options for viewing the hologram. Wireframe holograms were found to be more accurate to use for markings than shaded holograms, as the edges were clearer and easier to align. As each piece was overlayed on each offcut scrap, assistants could toggle on an option that would visualize where a particular piece was located in the overall aggregation.

Fig. 4. A holographic interface was used to update aggregation parts for a dynamic nesting in the offcut scraps

3.5 Jointing

Once an offcut piece had been marked and nested, the piece was cut on a drop saw set to a standardized angle of 30°. The standardization of the joint meant each piece could be cut quickly and efficiently. Approximately 2.5 mm length was lost in each part during the cutting process, so this length was adjusted for in the holographic overlays in the marking process to allow for accurate fabrication of each part. Two simple methods were used as joint systems in the assembly: a standardized domino joint system commonly found in furniture making and nail guns into planar facing pieces. A specialized jig (see Fig. 5) was setup alongside a domino machine to accurately and rapidly fabricate the joints for each part in the aggregation.

3.6 Fabrication and Fixing Methods

The overall aggregation was broken down into a series of chunk studies to enable multiple users to assemble the final timber aggregation simultaneously. Assistants made use of their interactive interface to control the number of parts they saw in each chunk at one time. This interface also contained information about the part numbers, so users could clearly see which part belonged in which location, and which parts it was connected to. Once each chunk was assembled, assistants used an overall holographic model to piece together each part in the aggregation. Offcut scrap pieces not used in the marking and

Fig. 5. A standardized timber domino system was used to fix the parts of the aggregation together

cutting process were used to stitch and brace these chunks together to enhance structural stability of the aggregation (Fig. 6).

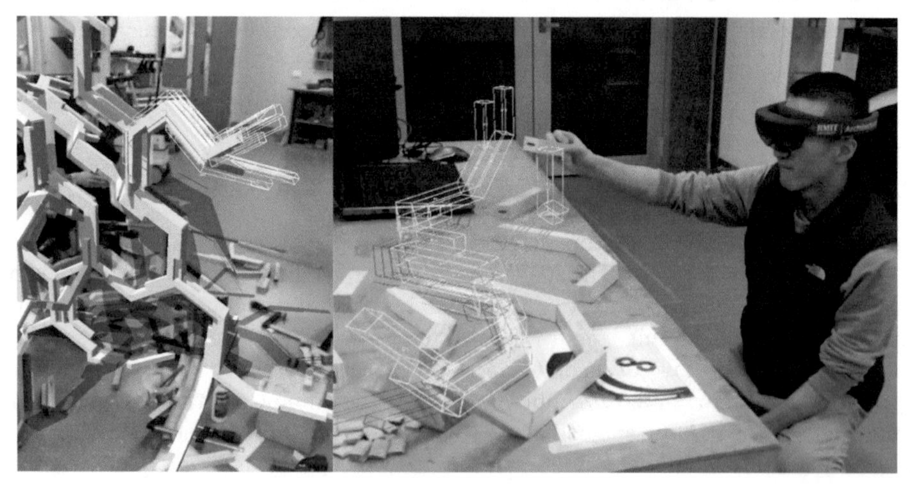

Fig. 6. The aggregation was assembles in chunks, and then together using a series of holographic guides

3.7 Method 2 | Working to a Flexible Digital Model

This second project focuses using non post processed recycled timber material and the development a flexible digital design model with the capacity to update in response to the digitised material inventory. While timber construction materials are to some degree standardised, one the challenges to using recycled materials is the variation in overall length and onsite modification made for during initial use phase.

3.7.1 The Digital Inventory

High resolution 3D scanning is often used to create 'digital twins' of physical objects that can then be used and referenced within 3D models. These 3D scans however are complex setups (especially for part scanning), and result in workflows that require noise removal and digital post processing workflows to produce clean, accurate models (Stojanovic et al. 2018). Despite being odd shapes and sizes, each offcut timber scrap as rectangular in its shape, with a defined length, width and height. Rather than using 3D scanning (high digital cost) or measuring with a tape (high manual cost), Aruco markers were used in conjunction with Fologram's marker tracking tools to create digital twins of each of these offcut timber scraps. Participants would align a timber scrap at a holographic XY plane (set-out in physical space by a QR Origin code), and align an Aruco marker at the opposite corner diagonally, and tap once inside the HoloLens (see Fig. 7 and 8). This marker would send a location plane to Rhino & Grasshopper and create a 2-point rectangular box of the dimensions of the timber scrap, thus creating an accurate digital twin of the offcut pieces with no post processing or manual measurement. The low-resolution digitised geometry becomes one of the library of parts that then informs the design.

Fig. 7. Individual Aruco markers can be placed on top of each timber offcut and aligned to a QR Origin code

3.7.2 Updating the Digital Design

To begin the process, a base design is modelled which is updated and recreated live as each of the pieces are scanned. In this case, a simple undulating vertical wall was the design object and contoured into a series of curves that perform as guidelines which provide the preliminary alignment for the digital parts to align with. Each of the curves is assessed on its curvature and length against the inventory of digital lengths. Genetic solvers within Grasshopper (Mirjalili 2018) were used to establish a best fit of each of the timber lengths to sectional curves through the wall. This technique avoids having to cut or additionally process the timber lengths, further increasing its efficient reuse (Fig. 9).

The genetic solver then breaks down the contoured curves from the wall and tries to best fit the length of timbers. This capacity of the model to absorb differentially

Fig. 8. Aruco marker tracking in Fologram & a holographic origin point created 2-point boxes over the timber parts

Fig. 9. Designed wall (left) contoured timber length guidelines (right)

dimensioned timber lengths encourages the emergence of highly textured 3D surfaces (Fig. 10).

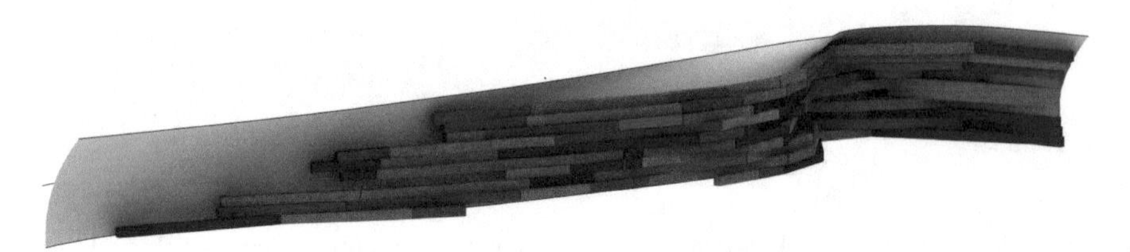

Fig. 10. Timber Lengths fitting to the original design

Some designs will be more efficient in how they use the timber lengths. As shown below in Fig. 8. The image on the left showing a shorted wall with greater curvature within the curve will not easily fit many of the pieces. In contrast, a longer less curvaceous wall will be more conducive with using more of the lengths of timber without the need for further cutting (Fig. 11).

3.8 Results

The method used for project 1 allowed for the use of uniquely shaped/mass customized timber pieces to assemble an aggregation made entirely of uniquely sized pieces. The nesting AR marking method reduced the wastage associated with each offcut timber piece. The end pieces meant participants could adjust their nesting dynamically to reduce offcut waste, or eliminate it entirely. Grouping the timber waste by cross section profiles

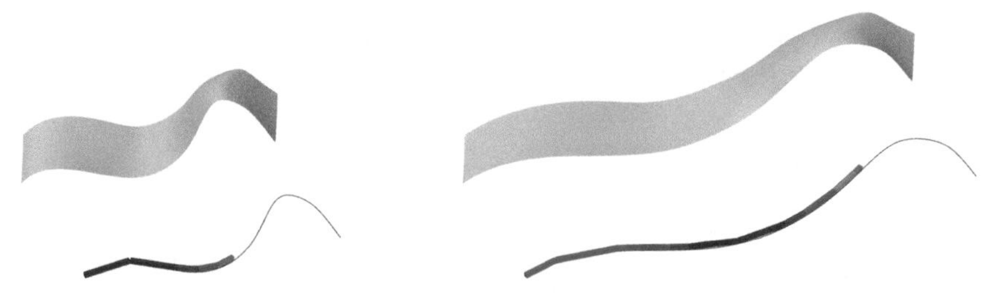

Fig. 11. Adjusting the digital model to increase efficient use of timber lengths

enabled the pieces to be located in the aggregation with a best fit method, which enabled the use of unique scrap material.

This method, however, came with complexities and problems. Many of the pieces did not have the required cross profile dimensions and were not used in the project. Some of the joints were poor and did not fit perfectly in the aggregation. The material still required machine manipulation through the cutting process needed to reach the aggregation piece lengths after marking, and still resulted in some (minor) wastage despite the nesting. This is compounded by the rigidness of the aggregation which did not respond to material input.

Project 2's method attempts to improve some of these processes of project 1 by creating an aggregation responsive to the material input. The digitization of each material scrap helped create a clear digital record of the pieces used in the aggregation

The capacity to update and experiment with the changing digital model meant that pieces could be used without any cutting or manipulation, which essentially removes all material wastage in the process and makes each scrap piece useable regardless of dimensions and cross sections.

4 Discussion

Providing a methodology to digitise and inventory land fill bound construction waste material has the potential to increase their reuse and recycling. Both of the projects discussed here demonstrate small non-structural objects designed and built from timber offcuts. The first project presents a more customised approach where augmented technology enables one to efficiently fit and cut the required parts to a static pre-designed object. The designed joint angles of the L system enabled nesting of the similar parts in an efficient manner and reduced the degree of cutting. This technique resulted in a monolithic well finished structure. In contrast to this, the second project had the capacity to absorb differentially sized and finished timber lengths creating a more textured outcome as the number of parts increase. The second project extended upon the first in establishing techniques for further reductions in material processing and inherent limitations of construction something predetermined from a randomised selection of timber length. The second project also provided the designer with a flexible relationship to the design and the digitised inventory of parts as opposed to the more traditional method of design then document. This method also has an underlying shift in the design paradigm where designers work from a blank slate of materials, where the only limiting factor is generally cost or structural competency.

4.1 Future Development

The process outlined in this paper offer two approaches to working with providing value to construction waste material. In the further evolution of this proof of concept it is intended to address more fully methods of geo-locating and stacking timber. Stacking the timber in such a way would take into account the order in which each piece is required in the construction process and then also guide the builder to the pieces whereabouts on site. Further work intends to establish methods to tag and link additional information to each scanned piece such as material type, structural/non-structural pieces or knots and blemishes in the material as a few examples. There is also more research to be done in creating a workflow that organizes the digital inventory and highlights the currents pieces location within the assembly in the augmented environment.

References

Jahn, G., Newnham, C., van den Berg, N., Iraheta, M., Wells, J.: Holographic construction. In: Gengnagel, C., Baverel, O., Burry, J., Ramsgaard Thomsen, M., Weinzierl, S. (eds.) DMSB 2019, pp. 314–324. Springer, Cham (2020)

Kaur, M.J., Mishra, V.P., Maheshwari, P.: The convergence of digital twin, IoT, and machine learning: transforming data into action. In: Farsi, M., Daneshkhah, A., Hosseinian-Far, A., Jahankhani, H. (eds.) Digital Twin Technologies and Smart Cities. IT, pp. 3–17. Springer, Cham (2020)

Krausmann, F., Wiedenhofer, D., Lauk, C., Haas, W., Hiroki, T., Fishman, T., Miato, A., Schandl, H., Haberl, H.: Global socioeconomic material stocks rise 23 fold over the 29th century and require half on annual resource use. In: Proceedings of the National Academy of Sciences, vol. 114, no. 8, pp. 1880 (2017)

Mirjalili, S.Z., Mirjalili, S., Saremi, S., Faris, H., Aljarah, I.: Grasshopper optimization algorithm for multi-objective optimization problems. Appl. Intell. **48**(4), 805–820 (2017). https://doi.org/10.1007/s10489-017-1019-8

Mohammadi, N, Taylor, J.E.: Smart city digital twins. In: IEEE Symposium Series on Computational Intelligence (SSCI), Honolulu, HI, pp. 1–5 (2017). https://doi.org/10.1109/SSCI.2017.8285439

Shooshtarian, S., Wong, S., Khalfan, M., Maqsood, T., Yang, J.: Green construction and construction and demolition waste management in Australia. In: Proceedings of the 43rd Annual Australasian University Building Educators Association Conference (AUBEA 2019), Noosa, Australia, 6–8 November 2019, pp. 18-25 (2019)

Stafford-Smith, M., Griggs, D., Gaffney, O., Ullah, F., Reyers, B., Kanie, N., Stigson, B., Shrivastava, P., Leach, M., O'Connell, D.: Integration: the key to implementing the sustainable development goals. Sustain. Sci. **12**(6), 911–919 (2016). https://doi.org/10.1007/s11625-016-0383-3

Stojanovic, V., Trapp, M., Richter, R., Hagedorn, B., Döllner, J.: Towards the generation of digital twins for facility management based on 3D point clouds In: Gorse, C., Neilson, C. J. (eds.) Proceeding of the 34th Annual ARCOM Conference, 3–5 September 2018, Belfast, UK, pp. 270–279. Association of Researchers in Construction Management (2018)

Zou, P., Hardy, R., Yang, J.: Barriers to building and construction waste reduction, reuse and recycling: a case study of the Australian Capital Region. In: Panko, M., Kestle, L. (ed.) Proceedings of the Sustainability In Construction and Deconstruction Conference, Auckland, New Zealand, 15–17 July 2015, pp. 27–35

Growing Shapes with a Generalised Model from Neural Correlates of Visual Discrimination

Pierre Cutellic(✉)

Chair of Digital Architectonics, Institute of Technology in Architecture, Faculty of Architecture, ETH Zurich, ETH Hönggerberg, Building HIB, Floor E 15, Stefano-Franscini-Platz 1, 8093 Zurich, Switzerland
cutellic@arch.ethz.ch

Abstract. This paper focuses on the application of visual Event-Related Potentials (ERP) in better generalisations for design and architectural modelling. It makes use of previously built techniques and trained models on EEG signals of a singular individual and observes the robustness of advanced classification models to initiate the development of presentation and classification techniques for enriched visual environments by developing an iterative and generative design process of growing shapes. The pursued interest is to observe if visual ERP as correlates of visual discrimination can hold in structurally similar, but semantically different, experiments and support the discrimination of meaningful design solutions. Following bayesian terms, we will coin this endeavour a *Design Belief* and elaborate a method to explore and exploit such features decoded from human visual cognition.

Keywords: Design computing and cognition · Brain-Computer interface · Machine learning · Generative design · Neurodesign

1 Introduction

Well known Event-Related Potentials (ERP) from neuropsychology [1] are widely studied and documented for reproducibility, and can serve the role of evaluating acquisition, pre-processing and classification methods. New applications seeking to involve known paradigms mean that new experiments need to be designed with these precedents in mind in order to compare meaningful results. In order to dissociate the question of acquiring, preprocessing and successfully decoding neural correlates of cognitive processes from their applications for CAAD purposes, one should refer to current EEG signal challenges and transferability of learned patterns across modalities [2].

Aims. While state-of-the-art research in cognitive science is actively dealing with that matter [3, 4], the present research focuses on the application of such potentials for better generalisations in future technologies for design and architecture. It engages with adapting known and generalised methods of acquisition, preprocessing, presentation, classification and exploitation from a P300 Visual Speller [5], for visual environments of increasing richness in information as commonly found in CAAD modelling interfaces.

P. F. Yuan et al. (Eds.): CDRF 2020, *Proceedings of the 2020 DigitalFUTURES*, pp. 68–78, 2021.
https://doi.org/10.1007/978-981-33-4400-6_7

It is known that, based on informational Bayesian models [6, 7], visual discrimination may occur in complex visual environments and their relevance for decision making rely on the degree of visual experience an individual may hold to construct prior beliefs upon which to infer [8, 9]. We will make use of previously built generalised techniques and trained models on a singular individual EEG signals and observe the robustness of advanced classification models to initiate the development of presentation and classification techniques for enriched visual environments by developing an iterative and generative design process of growing shapes. What is of interest is to observe if visual ERP as correlates of visual discrimination can hold in structurally similar, but semantically different, experiments and support the discrimination of meaningful design solutions. Following bayesian terms, we will coin this endeavour a Design Belief and elaborate a method to explore and exploit such features decoded from human visual cognition.

Significance. This research focuses on a generalisation and application of predefined Rapid Serial Visual Presentation of an Oddball Task (RSVP-OP) techniques and pre-trained classifiers to assess visual ERP as neural correlates of what we previously defined as design beliefs. Its goal is to advance research methods on related CAAD modelling applications.

2 Methods

The hereafter described methods are divided into sections concerning the necessary design of data flows from the generation of visual stimuli to the acquisition of EEG signals and their analysis to finally contribute in a generative process of shapes. While using the RSVP-OP as a basis, we will develop further on the generalisation of visual stimuli and their tokenisation, presentation for human visual cognition, and to which will be correlated acquired and processed aggregated EEG signals from a single person.

Tokenisation. The adapted visual stimuli use 3D metaballs rendered by a marching cubes algorithm [10, 11] in order to provide a generic and smooth visual flow in the continuous variation and presentation of generated shapes by the rendering of implicit functions of isosurface. Each flashing epoch, previously showing a row or a column in the reference case of the visual speller, is replaced by the uniform random position of a new metaball instance in spherical coordinates (Fig. 1).

The center of the spherical coordinates being either the origin of the rendered scene, or the center of one of the generated metaballs, if at least two already exist. In the case of none existing yet, a first instance will be placed at the origin for a second one to be generated from. Once the scene contains at least two instances of a metaball, the center point to generate a new one will be selected in a similar random fashion and produce the previously described relative coordinates for the new instance to be added for the rendering of the isosurface (Fig. 2).

As a result, each new metaball instance P is parametrised with its coordinates xyz, and two parameters of field strength St and substract Su related to the isosurface calculations. Ideally, the radius R of the sphere to be rendered as a metaball is $R = (St/Su)^{0.5}$, such that an instance of P can be parametrised as

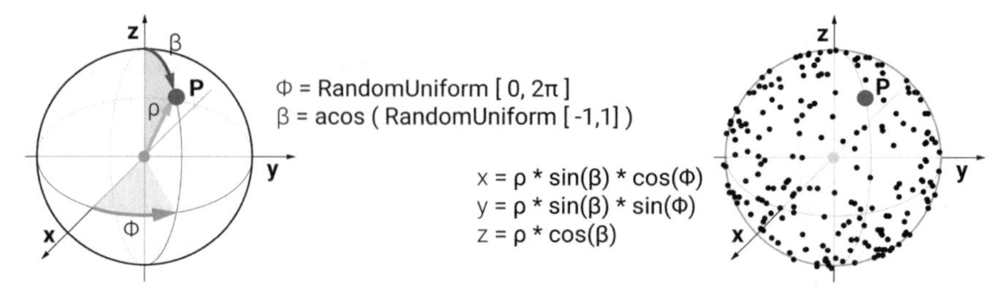

Fig. 1. Random uniform spherical distribution of solutions for new possible coordinates of a metaball instance being later presented as a new stimulus. From a given center point *O* (grey) of a sphere of radius *p* (green), one generates the new coordinates *x, y, z,* of a point *P* (pink) and relative to *O*.

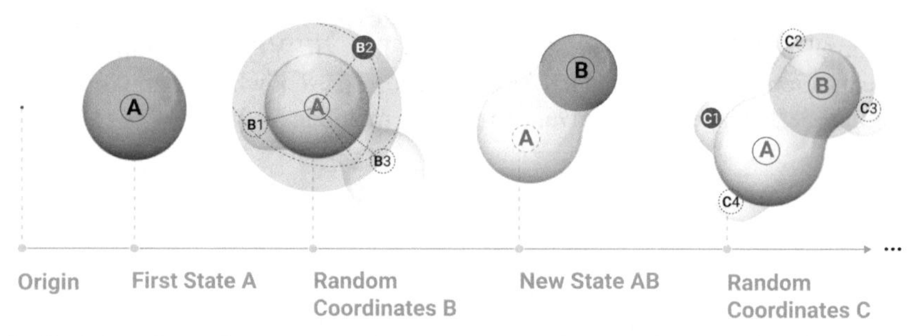

Fig. 2. Sequence for updating the rendered isosurface. From Left to Right: a first instance is placed at the *Origin* of the scene: *First State A*. New coordinates (i.e. *B1*, *B2*, *B3*, …, as examples of possible generations) are generated, given the center of *A* and its radius: *Random Coordinates B*. From *B2*, a new instance *B* is added to the scene: *New State AB*. Similalrly, new coordinates are generated (i.e. *C1*, *C2*, *C3*, *C4*, …, as examples) from a randomly selected preexisting instance *A* or *B: Random Coordinates C.*

$P : (Px, Py, Pz, Pst, Psu)$ and for an entire token T constituted of nP such that $T : \left[(P_0x, P_0y, P_0z, P_0st, P_0su), \ldots, (P_{n-1}x, P_{n-1}y, P_{n-1}z, P_{n-1}st, P_{n-1}su)\right]$. Eventually the final distance added to the coordinates of a new instance from the center of a previous one is equal to the radius of the later and the new resulting radius *R*. Each token can possibly have different distances between connected metaballs and each metaball can possibly have different radii (Fig. 3). One will consider these two configurations as two distinct classes *C1* (same distances and radii) and *C2* (random distances and radii).

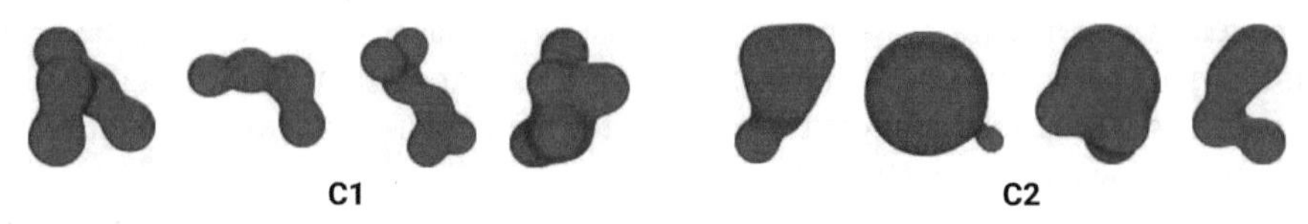

Fig. 3. Left: class *C1* samples with same radii and distances. Right: class *C2* samples with random radii and distances.

In addition, three main kind of shaders are applied to each tokens: *S1* - a plain white shader with no depth or shadow, *S2* - a Phong material shader with specularity and reflectance, *S3* - a black and white dot-patterned shader with no depth or shadow but applied on the uv coordinates of the shape (Fig. 4). These three shaders allow for three different kinds of visual distinction of the complex geometry, depth, silhouette and curvature being rendered. They all relate to a certain kind of basic information sent to the visual system for early processing and known as information of shape from texture and motion [12, 13]. The three applied shaders will be considered as three unrelated categories *Q1*, *Q2*, *Q3* for comparison of results, as providing different degrees of shape information.

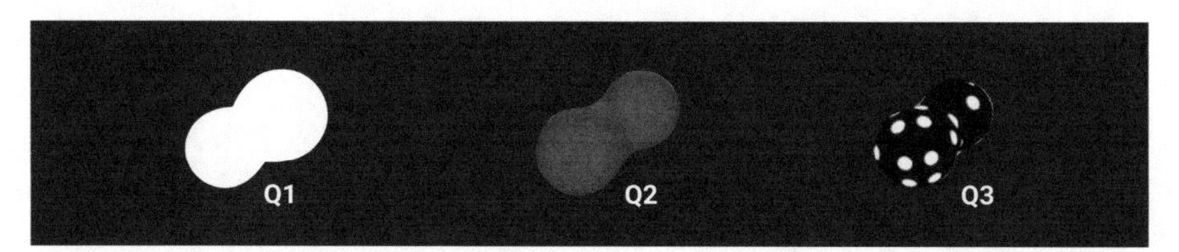

Fig. 4. Rendering of the three distinct shaders as visual categories. A black background is used during the recording sessions.

Visual Presentation. From the previous study of visual spelling with an ERP-BCI [5], the Rapid Serial Visual Presentation of the Oddball Paradigm Task (RSVP-OP) is preserved with a similar time and tokenisation structure. Each presentation contains a sequence of 12 tokens shuffled and shown 15 times so that each token would be viewed 15 times in a random order of appearance. An initial period, to ease-in the user's attention into the visual scene and show how the tokenisation will be presented, is set to 2.5 s. Similarly, a minimum of 2.5 s of a break period is set between presentation periods to avoid rapid fatigue and disengagement. Since the temporal method used for classification is *offline* learning and the next presentation period is dependent to the processing and the returned discriminated token by a pre-trained classifier (i.e. a new tokenisation can happen only if there is a new state returned), the break period is also extended until a value is returned (i.e. the index of one of the presented tokens or none in case of no discrimination found). Each token is presented on screen for a duration of 100 ms and followed by a blank screen for a duration of 75 ms while the standard refresh rate of the visual presentation is approximately 60 Hz. Each recording session has been kept under a maximum time of 18 s (excluding the break periods) and 6 discriminated tokens forming the overall shape. The main adaptation from the generalised RSVP-OP consists in augmenting its temporal structure (Fig. 5).

While the RSVP-OP occurs, data is acquired accordingly. And while the data is being processed during break periods, the current state of the shape is kept visible until a new value is returned and the new state of the shape is shown for a second before starting the new RSVP-OP and in order to generate the new tokens. Additionally, and since the complexity of visual scenes presented is more important than in the case of a word speller, the RSVP is adapted at every token flashed so that its silhouette appearing

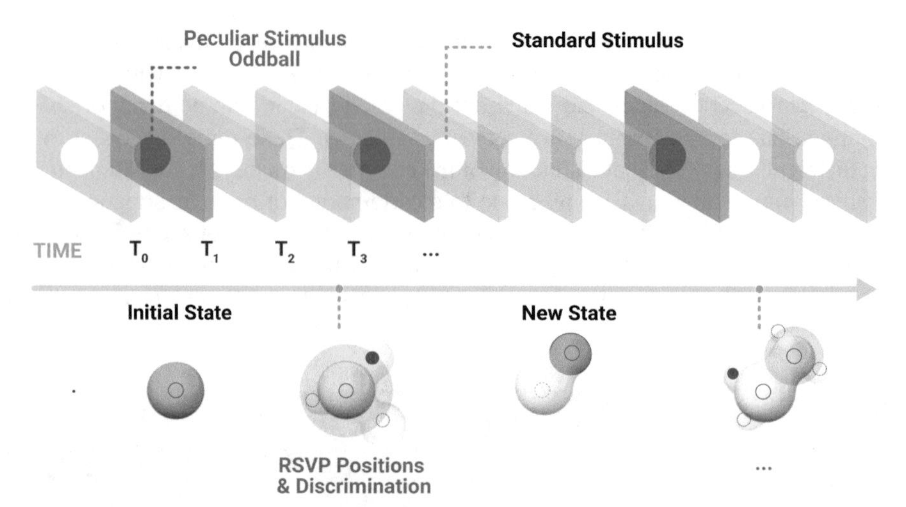

Fig. 5. Adaptation of the typical RSVP-OP task. Between stimulus presentation, another presentation period is introduced to show the new state of the overall shape formed by discriminated tokens and from which new tokens will be generated for stimulus presentation. This new period also plays the role of helping to reduce prolonged cognitive workload and await for discrimination and generation of new stimulus for the next RSVP-OP.

on screen is maximised. This effect is achieved by measuring the angle between: *a* - the line formed by the centroid of the shape and and the center of the presented token; *b* - the *X-axis* of the scene always horizontal and parallel to the camera *X-axis*. A rotation is then applied to the shape as in Fig. 6.

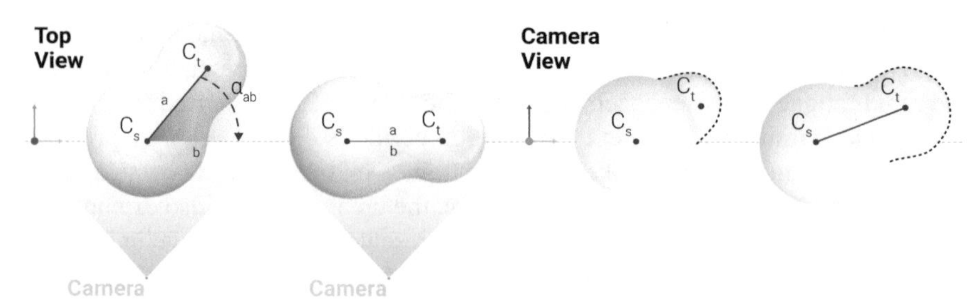

Fig. 6. Rotation applied to the shape around its X-axis during each token presentation of the RSVP-OP period in order to increase each presented token's visibility. Cs is the centroid of the shape in its current state and Ct is the centroid of the token being presented as a stimuli. Both are forming a line a from which is measured its angle ∂ab formed with a vector parallel to the camera's Y-axis and passing by Cs, noted b. The rotation is applied on the whole shape (including the token) around the vector parallel to the camera's X-axis and passing by Cs, so that $\partial ab = 0°$.

This method provides a new view angle of the shape at each RSVP and allows for novel information of the overall shape from motion and texture [12, 13], while the visibility of the token is emphasised to ease the discrimination. Additionally, a random rotation is constantly applied during break periods to show more information of the overall shape after and before RSVP. A re-centering and re-scaling of the camera occurs before every new RSVP to ensure that the centroid of the shape remains at the center of the scene and the whole shape is being contained and visible on the screen.

Signal Acquisition and Processing. The EEG data is acquired through a *Lab Streaming Layer protocol* [14], synchronised, and by a 16 channels *OpenBCI [15]* (i.e. *Daisy + Cyton* configuration) at a sampling frequency of 125 Hz with electrodes placement at *FC5, FC6, C3, Cz, C4, CP1, CP2, P3, Pz, P4, PO3, POz, PO4, O1, O2* and *Oz* positions of the *Modified Combinatorial Nomenclature* (MCN) of the *International 1020* placement [16]. Signals are digitised at their device's sampling frequency and then filtered with an eight-order bandpass filter with low and high cut- off frequencies of 0.1 and 20 Hz, to finally build epochs −0.100 to 0.700 ms onset visual stimulus and downsample each signal to the high pass limit since most ERP components can be found below 20 Hz [17]. No particular artifact rejection method is applied except for amplitudes superior at 75 μv to reject outliers from muscular movements. This allows for a minimisation of data points to process within the range of ERP detection.

For similar reasons explained in previous experiments [5] concerning challenging EEG signals features for stable classification (mainly signal-to-noise ratio and non-stationarity), the capacity for a given classifier to learn across different modes (different sessions, experiments and users) without calibration is a question of research on *Transfer Learning* itself [2] and can be approached by either *Information Geometry* [3] or *Deep Learning* [4] methods. Given the low amount of data and the user-based approach of the experiment, information geometry classifiers have been chosen and trained for a single person on multiple recording sessions of a P300 word speller, so that the assumed learning across-modalities would concern only the cross-experiments mode (i.e. From spelling words to growing shapes). The pre-trained classifier is a *riemannian* classification pipeline constituted of ERP covariance matrices and projection on the tangent space [18–20] with an AUC accuracy of 97.5% after 12 training sessions. Given previously mentioned EEG features and an increasing variance in the data when applying new experiments, the robustness of such method is evaluated by observing the difference of averaged discriminated samples recorded during the new experiment (Fig. 7). Though observed on less data amounts than for training, one can see that despite changes in the morphologies of signals and presence of noise, the classification accuracy across experiments for a single user can be maintained to a certain degree, although it may not provide for a continuous and fully robust adaptive classification across all mentioned modalities.

Shape Generation. During the developed RSVP-OP sessions and parameters, two types of data are recorded into a user's database: *a* - aggregated and processed time series used as input for the classifying pipeline, segmented by presentation period (i.e. one for each discriminated token). *b* - generated shapes directories, containing their *Q-C* shape labels (see section: *Tokenisation*), the mesh file and its associated material (Fig. 8.) computed by the programmed shaders (in **.obj* and **.mtl* formats), and a **.json* data file containing all parameters used to procedurally generate the given shape (Table 1.). The later is used for further understanding on the extent of the produced solution space

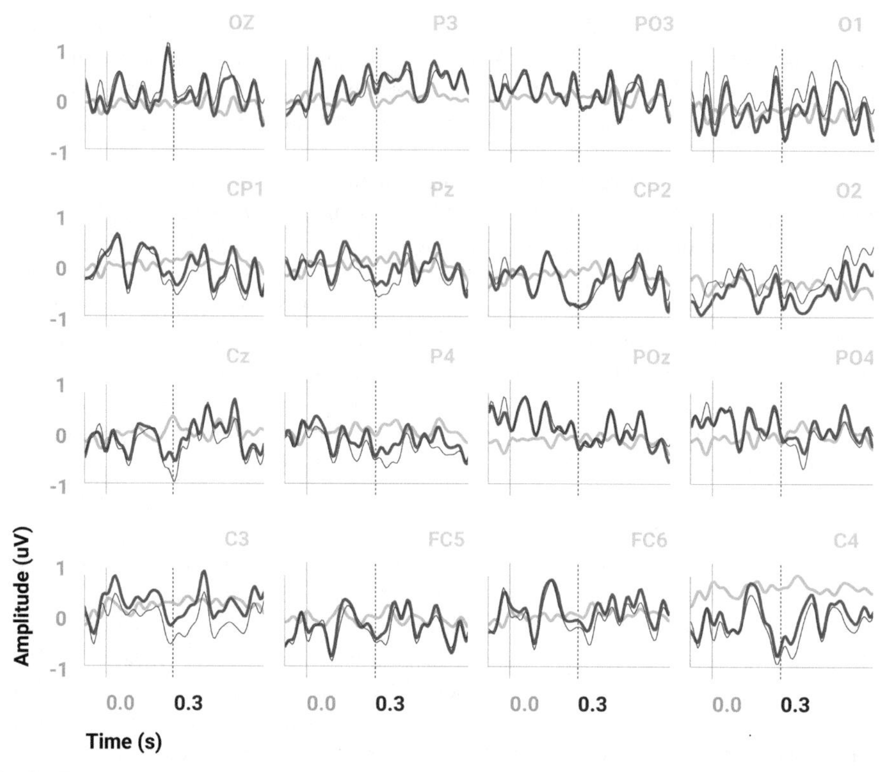

Fig. 7. ERP binary conditions of discriminated data with the shape generation experiment. Red plots show positive Target. Blue plots show negative ones. A difference waveform is plotted in dark blue.

and its features. Eventually, a similar method can be used to proceed from an inverse modelling fashion to generate such shapes given an adequate artificial generator.

3 Results

From all data files of *Q-C* shapes generated, a dimensionality reduction is applied from the initial 36 shape dimensions (6 instances X 6 params) to a 2d mapping using *T-SNE* [21] and *UMAP* [22] to evaluate the topology of the aggregated data and account for possible manifolds. In order to observe a differentiation between possibly random shape generations and otherwise meaningful ones, they are compared with randomly generated data using similar procedural methods and parameters for both *C1* and *C2* classes (Fig. 9).

Both T-SNE and UMAP methods show similar clusters and suggest that discriminated data correlate only in part with random data points. As some clusters appear outside the random ones in more compact topologies, they suggest a meaningful convergence for some generated shapes. Since visual ERP is clearly correlated with visual

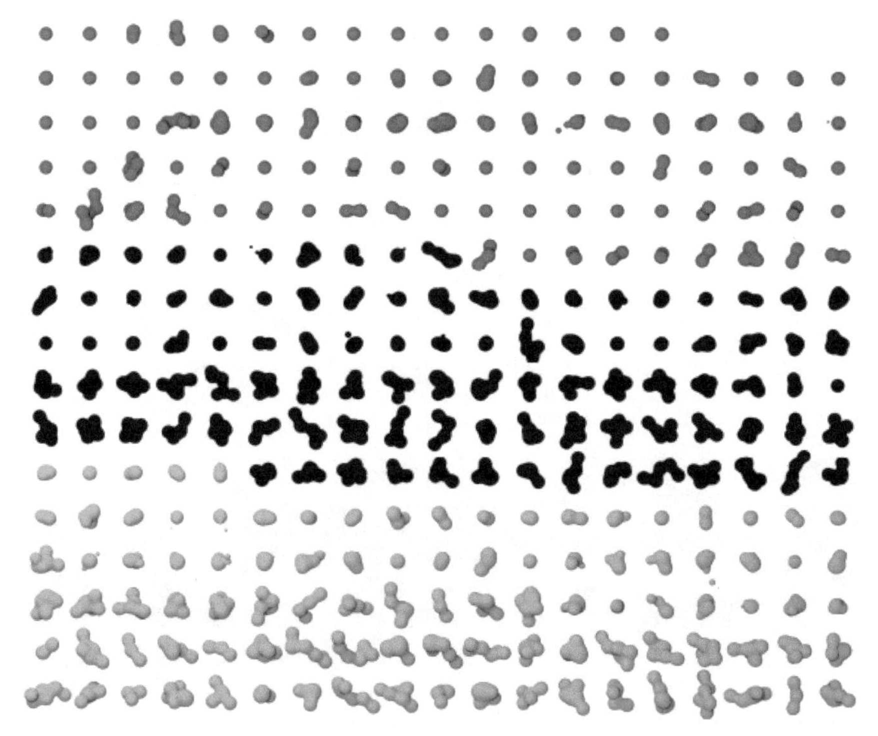

Fig. 8. Grid of all the generated shapes by *Q-C* order. Original shaders have been replaced here by colors for better overall visualisation. Bottom to top, Left to Right: *Q2* (pink) with 50 *C1* shapes and 50 *C2* shapes; similarly followed by *Q3* (black) and *Q1* (grey).

Table 1. A sample *.json file containing parameters *Px, Py, Pz, Pst* and *Psu* for each instance (0 to 5) necessary to procedurally generate its associated shape.

	Px	*Py*	*Pz*	*Pst*	*Psu*
0	0.5	0.5	0.5	0.61	15.6204993518133
1	0.556925825892438	0.495865511380644	0.645404113938775	0.61	15.6204993518133
2	0.454769577708112	0.495483133279051	0.763573292424071	0.61	15.6204993518133
3	0.511977763440335	0.644353703985817	0.663595549798839	0.61	15.6204993518133

attention [1], another index of engagement is added to help in visualising the relation of engagement with discriminated shapes (Fig. 10.). The index used for this is a commonly used *Beta to (Alpha + Theta)* index [23], where the mean relative band power of *Theta* (4–8 Hz), *Alpha* (8–12 Hz) and *Beta* (12–30 Hz) frequency bands are computed for each aggregated time series of a shape. Since the index *E* is computed on pre-processed EEG data which has been filtered and resampled to a maximum of 20 Hz, the *Beta* band is being cut by approx. 55%, a naive factor *k* is applied on the mean value of the *Beta* band such that k = 0.55 and: $index = (\beta + k\beta)/(\alpha + \theta)$.

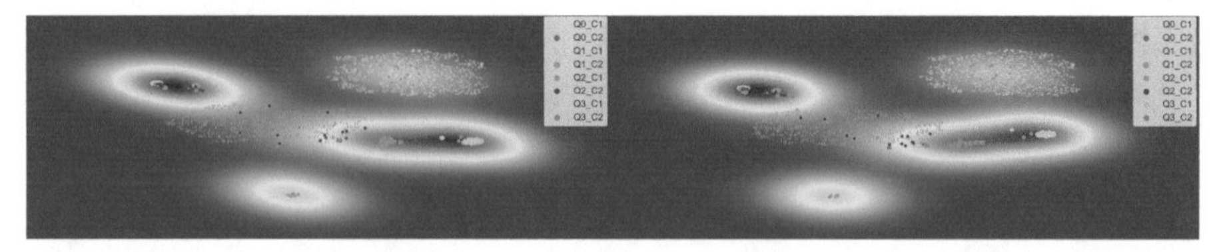

Fig. 9. T-SNE and UMAP 2D dimension reduction of the generated data. Both are run with several perplexity/proximity parameters (euclidean, 15 to 45 with step of 10) to observe the persistence of clusters topology across global/local structures. The map contains 20000 random data points and 300 discriminated ones. A gradient plot in the background shows the proximity of discriminated data points in clusters. Label *Q0* is a random generation with *C1* and *C2* parameters found in *Q1, Q2* and *Q3*. Sample 1–2: T-SNE (Left) and UMAP (Right) clusters with perplexity/proximity parameters 35. Similar results are found across all parameter sets.

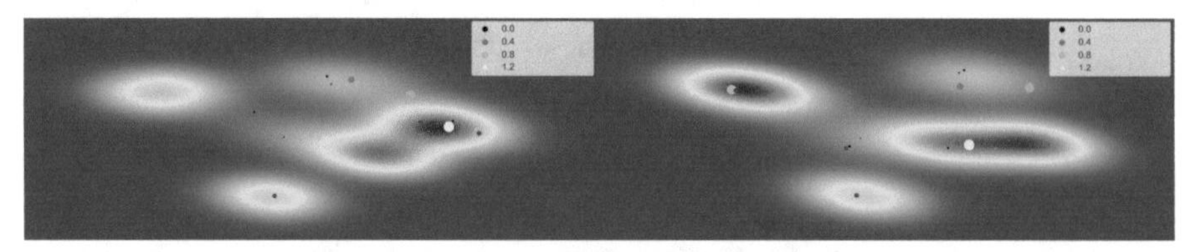

Fig. 10. Mapping of the Engagement index for each discriminated data points already projected on the 2-dimensional plane with T-SNE with perplexity parameter of 15, 25, 35 and 45. From Left to Right: sample plot with perplexity parameter of 15 and 35.

4 Conclusion

The mapping of engagement index on clustered discriminated data shows peaks of engagement both in specific clusters and random ones. It also shows that very few low peaks are present on the specific clusters. One can interpret such topology by summing that some meaningful clusters are formed but some data points outside of them might also be of interest and that such index would be helpful to adjust their meaningfulness. The robustness of generalising the acquisition and classification methods across experiments for a single user can be maintained to some extent and would greatly benefit from further adaptive research in stimulus presentation and transfer learning. We have engaged into modifying typical RSVP methods to the end of easing the rendering of complexified stimulus presentations towards design and architectural modelling purposes. Through the accumulation of generated shapes, we have shown that some meaningful clusters emerge to form what we can now call a Design Belief in the way they aggregate around regions in the latent space for certain design solutions and parameter ranges over time and based on typical informational bayesian prior beliefs. In addition, engagement indices of visual attention such as the one used in the present experiment can be purposed to value and ponder both formed design beliefs and episodic discrimination outside such regions but with high engagement index in order to notice other possible regions of interest. This should allow to further devise for a method to generate design solutions based on the discrimination of such design belief together with an exploitation/exploration ratio of the design space, in order to maintain variance over time in the generation of design solutions. Further experiments will develop this combined discriminative/generative method together with a better granularity of ERP classifications and stimulus presentations moving from the generation of shapes to the spatial articulation of parts for architectural modelling implementations.

References

1. Kutas, M., Kiang, M., Sweeney, K.: Potentials and paradigms: event-related brain potentials and neuropsychology. In: Faust, M. (ed.) The Handbook of the Neuropsychology of Language, pp. 543–564. Wiley, Oxford (2012)
2. Lotte, F., Bougrain, L., Cichocki, A., Clerc, M., Congedo, M., Rakotomamonjy, A., Yger, F.: A review of classification algorithms for EEG-based brain–computer interfaces: a 10 year update. J. Neural Eng. **15**(3), 031005 (2018)
3. Rodrigues, C., Luiz, P., Jutten, C., Congedo, M.: Riemannian procrustes analysis: transfer learning for brain-computer interfaces. IEEE Trans. Biomed. Eng. **66**(8), 2390–2401 (2018)
4. Tuleuov, A., Abibullaev, B.: Deep learning models for subject-independent ERP-based brain-computer interfaces. In: 9th International IEEE/EMBS on Neural Engineering, pp. 945–48 (2019)
5. Cutellic, P.: Towards encoding shape features with visual event-related potential based brain–computer interface for generative design. IJAC **17**(1), 88–102 (2019)
6. Bayes, T., Price, R.: Essai en vue de résoudre un problème de la doctrine des chances, vol. 18. Cahiers d'histoire et de philosophie des sciences. Paris (1763)
7. Pierce, J.R.: An Introduction to Information Theory: Symbols, Signals & Noise. 2nd, revised edn. Dover Publications, New York (1980)
8. Lindsay, P.H.: Human Information Processing: An Introduction to Psychology, 2nd edn. Academic Press, New York (1977)
9. Goldstein, E.: Bruce, and Thomson Learning (Firm). Sensation and Perception. Thomson Wadsworth, Belmont (2007)
10. Blinn, J.F.: A generalization of algebraic surface drawing. ACM Trans. Graph. **1**(3), 235–256 (1982)
11. Lorensen, W.E., Cline, H.E.: Marching cubes: a high resolution 3D surface construction algorithm. In: Proceedings of the 14th Annual Conference on Computer Graphics and Interactive Techniques, SIGGRAPH 1987, New York, pp. 163–169
12. Palmer, S.E.: Vision Science: Photons to Phenomenology 3rd printing. MIT Press, Cambridge (2002)
13. Stone, J.V.: Vision and Brain: How We Perceive the World. MIT Press, Cambridge (2012)
14. LSL is developed and hosted by The Swartz Center for Computation Neuroscience at UCSD San Diego. https://github.com/sccn/lsl_archived
15. OpenBCI specifications online
16. Sharbrough, F., Chatrian, G.-E., Lesser, R.P., Lüders, H., Nuwer, M., Picton, T.W.: American electroencephalographic society guidelines for standard electrode position nomenclature. J. Clin. Neurophysiol. **8**(2), 200 (1991)
17. Luck, S.J.: An Introduction to the Event-Related Potential Technique. A Bradford Book, 2nd edn. The MIT Press, Cambridge (2014)
18. Barachant, A., Congedo, M.: A Plug&Play P300 BCI Using Information Geometry. arXiv: 1409.0107 [Cs, Stat]. 30 August 2014
19. Congedo, M., Barachant, A., Bhatia, R.: Riemannian geometry for EEG-based BCI; a primer and a review. Brain-Comput. Inter. **4**(3), 155–174 (2017)

20. Barachant, A.: Python Package for Covariance Matrices Manipulation and Biosignal Classification with Application in BCI. see Alexandrebarachant/PyRiemann. Python (2019)
21. van der Maaten, L., Hinton, G.: Visualizing data using T-SNE. J. Mach. Learn. Res. **9**, 2579–2605 (2008)
22. McInnes, L., Healy, J.: UMAP: uniform manifold approximation and projection for dimension reduction. arXiv:1802.03426 [Cs, Stat]. 9 February 2018
23. Pope, A.T., Bogart, E.H., Bartolome, D.S.: Biocybernetic system evaluates indices of operator engagement in automated task. Biol. Psychol. **40**(1–2), 187–195 (1995)

Cyborgian Approach of Eco-interaction Design Based on Machine Intelligence and Embodied Experience

Guyi Yi[1(✉)] and Ilaria Di Carlo[2]

[1] Shanghai Tongji Urban Planning and Design Institute Co. Ltd., 1111 Zhongshangbeier Road, Shanghai, China
ucbqgyi@ucl.ac.uk

[2] The Bartlett School of Architecture, 22 Gordon Street, Bloomsbury, London, UK

Abstract. The proliferation of digital technology has swelled the amount of time people spent in cyberspace and weakened our sensibility of the physical world. Human beings in this digital era are already cyborgs as the smart devices have become an integral part of our life. Imagining a future where human totally give up mobile phones and embrace nature is neither realistic nor reasonable. What we should aim to explore is the opportunities and capabilities of digital technology in terms of fighting against its own negative effect - cyber addiction, and working as a catalyst that re-embeds human into outdoor world.

Cyborgian systems behave through embedded intelligence in the environment and discrete wearable devices for human. In this way, cyborgian approach enables designers to take advantages of digital technologies to achieve two objectives: one is to improve the quality of environment by enhancing our understanding of non-human creatures; the other is to encourage a proper level of human participation without disturbing eco-balance.

Finally, this paper proposed a cyborgian eco-interaction design model which combines top-down and bottom-up logics and is organized by the Internet of Things, so as to provide a possible solution to the concern that technologies are isolating human and nature.

Keywords: Cyborg · Embodied cognition · Embedded intelligence · Interactive environment · AR · User experience

1 Tracing Cyborgian Theory and Embodied Cognition

1.1 A Hybrid of Part Clock Part Swarm [10]

Since the concept of "cybernetics" was introduced by Norbert Wiener in the 1940s to deal with complex systems of communications and control in machines and animals [24], the word has been imparted the meaning of "cross-species communication". The term "Cyborgian" or "Cyborg", a combination of "Cybernetic" and "Organism", was coined in 1960 by NASA scientists [4]. Different from bio-robot or artificial intelligence,

P. F. Yuan et al. (Eds.): CDRF 2020, *Proceedings of the 2020 DigitalFUTURES*, pp. 79–90, 2021.
https://doi.org/10.1007/978-981-33-4400-6_8

"Cyborg" is a rejection of human-machine dualism by obscuring the rigid boundary and advocating the man-machine symbiosis.

By the end of 1980s, the concept of "Cyborg" has already widespread in science fictions such as *Ghost in Shell* and *Blade Runner*. The world of tomorrow predicted in these futurism works, with the highly hybrid cyber-organ relationship as an essential feature, is becoming today's reality.

Attempts and practices of cyborg individuals has been conducted in various areas, starting at a relatively basic and safe level - as the substitutes of lost or damaged body parts. Then, the continuous breakthrough of technology propelled the development and acceptance of "enhancement prosthetics" - for example, British artist Neil Harbisson has had a cyborg antenna implanted in his head that allows him to extend his perception of colors beyond the human visual spectrum [17]. However, implanted cyborg remains a controversial issue. Opponents concern that this technology would aggravate social polarization and impair social order and ethics.

In a broad sense, implant surgery is not necessary for becoming a cyborg. Everyone holding a smartphone is a cyborg. Because the external apparatus has become an integral part of us, as a cyber-extension of our organic corporeity. Embedded and external devices share the same purpose – to enhance the perception, communication, interconnection and control of everything (Fig. 1).

1.2 The Importance of the Presence and the Bodily Experience

Embodied cognition is a promising theory that has been developing rapidly since the "postcognitivism revolution" [3] in the middle of last century. Embodied cognition challenges traditional theories such as Connectionism and Computationalism, which hold the notion of Disembodiment and Mind–body Dualism [25]. It opens a new chapter of cognitive psychology with emphasis on the indispensability of human body in the process of cognition.

In 1945, phenomenological philosopher Maurice Merleau-Ponty claimed that "the body is our general medium for having a world" [14]. In 1979, James J. Gibson, who fathered the school of ecological psychology, also expressed similar idea that we acquire the information in the environment through our active body [8]. It is widely acknowledged that emotions influence behaviours, while according to the theory of embodied cognition, vice versa (Fig. 2).

"A designer and a cognitive scientist seem like an unlikely pair ... both are trying to decode how humans interact with the world" [13]. Relational art/aesthetics is a mode or tendency in fine art practice with embodied cognition as one of its theoretical foundations. It is defined as a set of artistic practices which depart from the concerns of human relations and social context, instead of independent and private spaces [2]. Relational art values the encounter between an audience and an artwork, and the encounters between people.

On the basis of these two theory, here comes a question: since the corporeity is the main source of knowing the world and the others, when our body is augmented by advancing cyber technologies, how would human cognitive abilities and experiences be improved with the help of machine intelligence, and how would this be beneficial for the construction of a user-experience-oriented interactive environment? (Fig. 3).

Fig. 1. Cyborg human. Author's own work.

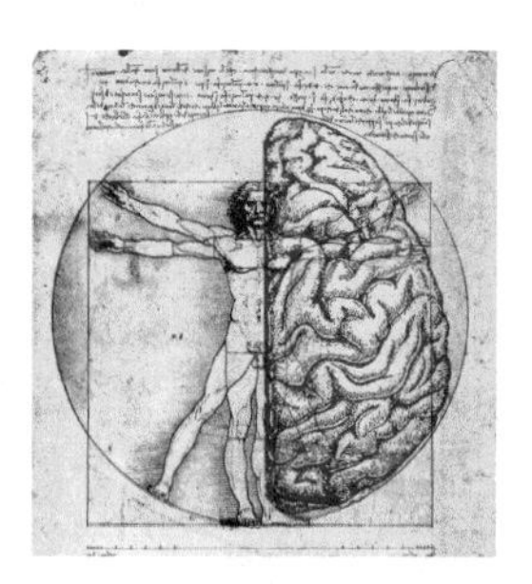

Fig. 2. Embodied mind. Author unknown.

Fig. 3. Relational art - geological memory. Image credit: teamLab

2 How Cyborgian Approach Activates Plants?

To encourage human-nature interaction, one way is to make plants more sensible, intelligent and interactive. This chapter is going to research on how cyborgian approach influences the procedures of plant's behaviour, namely sense (input), think (algorithm and feedback) and actuate (output).

2.1 How They Sense

We underestimated plants' sensing abilities and their feelings for a long time. Plants "are just very slow animals" [20]. In addition to the five senses that animals have, scientists believe plants have at least 15 other ways to feel the world. For example, they can perceive and calculate gravity, electromagnetic fields, moisture, and chemical substances [9]. Plant's sensing mechanisms are talented and exquisite enough, therefore, the question is how to make use of their powerful but implicit sensing capability for interactive functions. Focus on two aspects: the input and output of sensing process.

The former means to make them more sensitive to human behavioural inputs which is difficult for plants to understand without a cyborgian medium. It works in the way of exaggerating voice or motion signals or interpreting them into another type of signal that plants are more sensitive to, like electronic signals (Fig. 4).

The later means to externalize and visualize plants' internal biologic process. With the aid of cyber-devices, plants are augmented as sensors by visualizing their invisible sensing process. In the case of Cyborg Botany by MIT media lab, the electronics is transferred into plants. Internal wires are connected to sampling instrumentations, turning a plant into an inconspicuous sensor to detect motion and more [18]. These cyborg plant sensors are applied into many interesting scenarios, for example when a motion-sensitive rosebush senses that a cat run out of door, it will send an alert to its owner's computer.

2.2 How They Think

Recent research on plant intelligence has proved that plants are capable of certain levels of intelligent behaviours including communication, learning and memory.

Plants, trees in particular, communicate with each other relying on underground fungi networks [11]. Cyborgian plants, gifted by the cyber characteristics, become nodes in a more efficient communication network compared to biologic ones. These two forms of networks work cooperatively, sharing a large dataset of information, thus enhancing the adaptiveness of plants. Cyborgian plants are resonating and able to make predictions because they are in a network where they can talk to neighbours and understand what is happening at the other end of network. By simulating the process in the backend they can prepare themselves for the upcoming changes.

There's another project by MIT media lab - a pair of couple plants that can feel and respond to each other even when they're far apart. When one plant is gently poked, the other wiggles [18]. Cyber network enables them to "say hello" crossing distance. In the Resonating Forest at Jewel Changi Airport by teamLab, when someone passing by a tree, light color changes and a new tone resonates out. These information is transmitted to nearby plants, spreading continuously as if the plants are discussing the presences and locations of people.

As for learning and memory, it is proved by a team from Western Australia University that plants can build up classical conditioning through training and form memories through experiences [7]. The cyborgian approach imparts plants machine intelligence, featuring incredible capability of data processing and memory storage. By mass-analysing human behaviours as training inputs, cyborg plants learn about users' habits and preferences so as to better indulge them in nature.

2.3 How They Actuate

Plants are always sensing, thinking and responding to our voice and movements but in an extremely subtle way. Since we understand their sensing and thinking mechanisms, now we are able to guide and supervise their actuation by controlling what they sense and how they think.

By applying external stimulus, like changing light intensity and direction, human is able to guide the growth and movement of plants taking advantages of phototaxis or other biologic tropism. But the growth of plants is too inconspicuous to be noticed in a short time, therefore, their responds are usually transformed into other forms and are visualized through external devices which can be seen as their extended cyber-body. For example, MIT Media Lab created a robotic plant called Elowan. When there's light nearby, electronic signals within the leaves are detected by embedded wires, and the wheels of the robotic planter are triggered to move autonomously toward the light [19]. In the case of Breeze, an ambient robot inhabits the body of a Japanese maple, allowing her to sense and reach out to nearby people [5]. One difference between the two cases is: the bio-corporeity of Elowan itself acts as a sensor while in the project Breeze, there are embedded sensors around (Fig. 5). Anyway, all roads lead to the same purpose: to make human behaviours sensible for plants while make their responds visible for us.

In summary, cyborgian approach would be instrumental in the construction of a more interactive environment because:

1. Rather than augmenting the way plants sense, it augments the way how we understand and benefit from their gifted sensing abilities for interactive functions.

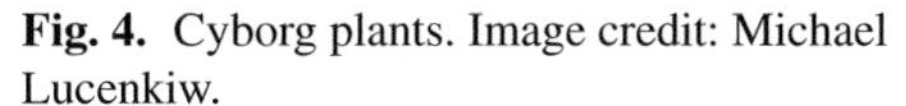

Fig. 4. Cyborg plants. Image credit: Michael Lucenkiw.

Fig. 5. Elowan - augmented plants as sensors, Displays and actuators. Image credit: MIT Media Lab.

2. On the basis of big data flowing in cyber-network, it augments plant intelligence including communication, learning and memory performance, thus improving capability of cross-species interaction.
3. It guides the growth and movement of plants by controlling input stimulus and even guides the process of evolution to a certain extent, towards the objective of the most appropriate level of interactivity.

3 How Cyborgian Approach Encourages Human Participation?

To encourage human-nature interaction, the other way is to improve bodily interaction experience. This chapter is trying to explore how would cyborgian system play a valuable role under the frame of experience hierarchy and assessment matrix (Fig. 6).

3.1 Experience Level

Rational Level. Communication is the most basic and rational need in the interaction with nature. Embedded intelligence makes plants cyborgs while wearable devices arm human as cyborgs, so we are able to communicate in a common language - binary machine language. Take self-caring planting pot linked with smart phones as an example, embedded technology provides plants the ability to respond to environmental changes in order to better survive, as well as enables human to understand their living conditions and feelings better.

Sensational Level. "Pleasure" and "the sense of alien" are the two dimensions to describe sensational stimulus [23]. Digital technologies expand the concept of reality. They allow us to explore the reality beyond human limits, to see the familiar world from an alien perspective, which can be quite interesting and motivating. For example, the VR project Mashmallow Laser Feast by B. C. Steel provides alien experience of discovering forests in the eyes of different animals.

Emotional Level. What eco-interaction aims to achieve is not only sensational pleasure, but more importantly, is emotional bonding and deep reflection. For example, in the project Talking tree by EOS magazine, anthropomorphic plants post their living conditions and feelings online. The purpose is to build up an empathetic bonding which would last longer and deeper than just rational or sensational memories, and would inspire reflects on the relationship with nature. This "reflective level of emotions" which Donald described as the supreme level can be achieved through aesthetic immersion as well (Fig. 7). Steel hopes to bridge the gap between science and art. For teamLab, their concept and purpose behind the aesthetic enjoyment is to cherish the balance between technology and nature and to create a global beauty culture.

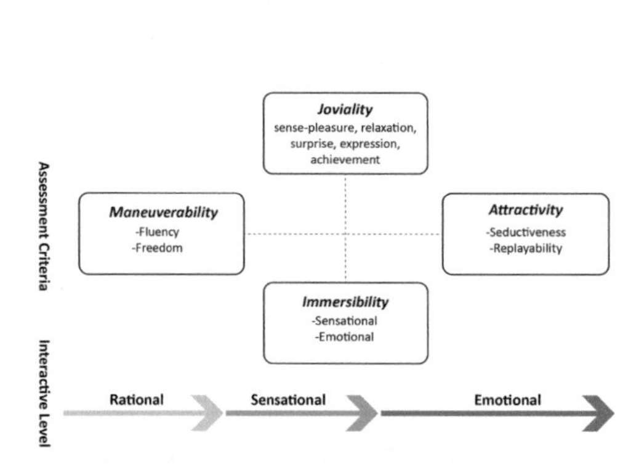

Fig. 6. Diagram of interaction assessment. Author's own work.

Fig. 7. Mashmallow laser feast. Image credit: Barnaby Churchill steel

3.2 Experience Assessment

"Interaction design is all about user experience" [1]. Experience of all users (not just human) should be the key assessment indicators for a design project. Cyborgian approach would be advantageous in human-nature bonding process. Its application and efficacy are investigated from four aspects: maneuverability, immersibility, joviality, and attractivity.

Maneuverability. It focuses on the quality and convenience of the operating process, or in other words, the pleasure of behavioural level [16]. Maneuverability includes fluency and freedom. Fluency concerns whether users operate naturally and smoothly with a clear logic. Freedom refers to the range and types of interactive inputs that users can play with. For example, in the interactive musical plants project Akousmaflore, by applying multi-touch interaction technology, every leaf becomes a tiny instrument. While in the project Forest Entrance by TeamLab, users get involved by walking around, as the project adopts somatosensory devices to capture and respond to users' position changes. The interaction mode in Forest Entrance is more fluent, but with less freedom of control, as there's only one type of corresponding input and output.

The development of human-machine interface (HMI) is a process with increasing naturalness, fluency and fitness in line with human cognitive and usage habits. Currently, somatosensory device is one of the most cutting edge HMI interface. Its innovation lies in the realization of contactless operation, which completely liberates people's hands and fingers and enables participants to freely control machines with their whole bodies. This development brings new opportunities for the interaction between man and nature. Interaction projects that respond to human bodily behaviour is more welcomed because it is the presence of human in outdoor environment that really matters.

Fig. 8. The history of HMI (Human-Machine Interface). Author's own work

Immersibility. Immersion is the engagement level of interactions. Immersibility is related to sensational experience and mental state. Immersive quality can be improved by enhancing multi-sensory experiences, balancing the relationship between "challenge" and "skill" based on feedbacks.

Sensationally. Although 83% of human's perception of the outside world comes from visual sense, the importance of other senses should not be neglected. For perceiving the world, the whole is greater than the sum of its parts [15]. High immersibility can be achieved not only by ensuring a wide scope of environment with minimalized visual distractions, but also by multi-sense activation including shape, smell, texture, softness and roughness with the help of perceptual augmentation devices.

Emotionally. According to Csikszentmihalyi, there are eight mental states in terms of the relationship between skill and challenge level: flow, control, relaxation, boredom, apathy, worry, anxiety and arousal (Fig. 10). These relationships can be constantly evaluated and adjusted with the aid of real-time concentration analysis and massive personal feedbacks, so as to reach the "Flow State" or "Peak Experience" [6], namely a status where fulfilment and enjoyment are coming out of high concentration and totally indulgence [12] (Fig. 9).

Joviality. The assessment of joviality is a composite index of all positive emotions in an interaction process. Electronic devices nowadays are not only able to monitor physical data such like heart rate and running pace, but are also capable of sensing and recognizing mental states (Fig. 8). Smart treadmills would automatically adjust their slope and speed in accordance with users' physical conditions. Similarly, an intelligent interaction system can also adjust in real time its difficulty level, surprise level, etc., according to users' emotional conditions in order to ensure constantly positive experience. In this way, emotion is no longer just the output or by-product of an interaction process. It participates in the process of feedback loop and becomes an input parameter that affects final output. There're already biometric sensors that can recognize micro emotions after deep learning. In the case of Emotional Design Language Orb, various emotions are visualized by different color and form of the orb (Fig. 11).

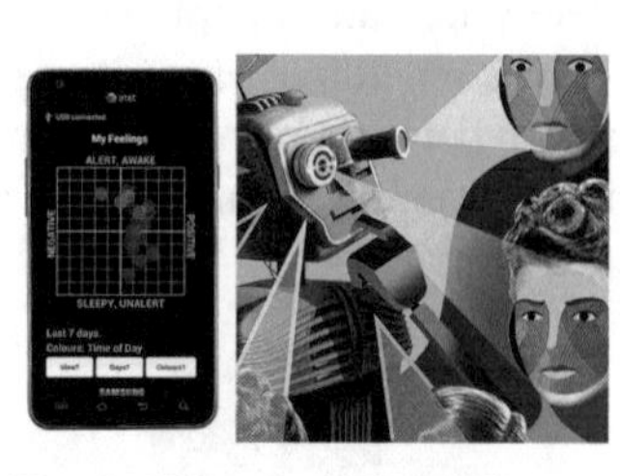

Fig. 9. Two ways of emotion sensing. Health data analysis & facial expression recognition. Image credit: Neurodata Lab, Москва

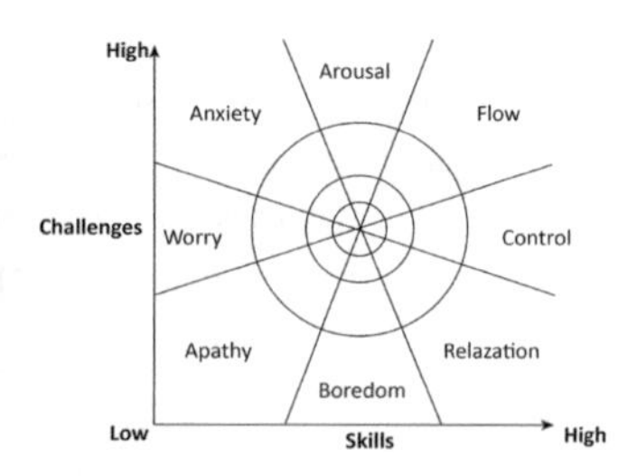

Fig. 10. Csikszentmihalyi mental states model. Author's own drawing.

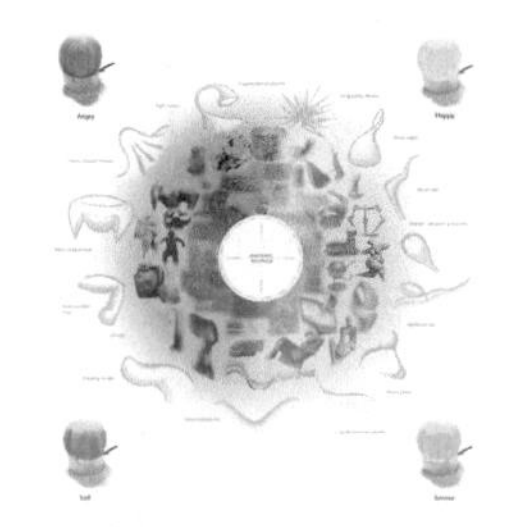

Fig. 11. Emotional design language Orb. Image source: MIT Media Lab.

Attractivity. It can be interpreted as "seductiveness" and "replayability." It evaluates the appeal for first-time users as well as the capability of encouraging re-participation without losing fun. Attractivity derives from beauty and creativity which can be enhanced by cyborgian approach. For instance, in the project Bio-responsive Garden, botanicals were fitted with microelectronics to make them physically animated. Creativity emerges from the unusual combination of "plants" + "dance". Another example is the Forest Entrance mentioned before. The resonating art work is rendered in real time in response to human behaviours. It is neither a pre-recorded nor imagery on loop [22]. Its attractiveness emerges from the non-repeatable organic aesthetics empowered by machine intelligence.

In summary, cyborgian approach would be instrumental as it allows us to better understand the physical world, improve bodily experience and add more joy and attractiveness to the interaction with nature. The analysis above provides anchor points for machine intelligence to intervene, facilitate and encourage human-nature interaction by making use of its sensing and computational capabilities.

4 Design an Interactive Outdoor Environment

As a matter of fact, many of the projects mentioned above are installed indoors. What hinders outdoor interactions? This chapter aims to figure out current constraints of outdoor eco-interaction projects and the opportunities provided by cyborgian technologies.

4.1 Challenges and Opportunities of Outdoor Interaction

One of the biggest technical obstacles of outdoor interactions is the more complex environment compared with indoor ones, which calls for better performance of input and output equipment. To address this problem, personal wearable devices which equip users as cyborgs would be a key tool. With fixed sensors only, it would be difficult to accurately capture motions and track positions of all users. Smart wearable devices greatly increase the accuracy of somatosensory input. Subtle movements and gestures are likely to be recognized, thus improving operational experience.

Besides, personal wearable devices increase the feasibility of multisensory enhancement in outdoor scenarios. As the name suggested, multisensory enhancement technologies aim to create alluring sensational experiences with the aid of various equipment, such as stereo, blower, perfume machine, tactile gloves, etc. It's like updating a 2D movie into a 3D (visually enhanced only) and even a 4D movie (multisensory enhanced). People need more attractive and comprehensive sensory experience to indulge themselves in a distractive outdoor environment.

Another challenge as well as principle of outdoor eco-interaction is to minimize the impact on voiceless ecological entities. Cyborgian approach would be helpful because: the living condition of plants can be always monitored, adjusted and guaranteed by their cyber apparatus; and our cyber extensions work as the medium for harmless virtual interactions that bring about genuine feelings. For plants, behavioural response in accordance with human inputs may be a negative interference. That's why in general, designers would apply non-material digital technologies such as light and sound which have no physical impact on the environment, turning nature into living art without harming it. However, due to the requirement of light and sound conditions, the suitable time for outdoor interaction is limited and unpredictable. The application of AR devices eliminates time limits in a way of mixing interactions happen in the cyberspace with reality in physical space in suitable lighting and volume settings.

Additionally, distributed wearable devices enable mass-customized multi-user interactions, which encourage not only physical but also social activities in natural environment. Although there're hundreds of users involved simultaneously, interactions can be tailor-made according to individually physical and mental state. Even the viewpoint can be set to be unique, for example, experiencing the world from the perspective of fish or birds, which brings alien experiences that can be attractive for potential users.

One weakness of cyborgian approach is that currently wearable devices are not light or user-friendly enough to be totally ignored while using, which may hinder users' movements and experience. There's always a gap between the real action (e.g., press the button) and the action executed in virtual world (e.g., pick up something). Similarly, the rationality of virtual viewpoint should be ensured. The sense of inconsistencies and the learning process keep reminding users the existence of a physical interface. Technologies are being optimized to reduce such inconsistencies of viewpoints and actions to minimize the perceptibility of external devices.

In summary, cyborgian approach, supplemented by multi-sensory enhancement technology, would work as a powerful toolkit to the challenges of outdoor interactions with enhanced immersibility, customized experience and minimal impact on the ecology.

4.2 A Cyborgian Eco-interaction Design Model

After all the researches and discussions above, this paper proposes a cyborgian eco-interaction design model depending on a distributed network of cyborgian intelligence.

On the one hand, non-human users are indispensable parts in an eco-interaction project, which means the ecological inputs that represent their status and interaction experience should not be neglected. Environmental data and plant living conditions that are monitored and evaluated by ubiquitous sensors have decisive impact on the system. Because the health and balance of ecosystem is the premise of human interaction.

On the other hand, for human users, their behavioural inputs including motions, gestures and voice commands etc. are recognised by wearable devices and stationary sensors. Besides, their interactive experience are also essential input parameters in an intelligent interactive system. As discussed before, users' feelings and experiences are no longer by-products of the interaction, but rather become inputs that trigger self-tuning mechanisms based on feedback loops.

The algorithms at the back end of interface are not fixed, instead, they are supposed to be determined by user experience. Users' experiences are evaluated from the four aspects mentioned before. The system automatically would customise its difficulty and intensity level according to individuals' feedback in order to help them reach and maintain "flow state". If the system notices that relying on bottom up self-tuning process only is not efficient enough to enhance user experiences, then human designers are required to get involved in order to modify the interactive design in a top down way. This procedure is similar to what Kevin Kelly called "control of control" [10]. In this way, the system provides customized and optimized experiences which in turn increase the attractivity (Fig. 12).

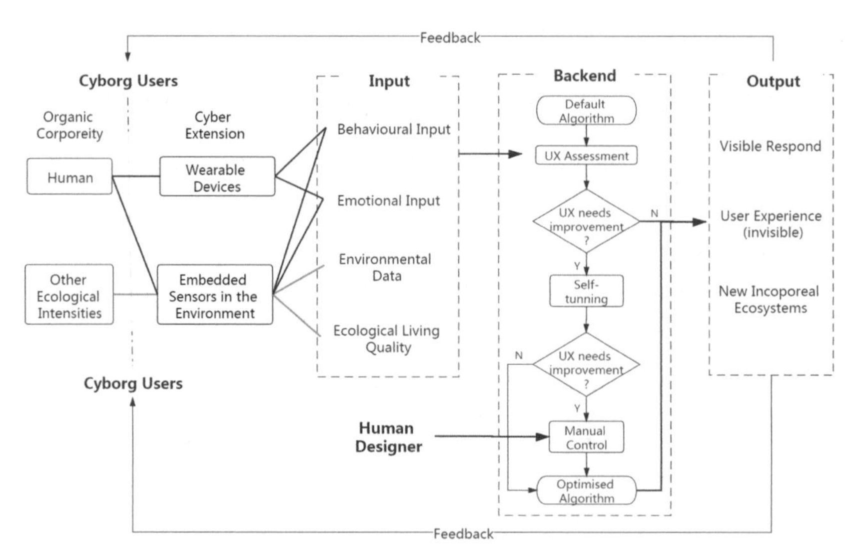

Fig. 12. Diagram of cyborgian eco-interaction model. Author's own work.

In sum, what distinguishes this model from the others is that it concerns more about the user experience at all levels for all ecological entities. The capability of enhancing user experience by applying a synesthetic approach to reality and the capability of mass-customising and self-evolving based on sensible user experience, are both gifted by a cyborgian methodology.

5 Conclusion and Outlook

The argument throughout this paper is to explore new opportunities and capabilities that are imparted by advancing digital technologies when it comes to establishing an empathetic relationship with the natural environment where we inhabit. The cyborgian theory supports the skeleton of the whole paper while the embodied cognition theory lays a solid foundation.

Cyborgian approach is a hybrid of machinery intelligence and biological intelligence as well as a mixture of top down and bottom up design methodologies. It allows designers to take advantages of both methods: the ability to secure efficiency and to manage complexity.

Embedded intelligence turns plants and other ecological entities into cyborgs while wearable devices transform human into cyborgs as well. Input data from human and non-human users are collected from the interface, processed at the backend and the output can be given back to any interface in this network. In this way human and environment are communicating and interacting in "the third space" [21] which is organized by the network of interface and augmented by digital technologies.

It is revealed by this paper that cyborgian approach would be instrumental in the construction of a more successful human-nature interactive system. This can be justified by the facts that cyborgian intelligence is gifted at: facilitating the whole process of cross-species interaction, namely sensing, processing, actuation and feedback; breaking the limitations of outdoor interaction; improving experiences of all participants which determines the quality of interaction.

The cyborgian eco-interaction design model this paper proposed may be far from satisfactory, but meaningfully, it is expected to work as a minnow that is thrown out to catch a whale, to broach the subject for further concerns and investigations.

Opportunities always come along with threats. Advancing technology is facilitating our life while challenging our relationship with nature. Insightful designers should be keenly aware of the pros and cons of technology, trying to come up with a new paradigm to rehabilitate this relationship, which would then be more than a design paradigm, but even become a new lifestyle in the upcoming digital future.

References

1. Anderson, S.P.: Seductive Interaction Design: Creating Playful, Fun, and Effective User Experiences (Voices that Matter). New Riders, Indianapolis (2011)
2. Bourriaud, N., Pleasance, S., Woods, F., Copeland, M.: Relational aesthetics, Les presses du réel Dijon (2002)
3. Calvo, P., Gomila, T.: Handbook of Cognitive Science: An Embodied Approach. Elsevier, München (2008)
4. Clynes, M.E., Kline, N.S.: Cyborgs and Space. The Cyborg Handbook, pp. 29–34. Routledge, New York (1995)
5. Coffin, J., Taylor, J., Bauen, D.: Breeze Interactive Animation (2006). http://www.danielbauen.com/robotany/
6. Csikszentmihalyi, M.: Flow: The Classic Work on How to Achieve Happiness. Random House, London (2002)
7. Gagliano, M., Vyazovskiy, V.V., Borbély, A.A., Grimonprez, M., Depczynski, M.: Learning by association in plants. Sci. Rep. **6**, 38427 (2016)
8. Gibson, J.J.: The Ecological Approach to Visual Perception. Psychology Press, Routledge (1979)

9. Hanel, S.: Plant intelligence – Our 5 senses + 15 more (2016). https://www.lindau-nobel.org/plant-intelligence-our-5-senses-15-more/
10. Kelly, K.: Out of control: The Rise of Neo-Biological Civilization. Addison-Wesley, Reading (1994)
11. Kirsh, D.: Embodied cognition and the magical future of interaction design. ACM Trans. Comput.-Hum. Interact. (TOCHI) **20**, 1–30 (2013)
12. Maslow, A.H.: Peak experiences as acute identity experiences. Am. J. Psychoanal. **21**, 254–262 (1961)
13. McNerney, S.: Embodied cognition and design: a new approach and vocabulary. Big Thinker (2013). https://bigthink.com/insights-of-genius/embodied-cognition-and-design-a-new-approach-and-vocabulary
14. Merleau-Ponty, M.: Phénoménologie de la perception. Paris (1945)
15. Negroponte, N.: Being Digital. Vintage Books, New York (1995)
16. Norman, D.A.: Emotional Design: Why We Love (or Hate) Everyday Things. Basic Civitas Books, New York (2004)
17. Ronchi, A.M.: eCulture: cultural content in the digital age. Springer Science, Dordrecht (2009)
18. Sareen, H.: Cyborg Botany: Augmented Plants as Sensors, Displays and Actuators (2019). https://www.youtube.com/watch?v=6gVrt37s9oY&feature=youtu.be
19. Sareen, H., Tiao, E.: Elowan: A plant-robot hybrid. MIT Media Lab (2018). https://www.media.mit.edu/projects/elowan-a-plant-robot-hybrid/overview/
20. Schultz, J.C.: Plants are just very slow animals. Presentations (Missouri Regional Life Sciences Summit 2010) Plant Sciences presentations (MU), University of Missouri (2010)
21. Soja, E.W., Chouinard, V.: Thirdspace: journeys to Los Angeles & other real & imagined places. Can. Geogr. **43**, 209 (1999)
22. teamLab: Abstract and Concrete - Forest Entrance (2018). https://art.team-lab.cn/w/forest-entrance/
23. van der Zee, D.: The complex relationship between landscape and recreation. Landscape Ecol. **4**, 225–236 (1990)
24. Wiener, N.: Cybernetics or Control and Communication in the Animal and the Machine. Technology Press, New York (1948)
25. Ye, H.: Embodied cognition: a new approach of recognition psychology. Adv. Psychol. Sci. **18**, 705–710 (2010)

Machine Seeing

A Large-Scale Measurement and Quantitative Analysis Method of Façade Color in the Urban Street Using Deep Learning

Jiaxin Zhang(✉), Tomohiro Fukuda, and Nobuyoshi Yabuki

Division of Sustainable Energy and Environmental Engineering, Graduate School of Engineering, Osaka University, Suita, Japan
xinxinxin266@gmail.com, {fukuda,yabuki}@see.eng.osaka-u.ac.jp

Abstract. Color planning has become a significant issue in urban development, and an overall cognition of the urban color identities will help to design a better urban environment. However, the previous measurement and analysis methods for the facade color in the urban street are limited to manual collection, which is challenging to carry out on a city scale. Recent emerging dataset street view image and deep learning have revealed the possibility to overcome the previous limits, thus bringing forward a research paradigm shift. In the experimental part, we disassemble the goal into three steps: firstly, capturing the street view images with coordinate information through the API provided by the street view service; then extracting facade images and cleaning up invalid data by using the deep-learning segmentation method; finally, calculating the dominant color based on the data on the Munsell Color System. Results can show whether the color status satisfies the requirements of its urban plan for façade color in the street. This method can help to realize the refined measurement of façade color using open source data, and has good universality in practice.

Keywords: Façade color measurement · Street view image · Deep learning · Quantitative analysis

1 Introduction

Color is people's first visual impression of the city, which can influence spatial perception in the human-made environment through different color compositions. Besides, color can reveal the history, culture, and the specific identities of regions [1]. Therefore, the city managers attached great importance to the planning of urban color and issued a series of rules. For example, in 1845, Turin published a guide to urban color, which served as a reference [2]. In the 1960s, Paris identified beige as the dominant color of the city. Tokyo requires that the color of building exterior walls should be light in the 1970s [6]. In China, mostly located in the eastern part of the country, conduct urban color planning which allows continuing the urban color context, in the way of historical heritage by the dominant color, but also to guide the development of new construction zones [3]. Nanjing, one of the most important cities on eastern China, organize experts

P. F. Yuan et al. (Eds.): CDRF 2020, *Proceedings of the 2020 DigitalFUTURES*, pp. 93–102, 2021.
https://doi.org/10.1007/978-981-33-4400-6_9

to discuss the urban color guidelines and identified light green as the dominant color of the city in 2004 [4]. Most of these urban color plans and regulations regard building color controlling as the most critical implementation program, because buildings are larger and more complex than the products created by human beings.

Currently, urban planners have established a complete workflow of facade color investigation [5]. First, researchers collect pictures and geographic coordinates of street facades by using digital cameras and GPS devices. Then, referring to the Munsell Color System to obtain color information such as the hue, saturation, and value of the façade image they collected. Finally, the results can show the facade color distribution and the dominant color of the city, meanwhile, planners make suggestions on the design or renovation of the city continuously. However, this workflow is limited to the manual collection at small quantity, so it usually takes some representative cases instead of the granularity analysis to ensure work efficiency in a large-scale practice [6]. Some urban designers and researchers were restricted to these empirical studies, getting refined results in the time specified by the project [7]. How to solve the large-scale, refined, and efficient acquisition of facade colors has become a significant obstacle to the further deepening and popularization of urban color planning.

2 Literature Review

Academic discussions and applications of large-scale measurement and quantitative analysis of building façade color have revolved around three categories.

2.1 Urban Color Planning

In the 1960s, Professor Lenclos Jean-Philippe proposed the color geography theory [8], in his points, the color of the city is formed by the specific geography, climate, and the relevant factors such as the national culture, history, and customs of the people living here. Therefore, the designers consider the characteristics of the city while planning the urban color. In this context, the first modern urban color planning was born in Tokyo in the 1970s. It mainly conducted a color survey across Tokyo [9], and while preserving the style of historical buildings, the planning guided the development of new buildings.

The planning of urban colors requires a quantitative definition of color system, and then the work can be carried out on a systematic and repeatable way. For example, Dinant suggest a standard color chart as a guidance for integral urban planning in Belgium [3]. In Greece, some researchers made studies of rehabilitation of contemporary Greek streetscape, by using regions of the NCS (Natural Color System) solid as color palettes [10]. In Tianjin (China), the authority used color charts based on the Munsell color classification system as a decision support tool [11], which is also a common practice in Chinese cities. These methods can investigate the current status of urban color and provide primary data for sustainable development of cities.

2.2 Façade Color Measurements

The current facade color measurement method uses digital cameras as the tools for collecting data. Investigators use digital cameras to obtain photographs of building façades in specific areas and distinguish whether the current status met the requirements of urban color planning. For example, Lu, et al. (2010) carried out field investigations on one avenue based on the color system of architecture design code (GB/T15608-1995) in Shenzhen to analysis the distribution of the dominant colors of the building façade [12]. Luan Nguyen (2017) presented a user-oriented protocol to characterize chromatic attributes of different class areas [3].

However, due to the consumption of time and labor in field investigations, almost all urban designers and researchers at the time could only conduct small-scale empirical research. They were unable to find a connection between theoretical study and practical operation and deepened their logical reasoning through induction and deduction, rather than exploring based on the principles of large-scale investigations.

2.3 Quantitative Analysis of Visual Quality in Urban Street

With the development of semantic segmentation algorithms and urban datasets represented by Cityscapes, it is no longer difficult to use a deep convolutional neural network architecture to achieve different feature dictation and high-precision segmentation results [13, 14]. The combination of street view data and machine learning in urban measurement can solve the problem in large scale with high refinement, and it is used in many aspects, such as identification of the neighborhood safety through street view pictures [15], the extraction of green street visibility [16], and the measurement of visual quality in street space [17]. The two considerations include in this study are the classic urban design theory and existing deep learning algorithms.

The combination of machine learning and street view images has changed the situation in which street data was difficult to obtain and challenging to use efficiently. Compared to traditional methods, we can quickly get large number of street view photos, then providing scientific suggestions to the visual quality of urban space. The method can solve the limitation of small samples by manual collection of building façade color analysis in the traditional way, which is possible to use in a wide range.

3 Methodology

3.1 Study Area and Workflow

Our research is conducted within the main districts of Nanjing, which is about 156.2 km^2. In Chinese history, Nanjing had long been a major center of culture, education, research, politics, economy, transport networks, and tourism being a multi-functional metropolis [18]. The study area located in the center of Nanjing [19]. The workflow of main measurement method can be seen in Fig. 1. The detail process in each section will be explained in the following section.

3.2 Street View Data Acquisition

The pictures in this study were extracted from the street view platform which can provide street view photos with a large coverage. The methods of street view download are the following four steps:

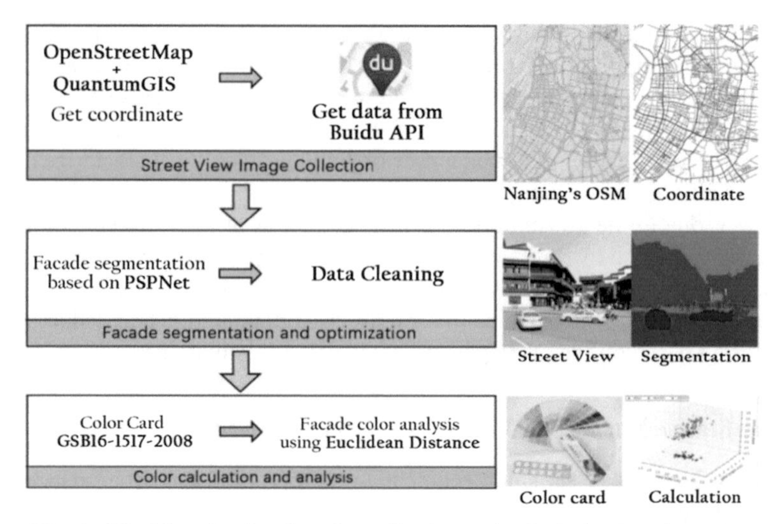

Fig. 1. Workflow for façade color collecting and calculating in urban street

1. The urban road network information is selected and obtained by using the rectangular frame within its scope through open street map [20].
2. The road network obtained in the previous step is simplified into single-line using QGIS, and the mean distance between adjacent points is 20 m which refer to the urban street design methodology of Gehl J [21].
3. We get sampling points with geographical coordinate information in QGIS. However, there are some sampling points in Street View API database without corresponding street view images.
4. In order to obtain building façade, we downloaded 2 pictures (both left and right) perpendicular to the road direction from street view service (viewing angle 90°, horizontal angle 0°, picture size 800 × 500 pixels) for each sampling points (Fig. 2).

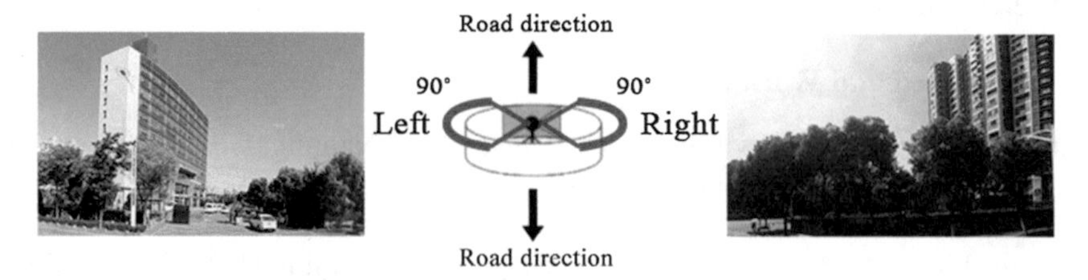

Fig. 2. Street view images acquisition at sampling point

3.3 Building Façade Segmentation and Data Cleaning

We used a deep neural network model, namely, PSPNet, to segment the building façade from street view images. Unlike the previous methods of extracting building element with RGB values, the network structure of the Pyramid Scene Parsing Network (PSPNet) was widely employed where spatial statistics provide a good descriptor for overall scene interpretation. A single PSPNet yields the new record of mIoU accuracy 85.4% on PASCAL VOC 2012 and accuracy 80.2% on Cityscapes [13]. The building segmentation results of Baidu street view images via the proposed trained PSPNet are shown in Fig. 3. Then, the proportions of building façade area can be measured by inputting street view images into the trained model to generate segmentation results, as Eq. (1) shows. However, there is a low proportion of buildings in some street view images, and these building images cannot reflect the color characteristics of the building. To improve accuracy of the experiment, we need to clean up the photos with a low proportion of the building.

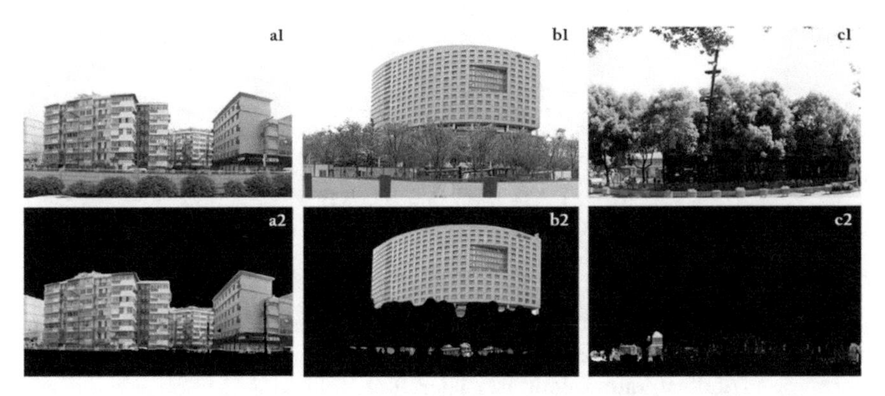

Fig. 3. Street view image segmentation through Pyramid Scene Parsing Network

$$P_{bldg} = \frac{\sum_{i=1}^{2} NW_i}{\Sigma_{i=1}^{2} N_i} \tag{1}$$

P_{bldg} is the proportion of building in a respective scene location, which is averaged by the values of street view images of the left and right side. N_i is the total number of pixels in image i, and NW_i is the number of building pixels in image i. By calculating the P_{bldg} of each sampling point, we delete the pictures, an example like Fig. 3(c2) shows, whose building proportion is less than 10%.

3.4 Façade Color Calculation

Due to the color deviation, not all the colors in street view images correspond to the colors of the building materials. Therefore, using the architectural standard color cards (GSB16-1517-2002) with 258 colors as a reference, we merge the raw color data to standard color. The merge method is to calculate the HSV value of the street view color (c_{siv}) and replace them with the closest architectural standard color (c_{std}) by calculating the Euclidean distance. The color distance is the closeness between two points in the HSV cone (the length of the hypotenuse is R, the radius of the bottom circle is r, and the height is h). In the HSV color space model (Fig. 4), the center of the ground is the origin and H = 0 is the positive direction of the x-axis. Then according to Eq. (2), we were able define the three-dimensional coordinates (x, y, z) of the color point (H, S, V) as:

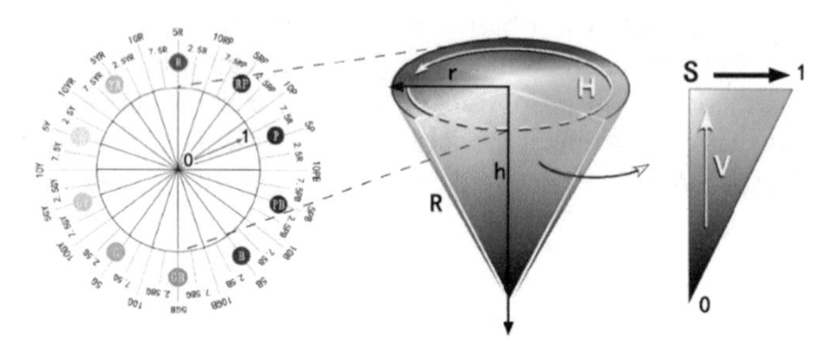

Fig. 4. HSV color space model

$$\begin{cases} x = r * V * S * \cos H \\ y = r * V * S * \sin H \\ z = h * (1 - V) \end{cases} \quad (2)$$

After calculating the color distance, we can count the proportion of each color in these pictures. Two of the most important factors affecting the dominant color are the proportion and the saturation contrast. Therefore, following is the selection way of dominant color: Firstly, the dominant color should be the largest part of the building facade. Secondly, when the color proportions are equal, the color with high saturation is used as the dominant color. Finally, when the perception of color is the same and the chroma is the same, the darker one is the dominant color.

4 Results

We finally obtained 77, 448 valid data in 110, 686 total images. Table 1 shows the dominant color statistics of street view pictures on both sides of the street at each sampling point, including the H (Hue), S (Saturation), V (Value), and geographic coordinate. By calculating and obtaining the quantitative color information of building facades at the city scale, we can not only do a large-scale analysis but also make targeted analysis for a specific area. Combined with the land use plan of the Nanjing City (2011–2020) (Fig. 5(a)) and the Nanjing Historical City Preservation Plan (2010–2020) (Fig. 5(b)), we can illustrate that the downtown blocks with different functions and the control of historic conservation.

Table 1. Valid street view data dominant color statistics

Picture ID	Dominant color	H	S	V	Latitude	Longitude
1	8.1GY6/1.4	192	10%	61%	32.0211	118.7632
2	8.1GY4.5/1.4	97	12%	41%	32.0213	118.7642
3	10YR9/1	37	10%	94%	32.0215	118.7636
……	10YR8.5/1	24	13%	86%	32.0605	118.7798
77,448	N7.25	240	1%	71%	32.0610	118.7803

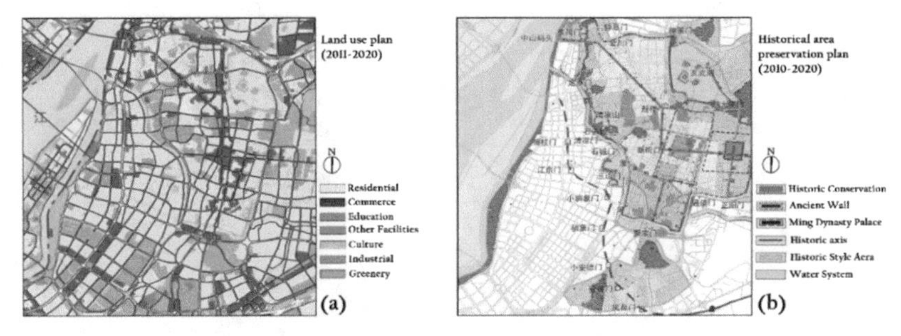

Fig. 5. (a) Land use plan of the Nanjing City, (b) Nanjing historical city preservation plan.

Figure 6(a) shows the distribution of the dominant color on both sides of the street network in the central city of Nanjing. We can find that the dominant colors are mainly grey, with dark grey (N6.25; N7.25) accounting for 33% of the total, greenish-grey (8.1GY4.5/1.4; 8.1GY6/1.4) accounting for 22%, and Moderate khaki (10YR9/1; 10YR8.5/1) accounting for 20% of the total.

Most proportions of the historic style area, the dominant color is low saturation (Colors with a saturation of less than 20% accounted for 73% of the total.), like Fig. 6(b) show. In the historic conservation area, the lightness of dominant color of façade usually higher than 60%, and such strict control are mainly distributed in *The Confucius Temple*, *The East Zhonghua Gate Historical Block*, *Sipailou Campus*, and other historic districts, like Fig. 6(c) show.

Compared with the traditional way. Street view images are obtained as open source, which has a large amount of acquisition and efficiency. In the building recognition stage, the semantic segmentation model we trained by using the Cityscapes Dataset and the IoU accuracy of the building is 90.6%. In the stage of color calculation and statistics, the color distribution of buildings calculated by the HSV model can be visually expressed on the spatial map.

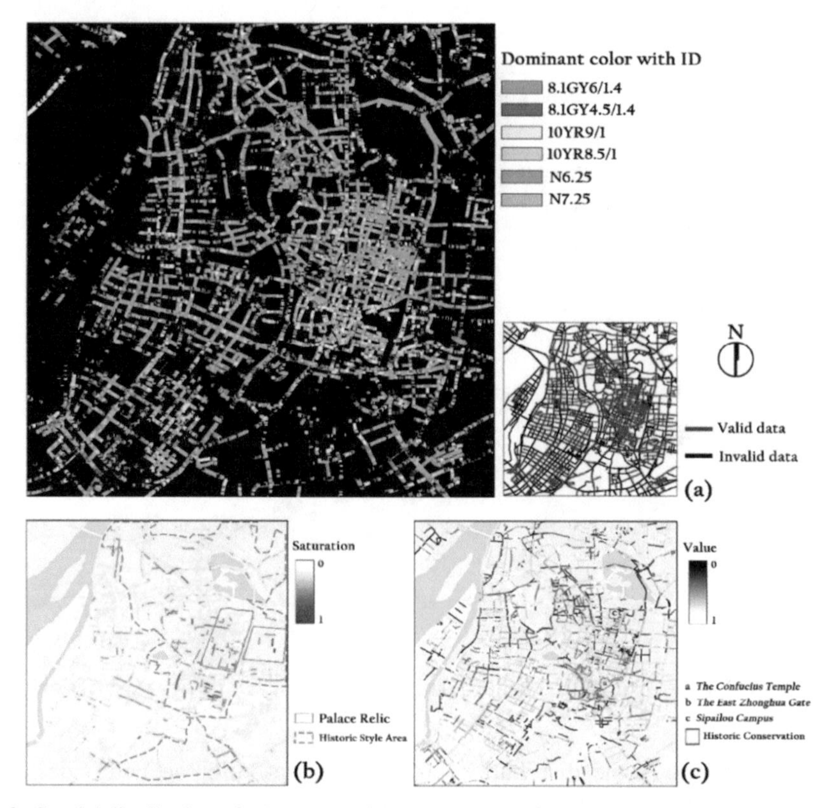

Fig. 6. Spatial distribution of urban street facade colors: (a) Color ID and valid data statistics, (b) Saturation distribution of dominant colors, (c) Lightness distribution of dominant colors

5 Discussion and Conclusion

This study proposed a framework for measurement and analysis of façade color in the urban street, including using open platforms to collect data and quantitatively analysis. The first was segmented the façades in street view images automatically with the support of PSPNet, the second was visualized in QGIS, while the last was achieved quantitatively through statistics. With façade color as the benchmark, the paper concludes the strengths and weaknesses of the large-scale measurement and quantitative analysis method of façade color in the urban street.

Empirical research on Nanjing's commercial and historical districts shows that Nanjing has gained a reputation for its cultural, historical, and architectural design values. The overall assessment is that color planning is highly consistent with the image of the city. In general, most of Nanjing's historic districts have retained their ancient form, structure, and color, despite being rebuilt in a modern way. Modernization challenges many traditional styles and characteristics of the inner city, and downtown is still a place that attracts the attention of developers.

Despite the contribution of this study in quantitatively measuring the façade color in large scale, it had several limitations that require further exploration. For example, street view photos are affected by outdoor light environment, causing the saturation and lightness to be very different from the actual data. The color calibration of street view images can improve the accuracy of the calculation results. In addition, how to use it to design a better visual environment is a significant problem in future work.

Conflicts of Interest Statement. Authors declare no potential conflicts of interest in relation with authorship, study and research conducted and/or publication of this article.

References

1. Ashihara, Y.: The Aesthetic Townscape. The MIT Press, Cambridge (1983)
2. Porter, T., Mikellides, B. (eds.): Color for Architecture Today. Taylor & Francis, London (2019)
3. Nguyen, L., Teller, J.: Color in the urban environment: a user-oriented protocol for chromatic characterization and the development of a parametric typology. Color Res. Appl. **42**(1), 131–142 (2017)
4. Jiang, H.: Research on quantitative control method of urban color main tone based on digital technology. Chongqing University (2013). (in Chinese)
5. Xu, L. et al.: Urban color survey and quantitative analysis: a case study of Shenzhen Shennan Avenue. Urban Planning, December 2010 (in Chinese)
6. O'connor, Z.: Bridging the gap: facade color, aesthetic response and planning policy. J. Urban Des. **11**(3), 335–345 (2006)
7. Liu, Y., Liu, X., Yuan, Y.: Collection and data analysis of suzhou classical garden color elements. Chinese Garden **32**(6), 46–51 (2016). (in Chinese)
8. Jean-Philippe, L., Lenclos, D.: Colors of the World: the Geography of Color. WW Norton & Company, New York (2004)
9. Sari, Y.: Case studies of color planning for urban renewal. In: AIC 2015 Tokyo Proceedings, pp. 113–118 (2015)
10. Tosca, T.F.: Environmental color design for the third millennium: an evolutionary standpoint. Color Res. Appl. **27**(6), 441–454 (2002)
11. Zhao, C., et al.: Urban color planning method: a case study of urban color planning of Tianjin. City Plan. Rev. S1 (2009). (in Chinese)
12. Xu, L., et al.: Urban color survey and quantitative analysis: a case study of Shenzhen Shennan Avenue. Urban Planning, 12 (2010). (in Chinese)
13. Zhao, H., et al.: Pyramid scene parsing network. In: Proceedings of the IEEE Conference on Computer Vision and Pattern Recognition, pp. 2881–2890 (2017)
14. Badrinarayanan, V., Kendall, A., Cipolla, R.: SegNet: a deep convolutional encoder-decoder architecture for image segmentation. IEEE Trans. Pattern Anal. Mach. Intell. **39**(12), 2481–2495 (2017)
15. Naik, N., et al.: Streetscore-predicting the perceived safety of one million streetscapes. In: Proceedings of the IEEE Conference on Computer Vision and Pattern Recognition Workshops (2014)
16. Li, X., et al.: Assessing street-level urban greenery using Google Street View and a modified green view index. Urban Forest. Urban Green. **14**(3), 675–685 (2015)
17. Tang, J., Long, Y.: Measuring visual quality of street space and its temporal variation: methodology and its application in the Hutong area in Beijing. Landscape Urban Plann. **191**, 103436 (2019)
18. Maier, C.S., et al.: The Nanjing Massacre in history and historiography. University of California Press, California (2000)

19. Wen, Z., Xiaolin, Z., Lili, X., Yahua, W.: Study on the characteristics of population spatial changes in Nanjing metropolitan area from 2000 to 2010. Geog. Sci. **36**(1), 81–89 (2016). (in Chinese)
20. Open Street MAP. https://www.openstreetmap.org. Accessed 24 Jul 2018
21. Gehl, J.: Cities for People. Island press, Washington, D.C. (2013)

Suggestive Site Planning with Conditional GAN and Urban GIS Data

Runjia Tian(✉)

Harvard University Graduate School of Design, 48 Quincy Street, Cambridge, MA 02138, USA
runjia_tian@gsd.harvard.edu

Abstract. In architecture, landscape architecture, and urban design, site planning refers to the organizational process of site layout. A fundamental step for site planning is the design of building layout across the site. This process is hard to automate due to its multi-modal nature: it takes multiple constraints such as street block shape, orientation, program, density, and plantation. The paper proposes a prototypical and extensive framework to generate building footprints as masterplan references for architects, landscape architects, and urban designers by learning from the existing built environment with Artificial Neural Networks. Pix2PixHD Conditional Generative Adversarial Neural Network is used to learn the mapping from a site boundary geometry represented with a pixelized image to that of an image containing building footprint color-coded to various programs. A dataset containing necessary information is collected from open source GIS (Geographic Information System) portals from the city of Boston, wrangled with geospatial analysis libraries in python, trained with the TensorFlow framework. The result is visualized in Rhinoceros and Grasshopper, for generating site plans interactively.

Keywords: Machine learning · Site planning · Generative adversarial network · GIS · Generative landscape design

1 Introduction

In 1984, Kevin Lynch formalized site planning as "the art of arranging buildings and other structures on the land in harmony with each other" [1]. Site planning has been crucial in the design process of architecture, landscape architecture, and urban planning.

Thus, site planning's primary goal is the generation and organizational process of the site layout. A fundamental step for site planning is the design of building layout across the site.

There have been various attempts over the past years to automate the site process. However, most approaches adopt rule-based generative design algorithms. Such approaches fail infidelity and diversity as this process is hard to automate due to its multi-modal nature: it takes in multiple constraints such as street orientation, program, density, plantation, and, most importantly, block shape.

In a broad sense, the site planning process in architecture, landscape architecture, urban design, and planning pedagogies refers to the organizational stage of the site planning proves. It involves the organization of land-use zoning, access, circulation, plantation design, and other procedures. In a narrow sense, site planning could be formalized as

P. F. Yuan et al. (Eds.): CDRF 2020, *Proceedings of the 2020 DigitalFUTURES*, pp. 103–113, 2021.
https://doi.org/10.1007/978-981-33-4400-6_10

a conditional generation problem solvable with state-of-the-art machine learning models such as Conditional Generative Adversarial Neural Networks (CGAN).

With the abundance of Geographic Information System (GIS) data, the existing urban environment proves a practical dataset for modelling and automating this design process with the statistical learning approach.

2 Related Works

In 2014, Ian Goodfellow proposed the structure of the Generative Adversarial Network [2]. In the paper, he proposed using two multilayer perceptron, a generator, and a discriminator to generate data and classify as true or false based on training data. In November 2014, Mehdi Mirza and Simon Osindero proposed Conditional GAN (CGAN) to place labels on training data for both generator and discriminator in a GAN structure [3]. In their work, they demonstrate how this approach can generate descriptive tags that are not part of training labels.

Isola et al. proposed image-to-image translation with their model Pix2Pix based on Conditional GAN in November 2018 [4] (Fig. 1).

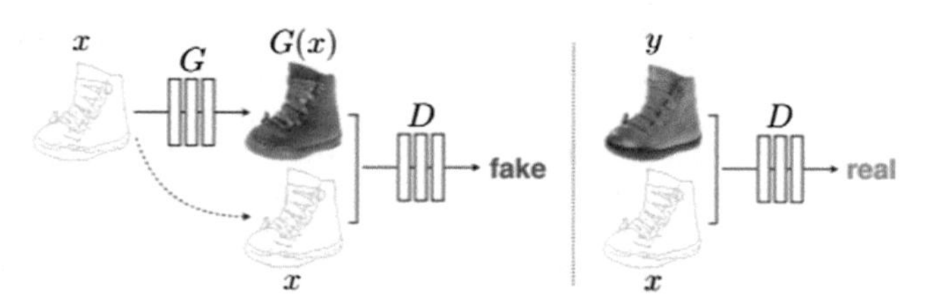

Fig. 1. Pix2Pix image translation, Isola et al.

The Pix2Pix model offers accessible interfaces to designers and has been proved useful in multiple scenarios, including generating architecture plan layouts and texture filling for sketches.

In 2019, Stanislas Chaillou proposed ArchiGAN: A Generative Stack for Apartment Building Design in his thesis [5]. Stanislas unpacked floor plan design into three steps: building footprint massing, program repartition, and furniture layout, and used a Pix2Pix GAN-model for each step for the three tasks correspondingly. The thesis of Stanislas laid a foundation for the generation of building footprint based on the parcel shape. However, his thesis is limited to apartment plan generation and does not incorporate into the complicated urban and landscape context of architecture design.

3 Methodology

The synthetic workflow of the project adopts the typical framework of a data science process. The core question for suggestive planning creative design tool can be divided into three steps (Fig. 2):

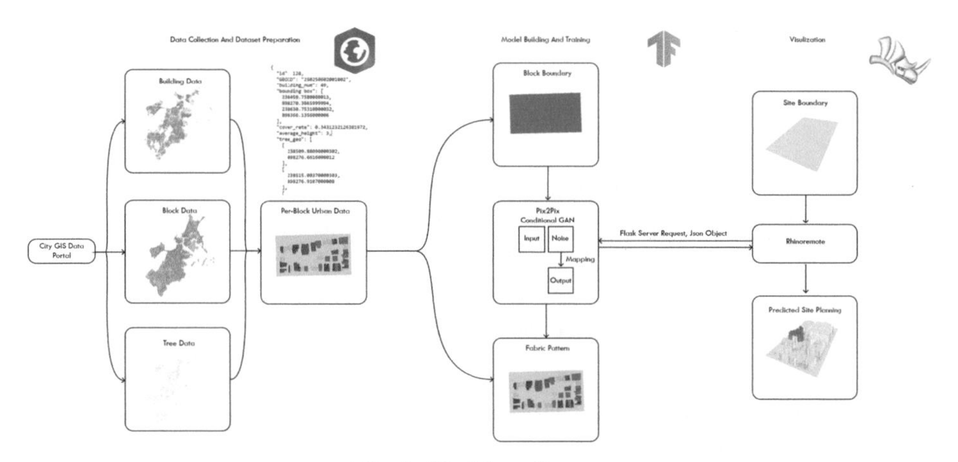

Fig. 2. Workflow diagram

1. **Data Acquisition and Feature Engineering:** A dataset containing necessary information is collected from open source GIS portals. The dataset was processed with geospatial analytical tools for standardizing and canonicalizing. The dataset of vector geospatial data is eventually rasterized into an image dataset for deep learning.
2. **Machine Learning:** Pix2PixHD Conditional GAN is used in this process to learn the mapping from a site boundary geometry represented with a pixelized image, to that of an image containing building footprint color-coded to various programs, trained with TensorFlow framework.
3. **Visualization:** The result is visualized in three-dimensional computer-aided design software with computational design access for generating site planning proposals interactively in real-time.

3.1 Data Acquisition and Feature Engineering

The data is obtained through various open-source GIS Databases. Therefore, the feature engineering and exploratory data analysis were crucial for this project as a spatial join is necessary to canonicalize and standardize the data across various data sources.

3.2 Machine Learning

The model uses the Pix2Pix pipeline for building reliable site boundary image to site planning image translation.

The Pix2Pix model consists of two parts, a generator, and a discriminator. The condition is concatenated with the Gaussian noise as input to the generator and is concatenated again with generator output as discriminator input (Fig. 3) [6]. The model's objective function is the sum of the GAN loss, a binary cross-entropy, and an l1 norm between the generated image and the ground truth.

For our research, the site boundary's input image is represented as a 256-pixel by 256-pixel image; the ground truth image is the same site with buildings color-coded

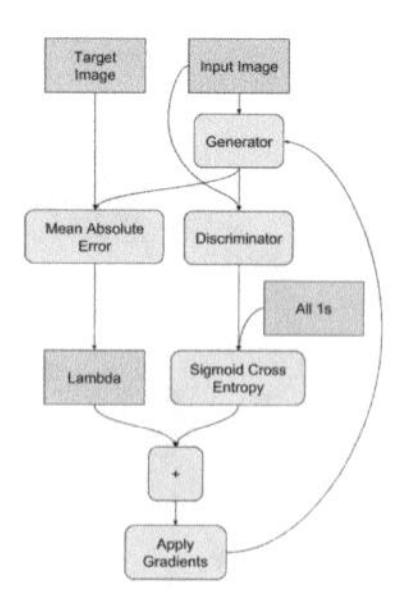

Fig. 3. Pix2Pix architecture

according to specific attributes. For this research, we use height data as a showcase, and similar approaches could be applied for other attributes.

3.3 Visualization

Specifically, the research utilizes Rhinoceros and Grasshopper as a platform for visualizing the result. The site boundary is represented as a closed curve in Rhinoceros. The Grasshopper platform rasterizes the curve geometry and sends the data as a raster image to the deep learning model. Flask framework is utilized for creating the inter-process communication between deep learning models and modelling software. The predicted site layout is then sent back to Grasshopper and visualized in Rhinoceros.

4 Case Study: Taking Boston as Example

The proposed workflow was tested in a course the author participated at Massachusetts Institute of Technology.[1] The objective of the course is to introduce foundational concepts and methodology of machine learning and its applications to design. The project was supported by industry partner Spacemaker.

4.1 Data Acquisition and Feature Engineering

The data for training the generative model is obtained through Boston GIS Data Open Portal, an open-source GIS Database. The author preprocessed GIS data with ArcGIS Pro. The raw GIS data was sparse and dirty. Moreover, it was hard to acquire a dataset that included building information and parcel information for Boston.

4.1.1 Data Acquisition

We acquire Boston Building GIS data [7], Boston Tree GIS [8] and data Massachusetts Street Block GIS data [9] from the public data portal of the City of Boston and Massachusetts State (Fig. 4). The GIS data is stored in .shp files and preprocessed with

[1] 4.453/4.S48 Creative Machine Learning for Design, Massachusetts Institute of Technology, Cambridge MA, U.S., 2020. Instructor: Caitlin Mueller, Renaud Danhaive.

ArcGIS Pro before performing any further analysis. Each object in the dataset has a .geojson geometry that indicates the geographical location for the object stored in polygon or point form. Geographical projection is canonicalized for spatial merging in the next steps.

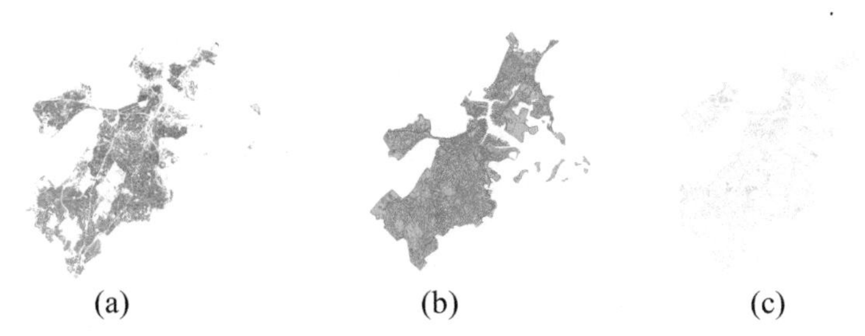

(a) (b) (c)

Fig. 4. (a) 90,308 building data acquired from Source [7] (b) 8,841 Street Block Data Acquired from Source [8] (c) 201,845 Tree Data Acquired from Source [9]

4.1.2 Data Merging and Wrangling in ArcGIS

The data wrangling consists of two parts: spatial joining and feature engineering. The spatial joining is accomplished in the ArcGIS Pro environment to speed up the geospatial data merging algorithm (Fig. 5). Feature engineering is accomplished in Anaconda environment with python geospatial analysis libraries include Geopandas. Then we split the dataset by parcel data so that we could bag the building to the closest street parcel that contains the building geometry.

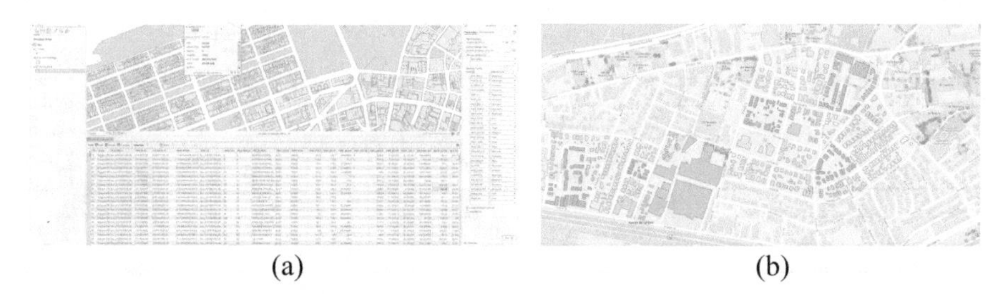

(a) (b)

Fig. 5. Spatial joining and profile merging in ArcGIS Pro

After we preprocessed the data, the building data frame and tree data frame, stored as .geojson objects, are spatially joined to the closest parcel that contains such objects. These files are ready for processing in python environments.

4.1.3 Feature Engineering

After this step, buildings within the same parcel are stored in the same tabular data frame object while maintaining the statistical numeric features such as elevation, building height, and land price. Feature engineering is accomplished in anaconda to calculate important design metrics such as coverage rate, plantation rate, and floor area ratio (Fig. 6).

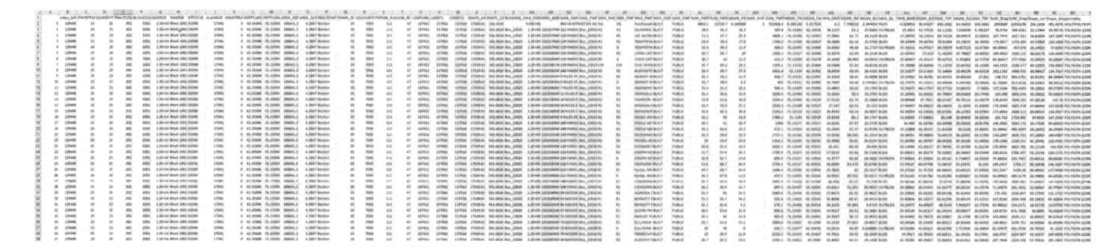

Fig. 6. Example data frame

After feature engineering, we can easily perform spatial query such as spatial join, which overlay two geospatial feature on top of each other according to selection metrics. The joined metric is stored in shapely geometry format, and we will further work with the geometry file parsed and aggregated this step as our cornerstone. For the program, each building in the site is color coded with a particular RGB value according to the categorical programs feature in the dataset. We use Matplotlib Tab 20 Colormap to maximize differences between various categories (Table 1).

Table 1. City of Boston GIS data land use code and dataset color coding

Use Code	Description	RGB triplet	Color
R1	Residential 1 Family	(44, 160, 44)	
R2	Residential 2 Family	(148, 103, 189)	
R3	Residential 3 Family	(227, 119, 194)	
R4	Residential 4-6 Family	(255, 127, 14)	
A	Residential 7 or More Units	(255, 187, 120)	
RL	Residential Lot	(152, 223, 138)	
CD	Condominium	(214, 39, 40)	
CC	Commercial Condominium	(255, 152, 150)	
CM	Condo Main	(31, 119, 180)	
C	Commercial	(140, 86, 75)	
RC	Mixed Residential Commercial	(196, 156, 148)	
CL	Commercial Land	(23, 190, 107)	
CP	Condo Parking	(199, 199, 199)	
I	Industrial	(127, 127, 127)	
E	Exempt	(188, 189, 34)	
Background	Background for the Site	(158, 218, 229)	

With the color-coding technique, we rasterize the GIS dataset into an image dataset ready for deep learning (Fig. 7). This research used the program matric as a show case for the research, as similar approaches could be applied to other data features such as height, price and plantation.

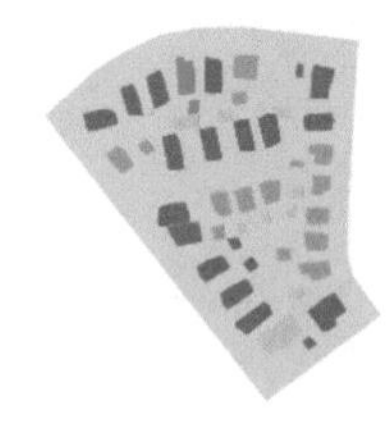

Fig. 7. Example data plotting

4.2 Model Building and Training

To prepare a dataset ready for Pix2Pix training, we need to prepare a dataset composed of training images and predicted images laid side-by-side (Fig. 8).

Fig. 8. Dataset samples

The model is trained in TensorFlow 2.1.0 environment with NVIDIA GeForce RTX 2080 Ti GPU acceleration. The training data consists of 4400 pairs of conditions and ground-truth site plans as training data and 400 validation data (Fig. 9).

4.3 Results and Visualization

The immediate result of the model is visualized in two-dimensional space as a masterplan generation in 2D (Fig. 10). As the training progress, the fidelity of the generation model gradually improves. However, a tendency of bias towards condo type is observed. This could be caused by the unbalanced number across types in the GIS Dataset.

Meanwhile, the color-coding for various program might also cause the bias in the generator. The generated site plans are blobby. This might be caused because various pattern of urban fabric coexist in the dataset, the density of urban setting could vary drastically from one area to another.

4.3.1 Visualization in Rhinoceros and Grasshopper

To further test and visualize the model, we use GH Python Remote [10] and flask server to request prediction and transmit data between the python server and the Rhinoceros-Grasshopper Visualization environment (Fig. 11).

For the current research, the extrusion height for building geometry uses the generic average building height of dataset, which is 16 m. In future work the height feature could be implemented using similar methodologies.

Various Site boundary shape is tested for site planning generation. The model performs best for sites with quadrilateral boundary shape, which might be a result because most city blocks in Boston are quadrilateral shapes. In future work, this aspect might be further enhanced with data augmentation.

Fig. 9. Training data batch example

4.3.2 Extended Usage: City Style Transfer

Generator essentially learns the mapping of a specific urban fabric pattern of a particular city in our case. Extensions of the model could be applied for urban analytic applications.

One extension is to develop a "city style-transfer" interface. The interface takes the urban parcel as input and output urban fabric. Thus we will be able to test how

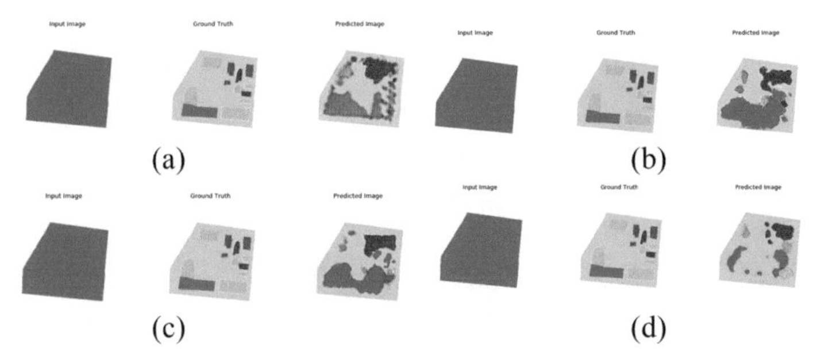

Fig. 10. Model generation after (a) 1 Epoch (b) 50 Epochs (c) 150 Epochs (d) 200 Epochs

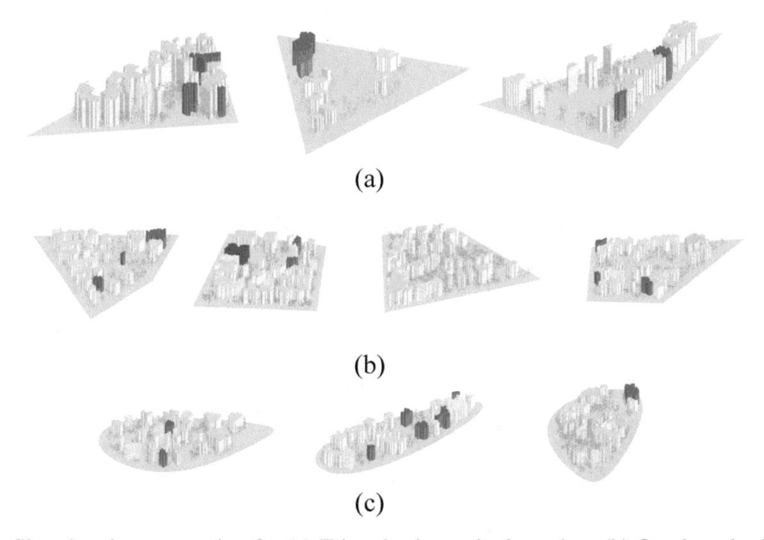

Fig. 11. Site planning suggestion for (a) Triangle shape site boundary (b) Quadrangle shape site boundary (c) Spline shape site boundary

the urban fabric of one city could be overplayed and juxtaposed onto another city. For demonstration, the generator trained on the GIS data of Boston is applied to the lower Manhattan of New York City (Fig. 12).

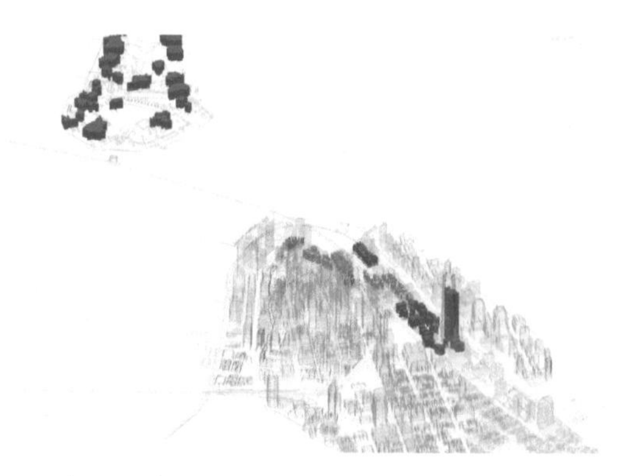

Fig. 12. City style transfer overlaying boston building footprint to lower Manhattan

5 Summary

The innovation for this project is that it proposes a generic workflow for building and fitting deep generative model that generate suggestive site planning in an accessible way. The toolkit could be utilized both in the context of site planning but for morphological analysis for city's urban fabric. Encoder-decoder network could be realized with transfer learning to the model's latent space. Here we provide some potential impact of the model and potential future improvements.

The project might have great impact for urban planners and urban designers, especially during the stage of concept design and massing studies. The project might also impact real estate industry much to a similar impact to urban designers. Real estate developers could use the tool for fast-prototyping design iterations at the very early stage of a real state financing project that involves complex site planning.

There are still several aspects that this model could be further improved. With the extensible framework, multiple GAN models could be trained to learn the mapping about multiple modalities of the data, and thus provides more comprehensive generation of building data.

Another possible improvement is the optimization of network structure. With structural Neural Networks and inference methods, we can achieve multimodal generation using a combination of conditional autoencoders and generative adversarial network such as CVAE-GANs.

Acknowledgements. The author would like to express great appreciation to the Professor Catlin Muller and Dr. Renaud Danhaive for their contribution in teaching. The also would also like to express deep gratitude to Ms. Karoline Skatteboe from Spacemaker who supported the research and offered insightful advice and feedback.

References

1. Lynch, K., Hack, G.: Site Planning, 3rd edn. MIT Press, Cambridge (1984)

2. Goodfellow, I., Pouget-Abadie, J., Mirza, M., Xu, B., Warde-Farley, D., Ozair, S., Bengio, Y.: Generative Adversarial Networks (2014). ArXiv.org, 10 June 2014
3. Mirza, M., Osindero, S.: Conditional Generative Adversarial Nets (2014)
4. Isola, P., Jun-Yan, Z., Zhou, T., Efros, A.: Image-to-Image Translation with Conditional Adversarial Networks (2017). ArXiv.org, 22 Nov 2017
5. Chaillou, S., Witt, A.: Harvard University. Graduate School of Design. Architecture for the 90%. The Iceberg (2019). http://stanislaschaillou.com/thesis/GAN/unit_program/
6. Pix2Pix Tutorial with Tensorflow. https://www.tensorflow.org/tutorials/generative/pix2pix
7. Boston Buildings - ArcGIS Online. https://www.arcgis.com/home/item.html?id=c423eda7a64b49c98a9ebdf5a6b7e135
8. MassGIS Data: Building Structures. https://data.boston.gov/dataset/trees
9. MassGIS Data: Trees. https://docs.digital.mass.gov/dataset/massgis-data-datalayers-2010-us-census
10. Pierre, C.: Massachusetts Institute of Technology, GH Python Remote. Food4Rhino (2017). https://www.food4rhino.com/app/gh-python-remote

Understanding and Analyzing the Characteristics of the Third Place in Urban Design: A Methodology for Discrete and Continuous Data in Environmental Design

Namju Lee[1,2(✉)]

[1] NJSTUDIO, Los Angeles, CA, USA
nj.namju@gmail.com
[2] Environmental Systems Research Institute (ESRI), Redlands, CA, USA

Abstract. With a rapid development of data-driven technologies, many opportunities have arisen to understand and characterize urban contexts. This paper addresses the methodology to understand a place in urban settings through the lens of third places and motility based on the walkable distance. To capture and process third-place data, fetched from Google Places, based on a given location, this paper discuses two data structures and process of discrete and continuous data. Representation of third places in a specific location of a city is characterized by representative queries. Its identified chart as a perspective of understanding a designated area could compare with other charts in different places. This method allows us to distinguish the constitution of third places based on the distance among places, enabling us to develop design strategies to differentiate or accord the sites based on mobility. The goal is to set up a method to process, interpolate, and visualize discrete and continuous urban data with representative queries of third places based on distance.

Keywords: Third place · Visualization · Continuous data · Discrete data

1 Background

Third place refers to the social environments between the two usual surroundings of the home as a first place and the workplace as second place. These places or settings such as cafes, clubs, parks, libraries, or restaurants, where we spend our daily life except for the first and second places. The diversity and density of the third place could become a barometer to characterize a place. According to the book "The third place", an engagement location, is where people consider it as a measurement of their sense of distinctiveness and wholeness. Third places are considered as sociability and nondiscursive symbolism. The benefits of third-place involvement are discussed regarding diversity and novelty, emotional expressiveness, color, and perspective. This means that the third-place shows a section of a city in many ways.

P. F. Yuan et al. (Eds.): CDRF 2020, *Proceedings of the 2020 DigitalFUTURES*, pp. 114–123, 2021.
https://doi.org/10.1007/978-981-33-4400-6_11

Energy is an essential resource in modern societies. Every major city on the earth consumes diverse types of energy, such as fossil fuel or electrical energy at the expense of our environment. Climate change has become a more critical issue in modern societies. However, in everyday life, we need a certain amount of energy to guarantee our experience. Among the significant energy consumption in cities, transportation is a significant issue and could be reduced, for instance: Fig. 1.

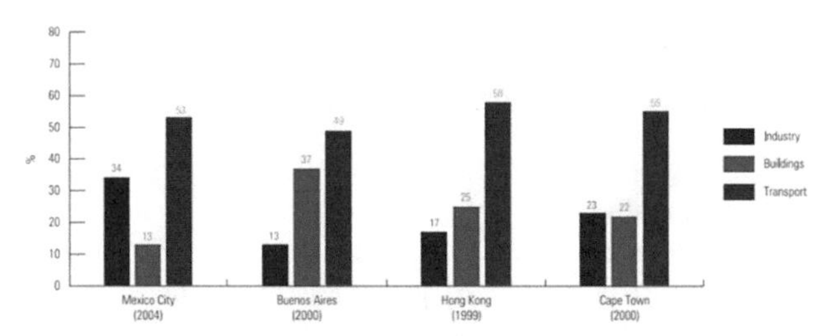

Fig. 1. Energy consumption in selected cities in middle-income countries

The urban population is more significant than rural areas, and the population increases the consumption of energy for transportation, Fig. 2. However, the energy use for transportations per person in a low density is higher than urbanized cities with high density because citizens need to use transportation to access third places in their daily life.

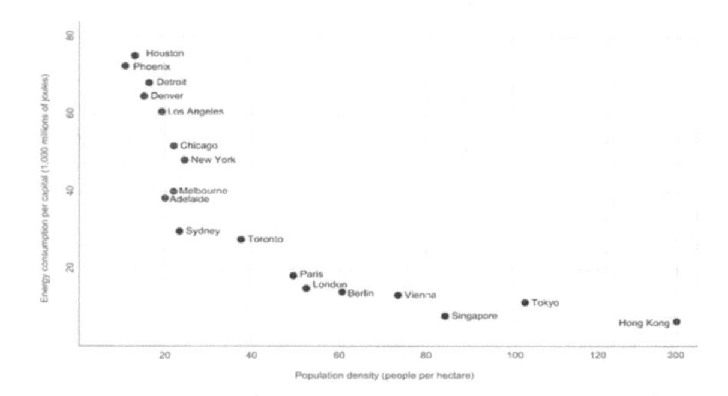

Fig. 2. Population density and energy consumption, selected World cities

The distribution of third place could be reflections from the population, sustainability, energy consumption, history, life, or style of a city. A certain place in a city could be a different setting from a place to the same city with zooming in. Even we could know the sequences of the differences between two particular points in the same cities based on the walkable distance. This lens is useful for understanding the context of an urban scale and characterizes the relationship at the architectural level.

Third place and walkable distance provides an important insight into accessible and energy for transportation in a city. Throughout the comparison among analyses in different urban settings, it becomes possible to adding removing and relocating third places in the city.

This paper discusses the mobility energy and third place which are considered as numerical data in the urban context and furthermore computes them to find a better solution of recognizing and distinguishing a position in an urban setting. With the measuring system of walkable distance, urban contexts would be reconsidered with the lens of third place.

2 Methodology

2.1 Data and Data Structure for Manipulation

The research methodology consists of two parts: 1. Data manipulation and tool making 2. Data structure for manipulation.

The workflow was developed in a custom component in Grasshopper of Rhino3d: Fig. 3. Google Places API was used to parse the third places in Cambridge, Massachusetts.

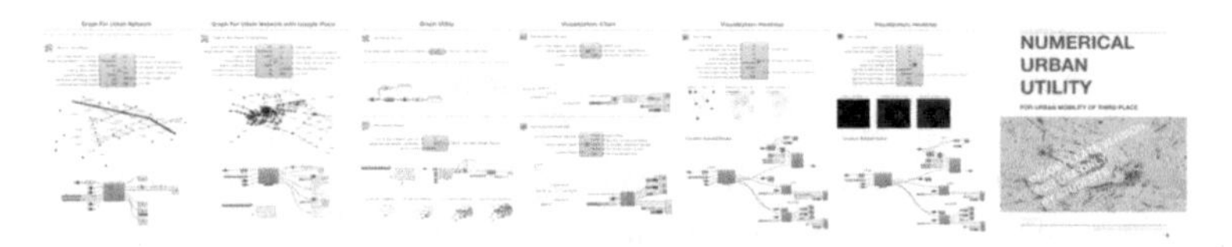

Fig. 3. Numerical Urban Analysis tool

The data structure for manipulation is unlike the well-known data array such SQL (Structured Query Language) as Tabular matrix-like CSV (Comma Separated Values), TSV (Tab Separated Values), or such NoSQL (No Structured Query Language) as JSON (JavaScript Object Notation) or graph structure. A data structure should be interactive and computable with its data sets and neighbors, children or connected data sets as efficient as possible to process more complex data.

There are two data structures for the experiment: (1) pixel and (2) graph data structure where individual data are populated and calculated. Pixel data structure is a two-dimensional matrix, interpolating a position data of an urban or district into continuous data as a finite setting for analysis, in which each pixel has the relationship with its neighbors, and each one computes its data based on neighbors' settings so that urban data can be naturally addressed and computed in the spatial context. Graph data structures enable the compute of discrete information such as distances from a particular place to third places to capture distance-based spatial urban data. Two data structures talk to each other to compute the final result as the character of the place.

2.2 Pixel Structure for Continuous Data and Blending Data with Neighbors

Pixel data structure based on a two-dimensional matrix consists of individual pixels that contain diverse data internally, for example: Fig. 4. As a parent of each pixel in the hierarchy, the pixel structure governs and controls computing and emerging the new data by processing not only its child pixel but also its neighbors. Like image processing and convolutional filters for feature extractions, the data affect their neighbors based on given algorithms so that the effect of data in the given relationships appears and emerges as a new pattern of data, for instance: Fig. 5. Thus, a point in a city becomes an area converting and visualizing the interpolated continuous data like a heat map.

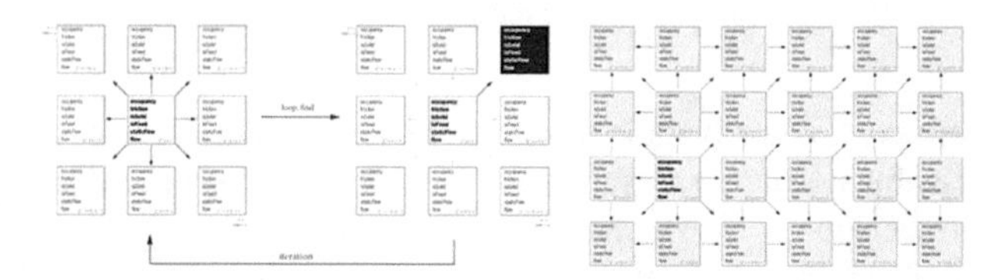

Fig. 4. The composition and relationship of a pixel map

Fig. 5. The interpolation of discrete data on the pixel map

The data interpolation could be expanded to a three-dimensional degree by the voxelization of multi-layers of the pixel map along the Z-axis. Based on the parameters of the interpolations (linear and nonlinear) for damping the driven data: Fig. 6. In the case of third-place data in three-dimensional space, captured by a graph data, the voxelized treatment could be applied as a pro-process, damping, and normalizing the interpolated data, for example: Fig. 7.

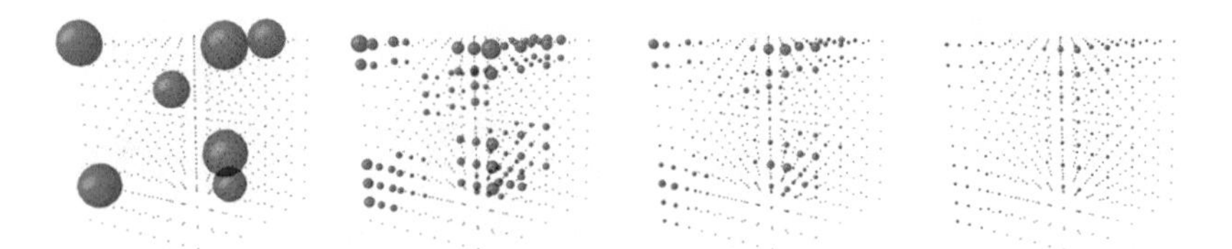

Fig. 6. Data blending in a voxel structure

For instance, a place could have a stronger attraction and others, and also a place push suppresses certain types of data. Locations of Police office could push out and minimize the density of data related to crimes. This type of assumption becomes an algorithm for interpolating and normalizing discrete data into continuous information.

2.3 Graph Structure for Discrete Data

As a different technique, A Graph data structure is utilized. The graph structure is a mathematical object that consists of nodes and edges and is widely used to represent

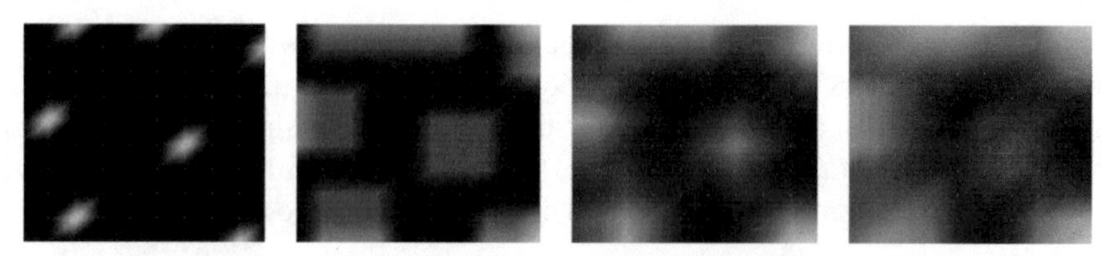

Fig. 7. Data blending in pixel by flatting the layer of voxel structure

relational data structures, for instance: Fig. 8. The street network of the urban, street, highway, or the subway map are examples of objects whose graphs closely resemble their physical form: Fig. 9. Thus, the structure is deployed to process urban data in spatial relationships to compute mobility energy consumption based on third place and the Pixel data structure.

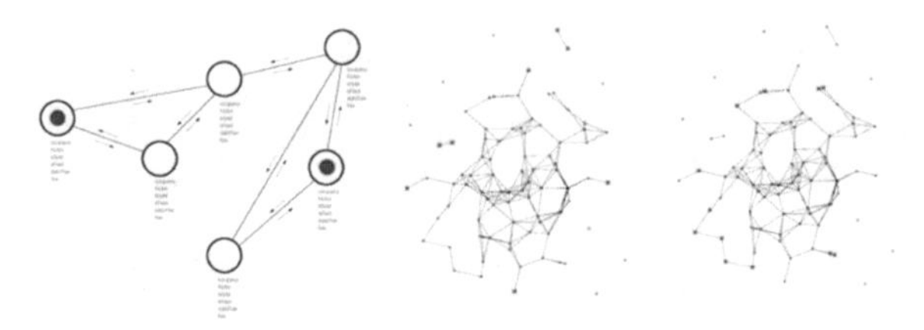

Fig. 8. The composition of nodes for parameters and edges for distances of a graph and its parameters

Fig. 9. Network analysis and data visualization

The graph data is constructed based on the pixel map, for instance: Fig. 5, and talk to each other in term of inspecting area information derived by position information. When considering the nature of the graph and performance issues, discrete information is needed. Thus, the pixel data map computes and sums up the weight and features of third-place data around the location, as numerical information, and feeds the information to the node of the graph. Therefore, to inspect the type and the density of third place based on walkable distance, a place could be selected in a city, and it finds the closest node of the graph to search the network of the graph based on the distance from the selected node. Walkable distance in a city

1 min (80 m, 0.0497097 mile), 5 min (400 m, 0.2485485 mile),
10 min (800 m, 0.497097 mile), 15 min (1,200 m, 0.497097 mile)
20 min (1,600 m 0.9941939 mile)

3 Case Study Implementation

3.1 Site Selection

There are four significant places which characterizes Cambridge, MA: Harvard Square Station, Kendall Station, Volpe Center, and MIT Building 7: Fig. 10. Each place has a unique place in terms of people in each site.

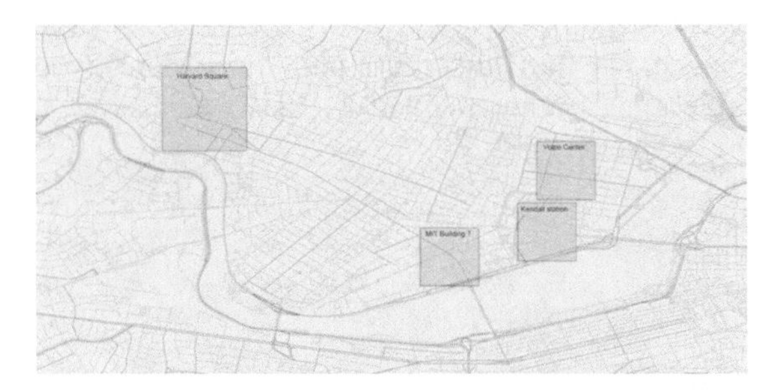

Fig. 10. Site selections in Cambridge

3.2 Parse Third Place Data and Visualization

Based on Google Places API, it parses third places around these sites. The API supports the types of queries below, and the data visualization of the places reveals patterns, highlighting certain streets and areas and contrasting the places: Fig. 11.

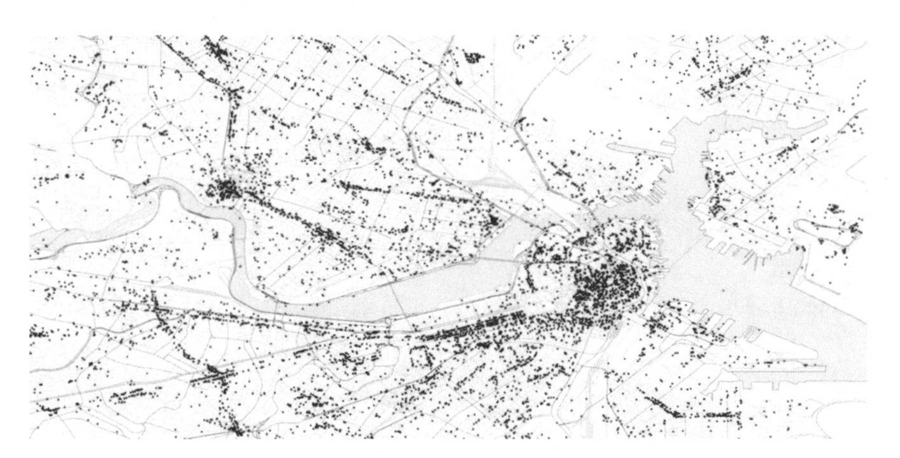

Fig. 11. Visualization: Third Place in Cambridge and Boston, Massachusetts

Type of query:

'parking', 'veterinary care', 'airport', 'plumber', 'roofing contractor', 'ATM', 'meal takeaway', 'hair care', 'insurance agency', 'school', 'synagogue', 'stadium', 'movie theatre', 'doctor', 'zoo', 'electrician', 'establishment', 'funeral home', 'spa', 'aquarium',

'storage', 'casino', 'park', 'courthouse', 'hospital', 'subway station', 'painter', 'moving company', 'movie rental', 'embassy', 'fire station', 'gym', 'bicycle store', 'local government office', 'book store', 'police', 'florist', 'museum', 'lawyer', 'car rental', 'real estate agency', 'physiotherapist', 'electronics store', 'hindu temple', 'car dealer', 'jewellery store', 'gas station', 'mosque', 'liquor store', 'campground', 'library', 'university', 'accounting', 'travel agency', 'finance', 'locksmith', 'bank', 'convenience store', 'health', 'church', 'bakery', 'lodging', 'laundry', 'shopping mall',

'dentist', 'store', 'cemetery'

Query reduction could be needed because multidimensionality tends to have much noise in terms of disclosing latent variables. In the six categories, the thirdplace data was processed, for example: Fig. 12.

Facilities : [' accounting', 'bank' , 'post_office' , 'library' , 'finance' , 'laundry'...]

Amusement : ['amusement_park' , 'zoo' , 'aquarium' , 'art_gallery' , 'spa' , 'stadium' , 'park' , 'museum','movie_theater', 'night_club...]

Health : ['doctor' , 'hair_care' , 'health' , 'hospital' , 'gym'...]

Store : ['clothing_store' , 'convenience_store' , 'department_store' , 'shoe_store' , 'store' , 'liquor_store' , 'hardware_store', ...]

Food : ['meal_delivery' , 'meal_takeaway' , 'food' , 'restaurant' , 'bakery' , 'cafe' , 'bar'...]

Transportation : ['taxi_stand' , 'subway_station' , 'train_station' , 'bus_station' , 'parking'...]

Fig. 12. The list of the queries of third places

3.3 Generate Data Structures and Inspect with Visualizations

The two data structures consume the data and process to examine a place to find and visualize the types and density of the third place: Fig. 13, Fig. 14, Fig. 15. Based on the selected node, it visualizes the places as a radial plot, which helps to compare it to others: Fig. 16, Fig. 17, Fig. 18.

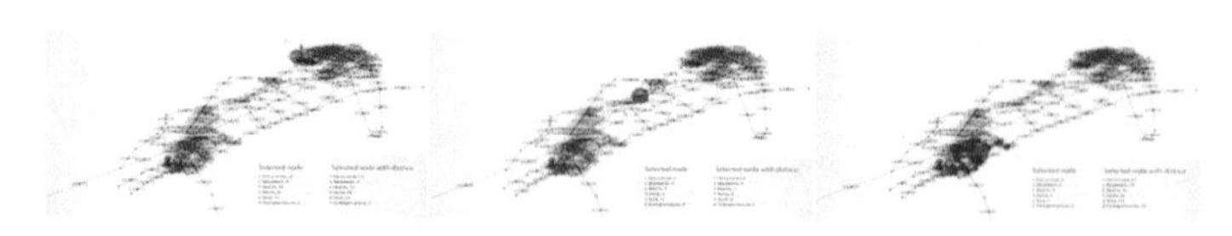

Fig. 13. Dynamic inspections and visualization the third places based on distance

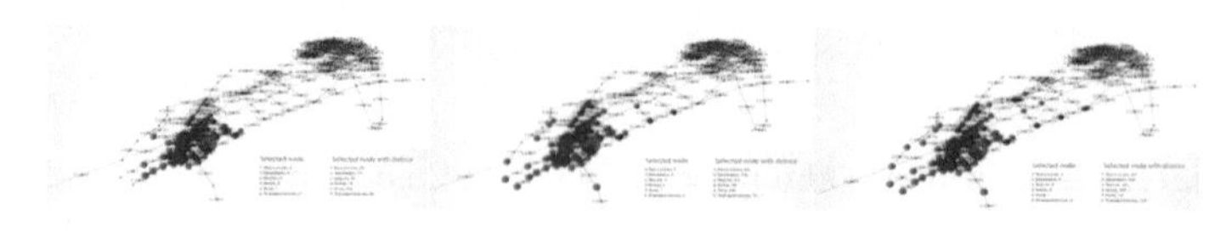

Fig. 14. Inspections for Harvard Square in 5, 15, 20 min' workable distances

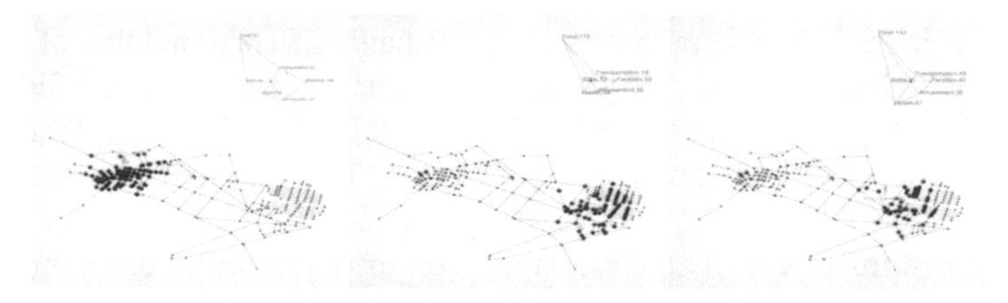

Fig. 15. Different places with the same distance and visualizing the third places.

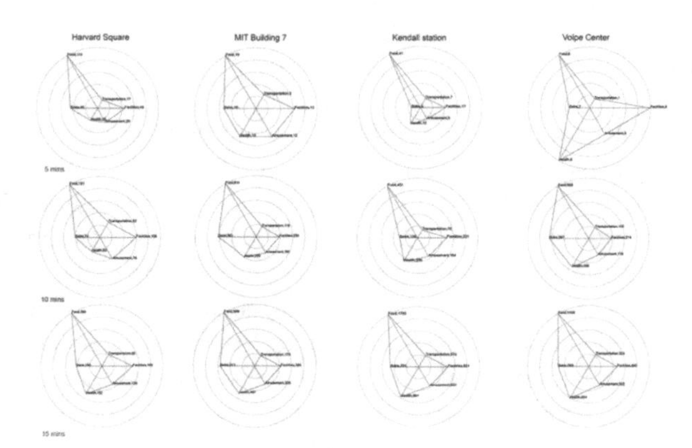

Fig. 16. Comparable charts at different distances

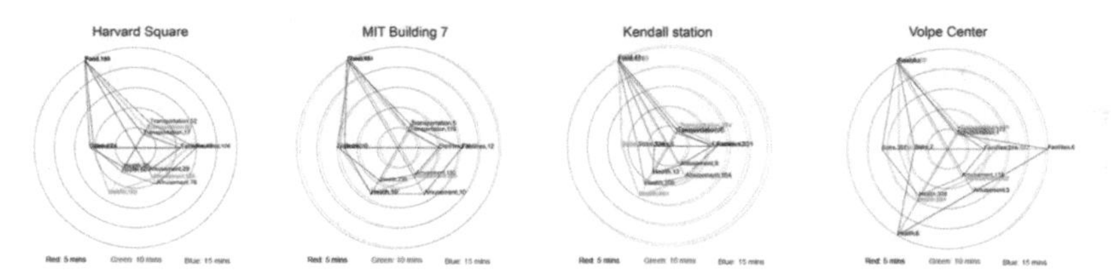

Fig. 17. Red: 5 min, Green: 10 min, Blue: 15 min

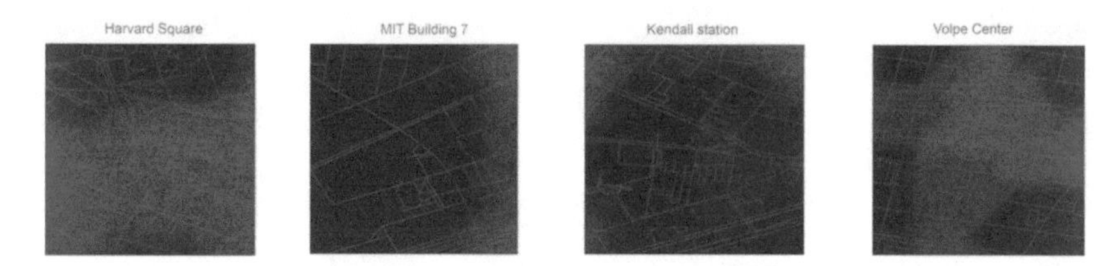

Fig. 18. Pixelated heat maps for the density of the third places in each site

3.4 Comparisons and Results

For example, Fig. 16, with many tourists and students, high density is shown compared to other places. There are 110 food places in 5-min walkable distances around Harvard

Square. For Volpe Center, where locate many companies, a small number of food places were inspected. However, it is rapidly increasing due to the Cambridge Side Galleria. In terms of mobility energy consumption, MIT Building 7 and Kendall Station are less efficient with the lack of diversity and density of third places. These places, except Volpe Center, have a high number of food areas, and Volpe Center has an imbalanced distribution of third places. This case study confirms certain assumptions about the distribution of places against people in the places based on distance. It reveals different perspectives to understand urban conditions by visualizing third places.

4 Discussion

Most of the urban data with location information are discrete data. In the real urban context, it is almost impossible to reinterpret a condition captured as data since the location always has visible or hidden relationships with their peripheries. The pixel data map allows us to interpolate discrete data as a continuous data set. Graph, a conventional and generic analysis tool, could be shaped as networks on top of the pixel map matrix to capture and compute third places data. Designers have better parameters to restore and set up urban modeling to process and characterize third place based on distances through the combination of two different data settings and processes.

This methodology makes it possible to compare and visualize a particular site with other different cities in understanding the context by third places and mobility or juxtapose a place with others in the same city. It reveals unique features and composition of third places by the comparison among designated places with the lens of the third place. As a decision-making process and visualization tool, designers could deploy it to differentiate the contrasts, or accord similarities by injecting or removing particular third place in the specific distance while developing their design in urban contexts.

5 Conclusion

Understanding the density of third place based on walkable distance could reveal the characteristic of cities which allow designers to develop their decision-making process by comparing with different locations or the past of the same location.

6 Future Work

Time-based data and other features such as the popularity of a place, race, population, or economic factors at different scales of time could reveal patterns or relationships with third places. It could provide a better understanding of the analysis at a different time, a day, and seasons.

By Deploying agents against the walkable-distance-based analysis, it makes it possible to develop diverse scenarios of modeling to simulate and understand urban contexts dynamically by parametrizing and controlling the follows and tendencies of agents with third places. Thus, we could compare the same agents in different urban contexts to reveal what we need to inject or remove third places to contrast or accord the cites.

References

1. Oldenburg, R., Brissett, D.: The third place. Qual. Sociol. **5**(4), 265–284 (1982)
2. UN-Habitat global obervertory 2008
3. Newman, P., Kenworthy, J.: Costs of automobile dependence: global survey of cities. Transp. Res. Rec. **1670**(1), 17–26 (1999). According to the Euclid's Definition, "A Pooint is that which has no part"
4. Numerical Urban Utility [Internet]. [updated 2016; cited 2020 May 23]. Available from: http://www.njstudio.co.kr/main/project/2016_MobilityEnergyConsumptionMITMediaLab/tool.html

Sensing the Environmental Neighborhoods

Mobile Urban Sensing Technologies (MUST) for High Spatial Resolution Urban Environmental Mapping

Maider Llaguno-Munitxa[1,2](✉) and Elie Bou-Zeid[2]

[1] Northeastern University, 11 Leon St, Boston, MA 02115, USA
m.llagunomunitxa@northeastern.edu, mllaguno@princeton.edu
[2] Princeton University, 59 Olden St, Princeton 08540, USA

Abstract. Given the benefits of fine mapping of large urban areas affordably, mobile environmental sensing technologies are becoming increasingly popular to complement the traditional stationary weather and air quality sensing stations. However the reliability and accuracy of low-cost mobile urban technologies is often questioned. This paper presents the design of a fast-response, autonomous and affordable Mobile Urban Sensing Technology (MUST) for the acquisition of high spatial resolution environmental data. Only when accurate neighborhood scale environmental data is affordable and accessible for architects, urban planners and policy makers, can design strategies to enhance urban health be effectively implemented. The results of an experimental air quality sensing campaign developed within Princeton University Campus is presented.

Keywords: Urban sensing · Environmental neighbourhood · Urban microclimate · Air quality

1 Introduction

Recent reports have demonstrated that intraurban and neighbourhood scale urban environmental gradients can be greater than between district or city-rural differences [1, 2], and these gradients have been associated with adverse health outcomes, including cardiovascular and respiratory disease, mortality and the exacerbation of asthma and chronic obstructive pulmonary disease [3–5]. In this context, the acquisition of urban environmental data at high spatial resolutions is critical to accurately characterize these urban environmental gradients at the neighbourhood scale and identify the areas of the city with most compromised urban environmental conditions. In the last decade, many municipalities have invested in extensive low-cost urban environmental sensing networks. That is the case with the C40 Cities air quality network led by the cities of London in the UK and Bengaluru in India [6], or the Array of Things (AoT) project in Chicago [7], amongst others. The quality control protocols required to guarantee an acceptable performance of low-cost sensing networks, however, make their maintenance challenging and costly. And most importantly, stationary sensors have proven incapable of identifying the spatial gradients present at the neighbourhood scale [1, 2, 8].

P. F. Yuan et al. (Eds.): CDRF 2020, *Proceedings of the 2020 DigitalFUTURES*, pp. 124–133, 2021.
https://doi.org/10.1007/978-981-33-4400-6_12

An increasingly popular alternative to low cost fixed environmental sensing networks, are mobile urban environmental sensing technologies. Mobile urban sensing is not a new sensing protocol [9–11]; however, in the last decade it has become increasingly popular [12, 13] given the limited investment required to cover a large urban area at a high spatial resolution. Recent mobile sensing experiments include campaigns developed in Berkeley, Houston and London [14–16] amongst others. As reported by Apte et al. [1], air quality concentrations can be 8 times larger within a street, and with a limited number of vehicle fleets a large urban area ~10 km^2 can be effectively mapped in a limited time frame. However, while mobile sensing technologies can potentially unveil the strong spatial gradients present at the neighbourhood scale, standard design criteria for low-cost mobile sensing technologies to accurately characterize these environmental gradients are yet to be defined.

This paper presents an autonomous, affordable, and fast-response mobile urban sensing technology designed to aid our research on the environmental neighbourhoods, defined as the surrounding urban area influencing the environmental quality of a given point in the city [2]. A prior generation of MUST sensing kits was deployed over the Seoul inter-city bus network as part of the 2017 Seoul Architecture and Urbanism Biennale [17, 18]. The current generation of MUST sensing kits, however, has evolved to focus on the acquisition of fine spatial environmental gradients to reveal the intimate relationship between local urban features and the surrounding environmental quality. On the one hand, the design criteria of the MUST sensing kit has focused on the definition of a mechanical design that guarantees resilience over adverse weather conditions such as overheating and heavy rainfall, while providing sufficient air exchanges to enable fast-response data acquisition. On the other hand, affordability and autonomy have also been two important criteria. Fast-response and low-cost readily available sensors have been used, and magnetic legs and solar panels have been incorporated into the design to guarantee power autonomy and an easy deployment over private vehicles and public transportation networks. The results of an experimental air quality sensing campaign developed within Princeton University Campus to test and demonstrate the capabilities of this technology are presented. MUST sensing kits were deployed over Princeton University printing and mailing service vehicles for mobile road air quality mapping. As displayed in Fig. 1, the collected mobile sensing datasets were stored in real-time over a cloud service and made available through web and mobile applications to wider audiences within campus.

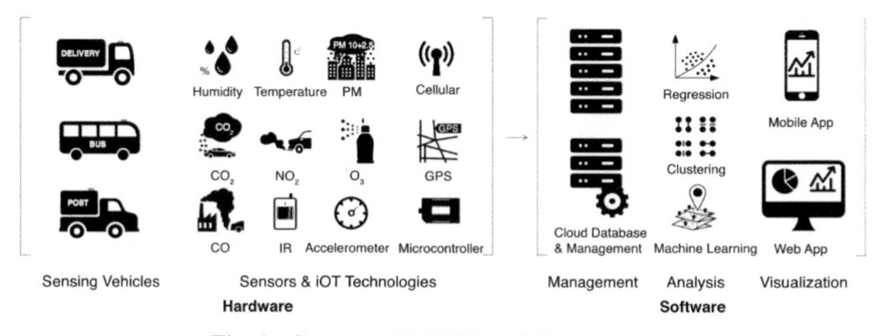

Fig. 1. Conceptual MUST workflow diagram.

1.1 Sensing Kit Design

The MUST sensing kit is an autonomous, affordable and fast-response sensing kit designed to enable low-cost, neighbourhood-scale environmental monitoring campaigns. The sensing kit relies on a solar panel to sustain the power consumption requirements of the system, and thus it can be placed on any vehicle without additional power supplies. To guarantee an easy deployment, strong rare-earth magnet legs have been introduced. Fast-response, readily-available sensors have been used to minimize the investment, and a focus on the design of the enclosure has been made to maximize the responsiveness of the sensing kit to capture the neighbourhood scale environmental gradients, while providing sufficient protection during adverse weather conditions (see Figs. 2 and 3).

Fig. 2. Deployment of MUST in NYC vehicle. Infrared (IR) video captured in NYC.

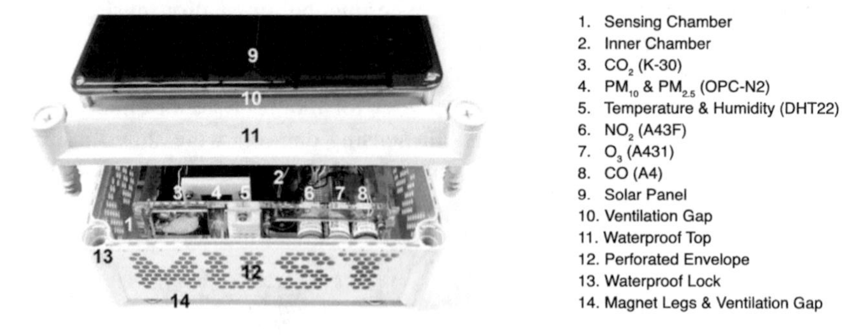

Fig. 3. MUST sensing kit configuration.

1.1.1 Hardware Design - Enclosure

The MUST sensing kit container has been designed aiming to protect the interior electronics from water, while enabling sufficient air exchange rate for the effective functioning of the sensors. Built in polyurethane, the exterior box of the MUST sensing kit has been perforated on $\approx$30% of its surface, with openings diameters ϕ ~0.5" to enable air flow to enter the sensing chamber while stopping the entry of large water droplets. An inner polyurethane shield faces the openings in the outer skin creating an inner chamber to collect the water droplets that enter the box and protect the microcontroller and electric boards, placed in the chamber, from the water. The sensing heads have been located such that their sensing surface protrude into the inner chamber, through the polyurethane shield and facing the side of the box toward the front of the vehicle, facing the incoming flow. A solar panel has been placed on the top of the box occupying the whole surface of the sensing kit to maximize its power collection capacity and guarantee that the sensing kit can operate autonomously.

Seeking a qualitative visualization to understand the ventilation potential of the box and the air exchanges that take place in the sensing chamber located between the perforated outer skin and the inner chamber, fog visualization wind tunnel experiments have been performed in the closed loop wind tunnel located in Princeton University's Instructional Fluid Dynamics Laboratory (see Fig. 4). The tunnel is a modified version of a wind tunnel model 407-A manufactured by Engineering Laboratory Design, Inc (for further details on the tunnel characteristics see [19]). The MUST kit sensing wall was positioned perpendicular to the inlet of the wind tunnel and the tunnel was set to very low wind speed velocities <1 m s^{-1} to minimize the presence of turbulence in the visualization and mimic slow vehicle movement where air exchange is weakest. Two setups were tested. In the first setup, the fog source was positioned upstream of the MUST sensing kit and was released through a sequence of $\phi \approx 0.25$" diameter nozzles pointing in the direction of the incoming flow. The nozzles are connected to the fog generator, which keeps the pipe full of fog and is released by the suction force induced by the incoming air flow. For the second setup, the fog was released through a small tube of ϕ ~ 0.35" located in the bottom of sensing chamber area. In this case a single release nozzle was used to fill the chamber space with smoke given that the goal was to visualize the efficiency of the sensing area to be cleared of smoke. The top sequence of images included in Fig. 4 shows that once the wind tunnel speed is setup to $\approx$1 m s $^{-1}$, in <4 s, the sensing chamber is clear of fog. The second row shows the nozzle configuration located in front of the box. The fog generator is started and once the sensing chamber is filled, the wind tunnel is set to ~1 m s $^{-1}$. As with the first configuration, in <4 s, the chamber is free of fog. Given the low speeds tested in the tunnel, with velocities more representative of a stationary condition than a mobile kit, this study guarantees that when the vehicle is on the move, the air exchange rates will be sufficient for data collection at a rate of $\approx$2 s (frequency of $\approx$0.5 Hz). (see Table 1). Given that the movement of cars will take place in an average intraurban velocity of about 20 km h^{-1} or 5 m s^{-1}, the spatial resolution of the measurement will be on the order of 10 m. Therefore, the observed air exchange rate has been considered sufficient.

To evaluate the effect a prolonged exposure to direct sunlight can have on sensor stability and temperature and humidity measurements, the offset in temperature within and outside of the sensing kit was measured in a mobile and stationary vehicle during a summer day within Princeton University campus. In the case of the mobile vehicle, a maximum Temperature $\Delta \approx 1$ °C was observed. In order to enhance heat transfer and avoid overheating, following the studies ventilation gaps were introduced between the

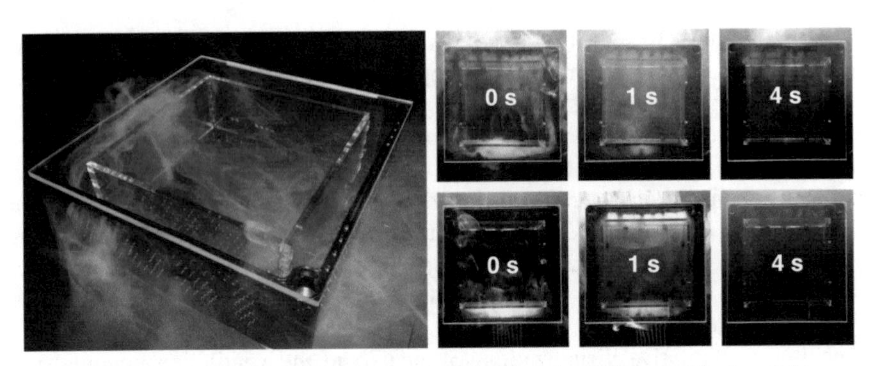

Fig. 4. MUST fog visualization. The image to the left shows the WT setup. The top right sequence shows the setup when the fog emitter is located in the inner sensing chamber. The bottom row shows the setup when the fog emitter nozzles are located in front of the box.

Table 1. MUST sensing kit sensor list, sampling intervals and calibration instruments.

	Commercial Brand	Sampling frequency	Calibration reference sensor
Temperature & Humidity	DHT 22	0.5 Hz	Rotronic MP101A & Vaisala HMS110
CO_2	CO2Meter K-30	0.5 Hz	LI-COR7500
PM_{10}, $PM_{2.5}$	Alphasense OPC-N2	0.1 Hz	Thermo Beta 5014i
NO_2	Alphasense NO2-A43F	0.5 Hz	Teledyne T500U
O_3	Alphasense OX-A431	0.5 Hz	Teledyne T400
CO	Alphasense CO-A4	0.5 Hz	Under development

vehicle and the bottom of the sensing kit, and between the solar panel and the cap of the sensing kit (see Fig. 3).

1.1.2 Hardware Design - Sensors

The sensing kit contains sensors for carbon monoxide, ozone, particulate matter (PM_{10}, $PM_{2.5}$, PM_1) nitrogen dioxide, carbon dioxide, temperature and humidity. As described in Table 1, low-cost and readily available sensors that provide a reasonable response time to develop neighborhod scale environmental sensing were chosen. The sensing kits are also equipped with a GPS shield that enables the kits to be geo-localized in real-time. Furthermore, the kit also comprises a cellular antenna that enables the acquired data to be transmitted in near real-time to an online database.

The sensor calibration has been performed following 3 methodologies. The MUST sensing kits were calibrated against i) a reference EPA air quality station located at the Rutgers University campus, ii) in the laboratory at the Princeton University Environmental Fluid Mechanics Lab, iii) and during the measuring campaign through cross-calibration between different MUST sensing kits (see Table 1 for instrument details).

Figure 5 shows the comparison between the MUST measurements of CO_2 and temperature against the reference sensors. The CO_2 calibration consistently showed an offset for the different K-30 sensors. The relative humidity readings have been converted to specific humidity for comparability purposes. It is also noted that for the specific humidity, there is an offset between the MUST sensors and the Vaisala HMS110. Based on these comparison, linear recalibrations of the MUST sensors are incorporated into the code to improve data quality. The recalibration for temperature on the other hand required a second-order fit given the distinct trends observed in the data. Finally, in order to enable sensor cross-calibration, for every measuring trip, two MUST sensing boxes were placed next each other. Additionally, an infrared camera was coupled to the MUST sensing boxes to evaluate the impact of the presence of neighboring vehicular exhaust or greenery, or exposure to direct sunlight in the data peaks and valleys.

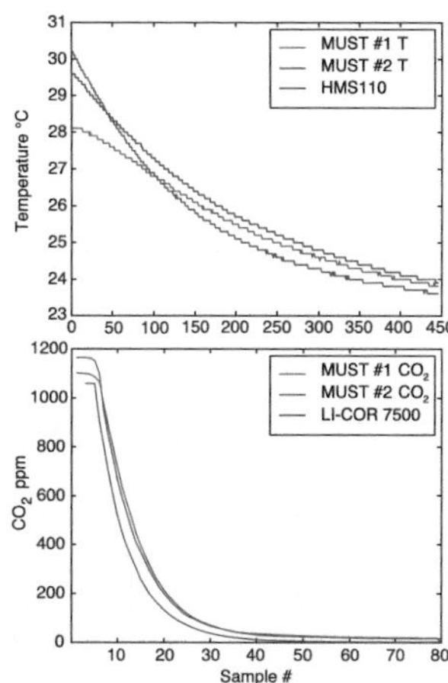

Fig. 5. To the left, the calibration setup against the Rutgers New Brunswick air quality station. To the right, the calibration results in the lab for temperature, and CO_2 are shown.

One of the main challenges of the system was to optimize the power consumption to minimize operation maintenance. The code includes a hardware sleep routine which disconnects the cellular connectivity to minimize battery consumption when the battery level reaches a given threshold, and the highest consuming sensors are disconnected through a power relay to minimize the consumption of the kit when the system is sent to sleep. The sleep routine can also be remotely activated if the weather conditions are not desirable, or the vehicle is stationary. A multicell rechargeable lithium battery of 20.000 mAh, guarantees that the system can remain awake for 3 days without being recharged by the solar panel, that is with no daylight.

1.2 Case Study

MUST sensing kits were deployed over Princeton university mailing and printing service logistic vans during summer 2019 for the duration of a month. In every measuring day, a daily average of ~3–4 h data was collected. In every occasion, a similar route was followed at least twice, and the collection time was performed from 3 pm onwards to guarantee that the rush hour peaks were also captured. As mentioned in Sect. 1.1.2, two MUST sensing kits where located on the same vehicle for cross-calibration purposes. The data collection was performed at 0.5 Hz frequency for all sensors, but the PM data were collected at 0.1 Hz. The data were sent to an online and publicly accessible server in real time. When the cellular connectivity of GPS fix was lost, the data were not sent to the server. However, such instances never exceeded 3% of the collected data volume and thus did not amount to an important loss of information.

In Fig. 6 the air quality data collected during a day on July 2019 within Princeton University central and Forrestal campus are shown. With the colormap, the median values and concentrations for the distinct collected parameters are plotted over the road network. Utilizing GIS road network datasets, the collected geo-localized datasets were mapped over the shapefiles binning based on proximity. In the plots shown above, a binning of ~100 m was performed. That is, the median of the data points contained within a ~50 m radius circle were computed. Overall, as shown in the displayed data, the highest concentrations were observed in the main intersections in the city and in the main traffic arteries that connect the campus to US Route 1. The presence of the lake and its influence on specific humidity and temperature were also consistent.

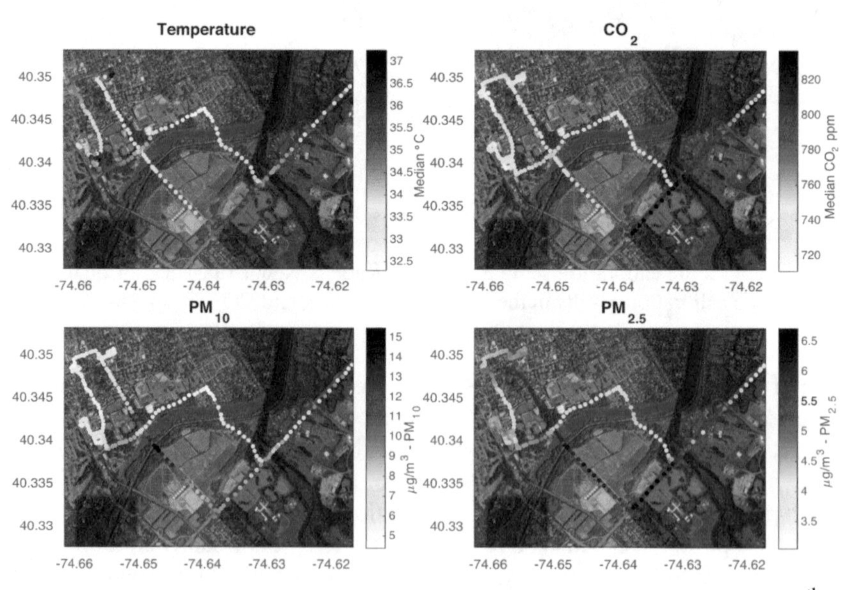

Fig. 6. Averaged Temperature, CO_2, $PM_{2.5}$ and PM_{10} data for the afternoon of July 17th 2019.

In Fig. 7, a 3-dimensional visualization of the CO_2 data collected in one of the measuring trips is displayed. The purpose of this visualization is to enable an intuitive understanding of the correlation between building density and concentration levels as well as the distinct road characteristics within the trajectories. The model comprises the whole extent of the campus until US Route 1.

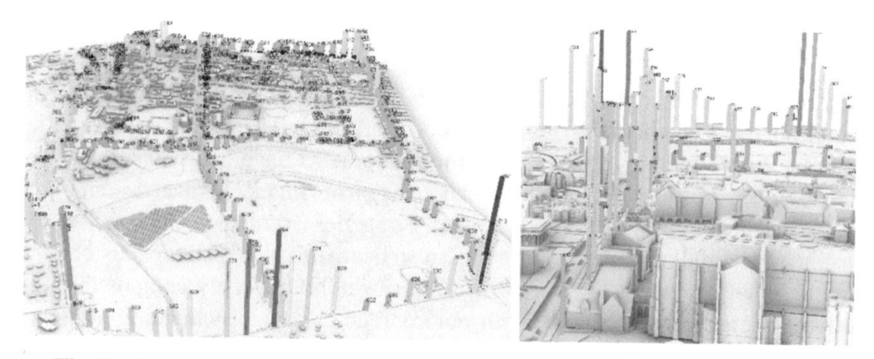

Fig. 7. CO_2 concentration visualization within the Princeton Uiversity Campus.

As shown in Fig. 8 the collected data were also made available in real time in the Princeton University website *must.princeton.edu.* Utilizing the google maps API, the environmental data are displayed together with the google traffic information, and map or satellite view. In the web app, a scatter plot displaying the data points collected by the sensing kits or an interpolated plot using a kriging interpolation is enabled. As shown in Fig. 8, a mobile app was also developed to display the environmental quality information as an interpolated gradient, with specific information on every road intersection.

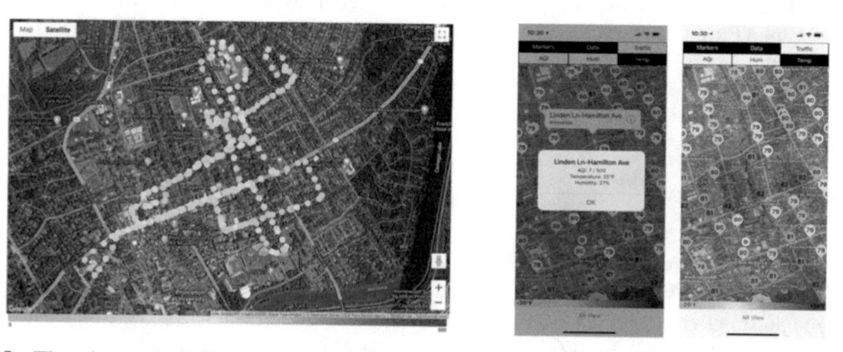

Fig. 8. The plot to the left displays the web application showing the scatter plot of an hour of measuring trip trajectory for the air quality parameter PM2.5. The plots to the right show the mobile application showing the environmental quality within Princeton University Campus (Built with google maps API).

1.3 Summary

An autonomous and affordable Mobile Urban Sensing Technology (MUST) to capture the urban spatial gradients at the neighborhood scale has been presented. The challenges to obtain accurate high spatial resolution measurements have been discussed, and the design criteria followed to enable a fast-response sensing kit mechanical design that overcomes these challenges have been described. Strategies to enhance enclosure resilience against heavy rainfall and overheating have also been discussed. The results obtained in an experimental campaign developed within Princeton University campus during July 2019 have been presented, along with environmental data visualization strategies to communicate the data to wider audiences. The study has shown that with a small investment and maintenance control, urban planners, policy makers and municipalities could make use of high spatial resolution urban environmental data to aim for environmental justice in the most compromised neighborhoods in our cities. Given the recent advances in computer vision technologies, it is becoming increasingly more plausible to couple affordable visible and IR cameras and environmental sensing kits to train a computer vision algorithm to emulate air quality information, or to develop a real-time data quality control mechanism, citizen exposure diagnosis, or pollution source detection method. Such technologies are being incorporated in the new generation of MUST sensing kits.

Acknowledgements. Authors would like to acknowledge the support of the Princeton University Campus as Lab grant, and the Army Research Office under award # W911NF2010216 (Program Manager Julia Barzyk). We would also like to acknowledge the contribution of Traci Mathieu who developed the mobile app.

References

1. Apte, J.S., Messier, K.P., Gani, S., Brauer, M., Kirchstetter, T.W., Lunden, M.M., Marshall, J.D., Portier, C., Vermeulen, R.C., Hamburg, S.P.: High-resolution air pollution mapping with Google street view cars: exploiting big data. Environ. Sci. Technol. **51**, 6999–7008 (2017)
2. Llaguno-Munitxa, M., Bou-Zeid, E.: Environmental Neighborhoods and Their Spatial Extent. IOP Publishing, Environmental Rersearch Letters (2020)
3. Brugge, D., Durant, J.L., Rioux, C.: Near-highway pollutants in motor vehicle exhaust: a review of epidemiologic evidence of cardiac and pulmonary health risks. Environ Health **6**, 23 (2007)
4. Cramer, J., Jorgensen, J.T., Hoffmann, B., Loft, S., Brauner, E.V., Prescott, E., Ketzel, M., Hertel, O., Brandt, J., Jensen, S.S., Backalarz, C., Simonsen, M.K., Andersen, Z.J.: Long-Term exposure to air pollution and incidence of myocardial infarction: a Danish nurse cohort study. Environ. Health Perspect. **128**, 5 (2020)
5. Environmental Defense Fund: Why new sensor tech is critical for tackling air pollution around the globe (2020). https://www.edf.org/airqualitymaps
6. C40 Cities: Air Quality Initiative (2018). https://www.c40.org/networks/air-quality
7. Chen, L.J., Hsu, W., Cheng, M., Lee, H.C.: DEMO: LASS: a location-aware sensing system for participatory PM2.5 monitoring MobiSys' Companion. In: Proceedings of the 14th Annual International. Conference on Mobile Systems, Applications, and Services Companion, pp. 98–98 (2016)

8. Environmental Defense Fund: Clean Air Innovation (2019). https://business.edf.org/insights/future-fleets-collecting-air-pollution-data/
9. Schmidt, W.: Die Vertilung der Minimurntemperaturen in derl Frostnacht des 12. 5.19.27 im Gemeindegebiet von Wien. In Fortschritte der Landwirtschaft (1927)
10. Peppler, A.: Die Temperaturverhaltnise von Karlsruhe an heissen Sommertagen. Deutsche Meteorologisches (1929)
11. Budel, H., Wolf, G.: Sonderklima der stadte geog. Wochenschr **1**, 25–31 (1933)
12. Van Poppel, M., Peters, J., Bleux, N.: Methodology for setup and data processing of mobile air quality measurements to assess the spatial variability of concentrations in urban environments. Environ. Pollut. **183**, 224–233 (2013)
13. Levy, I., Mihele, C., Lu, G., Narayan, J., Brook, J.R.: Evaluating multipollutant exposure and urban air quality: pollutant interrelationships, neighborhood variability, and nitrogen dioxide as a proxy pollutant. Environ. Health Perspect. **122**, 65–72 (2014)
14. BBC News: London air pollution: google Street View cars to carry monitors. (2019). https://www.bbc.com/news/uk-england-london-46878067
15. Messier, K.P., et al.: Mapping air pollution with google street view cars: efficient approaches with mobile monitoring and land use regression. Environ. Sci. Technol. **52**, 12563–12572 (2018)
16. Miller, D.J., Actkinson, B., Padilla, L., Griffin, R.J., Moore, K., Lewis, P.G.T., Gardner-Frolick, R., Craft, E., Portier, C.J., Hamburg, S.P., Alvarez, R.A.: Characterizing elevated urban air pollutant patterns with mobile monitoring in Houston, Texas. Environ. Sci. Technol. **54–4**, 2133–2142 (2020)
17. Llaguno Munitxa, M., Bou-Zeid, E., Bogosian, B., Al Tair, A., Radcliff, D., Fisher, S., Ryu, Y.: Sensing and information technologies for the environment (SITE); hardware and software innovations in mobile sensing applications. In: International Building Physics Conference 2018, Syracuse, NY, USA (2018)
18. Llaguno-Munitxa, M., Bogosian, M., Bou-Zeid, E., Fisher, S., Al-Tair, A., Radcliff, D.: Seoul On-Air: Augmented Environments for Urban Activism, Imminent Commons: The Expanded City. Seoul Biennale of Architecture and Urbanism 2017 (2017)
19. Jimenez, J.M.: High Reynolds Number Flows about Bodies of Revolution with Application to Submarines and Torpedoes. PhD, Deparment of Engineering, Princeton University (2007)

A Performance-Based Urban Block Generative Design Using Deep Reinforcement Learning and Computer Vision

Zhen Han[1], Wei Yan[2], and Gang Liu[1](✉)

[1] Tianjin Key Laboratory of Architectural Physical Environment and Ecological Technologies, Tianjin University, Tianjin, China
dkhertz@163.com, lglgmike@163.com

[2] Department of Architecture, College of Architecture, Texas A&M University, College Station, TX, USA
wyan@tamu.edu

Abstract. In recent years, generative design methods are widely used to guide urban or architectural design. Some performance-based generative design methods also combine simulation and optimization algorithms to obtain optimal solutions. In this paper, a performance-based automatic generative design method was proposed to incorporate deep reinforcement learning (DRL) and computer vision for urban planning through a case study to generate an urban block based on its direct sunlight hours, solar heat gains as well as the aesthetics of the layout. The method was tested on the redesign of an old industrial district located in Shenyang, Liaoning Province, China. A DRL agent - deep deterministic policy gradient (DDPG) agent - was trained to guide the generation of the schemes. The agent arranges one building in the site at one time in a training episode according to the observation. Rhino/Grasshopper and a computer vision algorithm, Hough Transform, were used to evaluate the performance and aesthetics, respectively. After about 150 h of training, the proposed method generated 2179 satisfactory design solutions. Episode 1936 which had the highest reward has been chosen as the final solution after manual adjustment. The test results have proven that the method is a potentially effective way for assisting urban design.

Keywords: Generative design · Deep reinforcement learning · Urban design · Performance · Hough transform

1 Introduction

Generative design was proposed first in the 1970s and was used in architectural design in 1974 (Frazer 2002). Since then, many research projects utilized different approaches, like cellular automata (CA) and shape grammar (SG), to help designers with their designs. Generative design methods are developed in order to automatically create new design schemes based on the rules or constraints set by designers. In some cases, the performance evaluation is embedded into the generative design methods to drive the creation of

P. F. Yuan et al. (Eds.): CDRF 2020, *Proceedings of the 2020 DigitalFUTURES*, pp. 134–143, 2021.
https://doi.org/10.1007/978-981-33-4400-6_13

schemes. Designers will choose an optimal solution from a large number of generated design alternatives.

Lambe and Dongre (2019) proposed a SG method to create an architectural design scheme based on the style of the existing architecture. Contextualism was used in their work to represent the relationship between new designs and the existing surroundings. Ozdemir and Ozdemir (2018) proposed a novel generation method with multi-criteria decision making (MCDM) techniques to generate alternatives for specific architectural models. Li et al. (2018) introduced the concept of circulation into shape grammar. Circulation is a design method used in architectural design that is formed by connecting the points left by indoor or outdoor space movements of human. In that research, the proposed method was tested on a commercial building, and different alternatives of circulation were generated successfully. Eilouti (2019) introduced a reverse engineering technique into generative design method and proposed a parsing tool to decode the morphogenesis in architecture. The method is synthetic, predictive, and generative. Lee et al. (2018) developed a generic Justified Plan Graph (g-JPG) grammar and proposed a hybrid method that combined Space Syntax and shape grammar to find out both the syntactical and grammatical genotypes of designs.

In addition to aesthetics, performance should also be considered in urban and architectural designs. For example, an appropriate design with superior performance will reduce energy consumption and improve human comfort as well. Many researchers have already combined the generative design methods with stochastic optimization algorithms like genetic algorithm (GA) and particle swarm optimization (PSO). Rodrigues et al. (2019) proposed a methodology of performance-based automated architectural design. This method took into consideration urban geometric and energy consumption and it can be used at the early design stages to explore a concept model. Chang et al. (2019) established some building prototypes and used deep reinforcement learning (DRL) to control the arrangement of buildings. All the schemes created by the DRL algorithm were then evaluated by their performance criteria such as energy consumption, sky openings and solar radiation. The best performance scheme under multi-constraints was chosen as the final design. Youssef et al. (2018) proposed a new method on generating the shape of building integrated photovoltaics (BIPV). This method adjusts the shapes or envelopes of the input buildings in order to generate a series of better BIPV shape alternatives. The optimal placement of BIPV for the optimized building is determined. Yavuz et al. (2018) proposed a novel shape grammar to guide the generation of acoustic panels in order to create an optimal indoor acoustic environment, from the generation of 2D geometric to the evolution of 3D acoustic panels. Rodrigues et al. (2018) proposed a step-by-step method to generate and evaluate schemes. An evolutionary program for the Space Allocation Problem (EPSAP) algorithm was introduced into the step-by-step method to create buildings, with an optimization algorithm used to find the optimal solutions. Sun and Rao (2020) proposed a performance-based generative design framework. The Grasshopper plugins Penguin, Butterfly, and Octopus were used to generate schemes, evaluate the performance, and optimize the designs, respectively.

With the help of the simulation and optimization tools, generative design methods can create the alternatives of high performance. Such an automatic generation algorithm is more effective and time-saving than a manual design approach. However, due to the restrictions of the optimization algorithms and rule-based grammars, the performance-based generative design approaches still have room for improvement, for example:

- The number of alternatives is limited in the rule-based generative design approach. The conventional approaches create the schemes according to rules or laws previously set, and that will influence the diversity of the alternatives.
- The number of design variables must be fixed during the optimization process. For example, the length of the genes in GA and the dimensions of the search space in PSO are constant till the algorithm meets the stop criterion. This means the designers need to determine the design variables at the beginning. However, some variables like the number of the buildings (in an urban design case) or the number of the lamps (in an indoor lighting design case) are very difficult to be determined in the beginning of the optimization algorithm.

In this study, a novel generative design approach using deep reinforcement learning and computer vision was proposed. A DRL agent, the deep deterministic policy gradient (DDPG) agent, is used to observe the site and generate a scheme with high-performance.

2 Methodology

Reinforcement learning is a branch of machine learning where an agent learns to handle an unknown environment based on rewards. DRL is a combination of traditional reinforcement learning and deep learning. Compared with the conventional rule-based generative design method, like CA and GA, methods using DRL can train agents to observe the environment and generate an action by themselves. During the training process, the agent will optimize the parameters to take better actions according to the rewards. Without human rules or laws, the agent can conduct a trial-and-error process automatically. Its end-to-end training doesn't need to determine design variables in the optimization process. In other words, this approach only needs the initial condition (the site information) and the goals (1. to generate an urban block which has a certain total building area; and 2. to calculate building performance by simulation tools like Honeybee in Rhino, as accurate as possible). There is no need to provide the algorithm with other information such as the number of buildings or the shape grammar rules to guide the generative process.

2.1 DRL Based Generative Design Framework

The DRL agent contains two parts: a policy and an algorithm. As shown in Fig. 1, a DRL based generative design framework was established using co-simulation with MATLAB and Rhino/Grasshopper, which includes three steps:

STEP 1: At time t of an episode, the agent observes the environment (*Observation,* S_t) and the policy takes an optimal action (*Action,* a_t) according to the observation.
STEP 2: According to the action from the agent, the environment will evaluate how successful the action is to achieve the task goal and send a reward (*Reward,* r_t) back

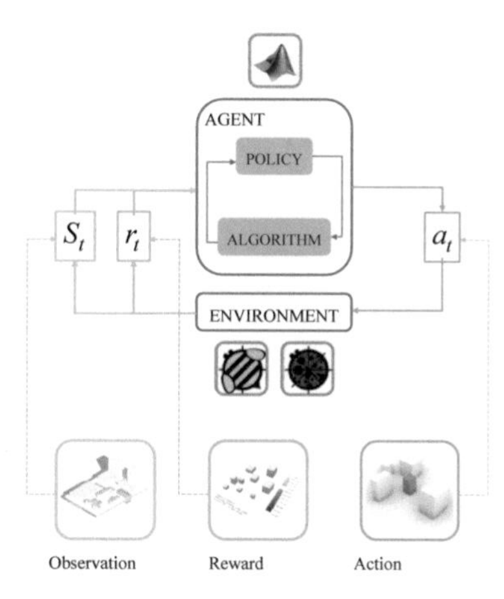

Fig. 1. The framework of the DRL based generative design approach

to the agent. At the same time, the environment will also update its state and send the observation back to the agent.
STEP 3: The algorithm will update the parameters of the policy based on the action a_t, observation S_t and reward r_t. The agent will generate a new action a_{t+1} according to the updated environment S_{t+1}.

The above three steps will repeat in each episode until S_t is a terminal observation. The training process will stop until the maximum episode iterations is reached or the other terminal criterions are met.

2.2 DDPG Agent

The goal of the DRL is to train an agent to take optimal actions to deal with changing of an unknown environment. In this research, the agent was trained using the DDPG algorithm, which is an off-policy, model-free and online DRL approach. The agent will calculate an optimal policy to maximize the long-term reward using actors and critics.

The actor and critic are function approximators used to evaluate the policy and value function. The DDPG agent includes the following four function approximators: an Actor $\mu(S)$; a Target Actor $\mu'(S)$, a Critic $Q(S, a)$ and a Target Critic $Q'(S, a)$. $\mu(S)$ accepts the observation S_t and outputs the optimal action a_t that maximizes the long-term reward; $Q(S, a)$ accepts the observation S_t and action a_t and outputs the prediction of the long-term reward. Both $\mu(S)$ and $\mu'(S)$ and $Q(S, a)$ and $Q'(S, a)$ have the same structure and parameterization. To improve the stability of the DDPG algorithm, $\mu'(S)$ and $Q'(S, a)$ will be updated periodically according to the newest $\mu(S)$ and $Q(S, a)$ parameter values, respectively (Lillicrap et al. 2015). In this research, the $\mu(S)$ and $Q(S, a)$ were established by two deep neural networks based on the observation and action (shown in Figs. 2 and 3, respectively).

As shown in Fig. 2, the actor only receives observation as input, which includes a Site Path and an Index Path. (The details of observation, action and reward is explained

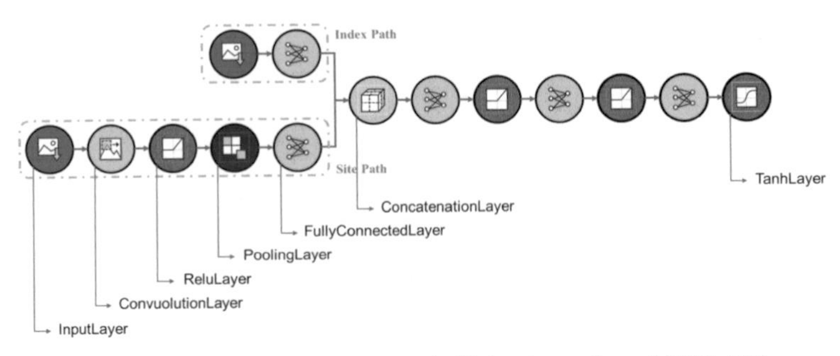

Fig. 2. Actor network structure (utilizing icons from MATLAB)

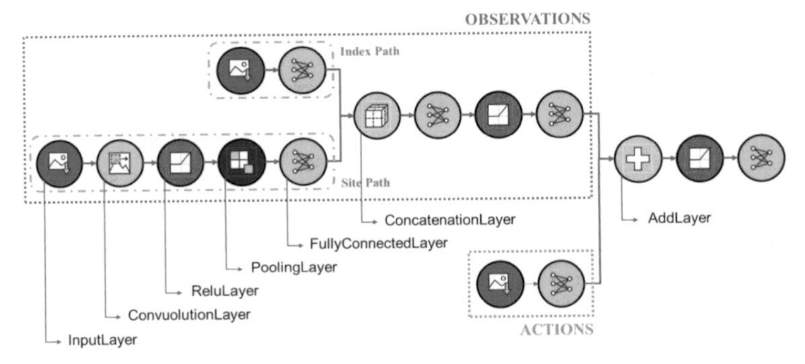

Fig. 3. Critic network structure (utilizing icons from MATLAB)

in Sect. 3.1). The inputs of the Site Path and Index Path are an image and a vector, respectively. A convolution neural network (CNN) is used in the Site Path. As shown in Fig. 3, the critic receives observation and action as well.

2.3 Hough Transform

In computer vision, Hough transform is used to detected lines or curves in an image (Duda and Hart 1972). The Hough transform algorithm can represent a line in the Cartesian space as a point in the Hough space. As shown in Fig. 4, lines in Cartesian space which go through the same point can be described as a curve in Hough space. So, the points on the same line (like Point A, B and C) in Cartesian space must intersect at one point in Hough space. The line in Cartesian space can be described as $r = x \cos\theta + y \sin\theta$, and the coordinate of the intersection point should be (r_0, θ_0) in Hough space.

Gap Distance (GD) is used to describe least distance between two line segments associated with the same Hough transform line. When the distance between the line segments is less than GD, the algorithm will merge the line segments into a single line segment. As shown in Fig. 5(b), five line segments (two blue and three orange) of the detected line segments were used as an example. The Hough transform algorithm will merge them into two line segments when GD was specified as infinity (Fig. 5(c)).

Considering the aesthetics of urban design, this research used Hough transform to evaluate an urban geometric design objective to make sure as many buildings aligned

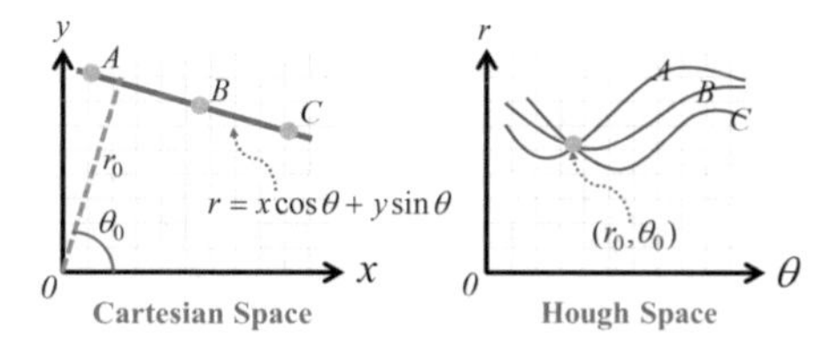

Fig. 4. The principle of Hough transform

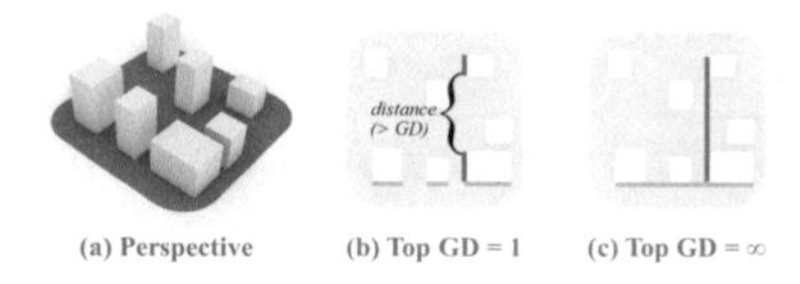

Fig. 5. The diagram of Gap Distance

as possible (as an example objective). Thus, after making GD to infinity, all the line segments in a same line will be merged to one. The fewer lines found after Hough transform means the more buildings are aligned with each others.

3 Case Study

3.1 Observation, Action and Reward

The observation in this research consists a 150 pixel-by-150 pixel-by-3 channel image representing solar radiation performance and a 3-by-1 vector representing building configurations. As shown in Fig. 6(a), the direct sunlight hours nephogram of the site calculated by a Grasshopper plugin, Honeybee, will be resized to $150 \times 150 \times 3$ and sent to the DDPG agent as one part of the observation. Another part of the observation is a vector which consists three elements: total building area, building coverage and floor area ratio (FAR), respectively.

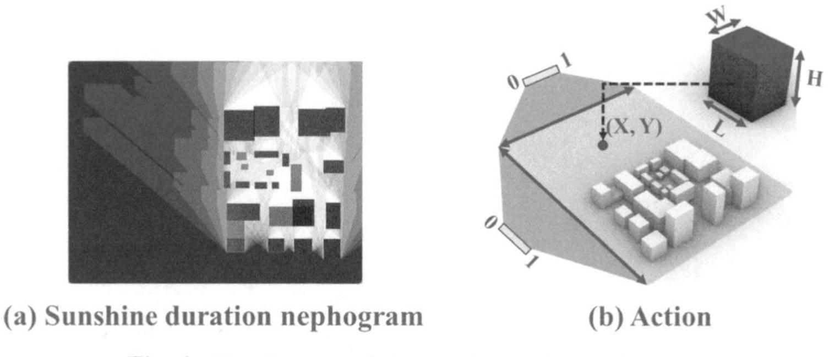

Fig. 6. The diagram of observation, action and reward

In one episode, the agent will arrange one building at one time until the episode is terminated. The action in this research is a 5-by-1 vector consisting of building location

X, location Y, length L, width W and height H. As shown in Fig. 6(b), the location X and location Y are two parameters normalized to a range of 0–1.

The reward function described in Formula (1) consists of six terms: (1) a solar heat gains reward R_{SHG} which is the average solar heat gain of the buildings in winter (*kW/h*); (2) a direct sunlight hours reward R_{SD} which is the average direct sunlight hours of the block on winter solstice (*h*); (3) an aesthetics reward $R_a = 4n - N$ where n is the number of buildings; N is the number of lines determined by Hough transform; (4) a constant reward $R_c = 10$ which encourages the agent to avoid termination; (5) a collision punishment $R_{cp} = -0.5$ and (6) a collision termination punishment $R_{ctp} = -30$.

$$R = \sqrt{R_{SHG}} + 1.5R_{SD} + 5R_a + R_c + R_{cp} + R_{ctp} \tag{1}$$

The coefficients and constants in Formula (1) were determined by a significant volume of tests that can make the agent performs best and they are used to make sure each item have the same order of magnitude (range between 0 to 30).

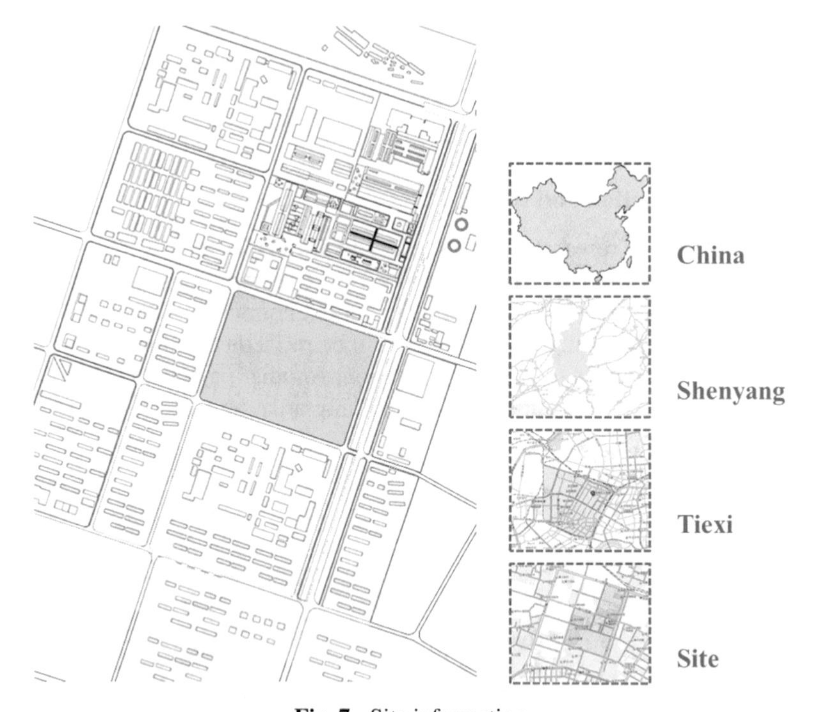

Fig. 7. Site information

In each episode, the agent will generate the urban block step by step. One building will be created at each step according to the environment until the agent meet the following terminal criterion (1) overlap of two buildings exceeds 50%; and (2) the FAR is over 3. And the environment will be reset to start a new episode until the training process is over.

3.2 Site Information

With the acceleration of urbanization in China, the old industrial districts in cities are being rebuilt. In this research, an urban design case located in Tiexi District, Shenyang, China was experimented to verify our approach. To simplify the calculation, the site only consists of one block (in blue) which is an old industrial area of about 60000 m^2 (shown in Fig. 7).

4 Results

In this research, the agent was trained using co-simulation with MATLAB and Rhino/Grasshopper. MATLAB was used to code the algorithm and Rhino/Grasshopper were used to establish the model and simulate the direct sunlight hours, solar heat gains, etc. After about 150 h of training (2179 episodes, Intel(R) Core (TM) i7-7700HQ CPU @ 2.80 GHz), the agent finally generated a series of alternatives. According to the results shown in Fig. 8, there was an upward trend from Episode 1 to Episode 2000. The last group at the lower right corner was manual adjusted according to Episode 1936 which had the highest reward according to Formula (1) among all the episodes.

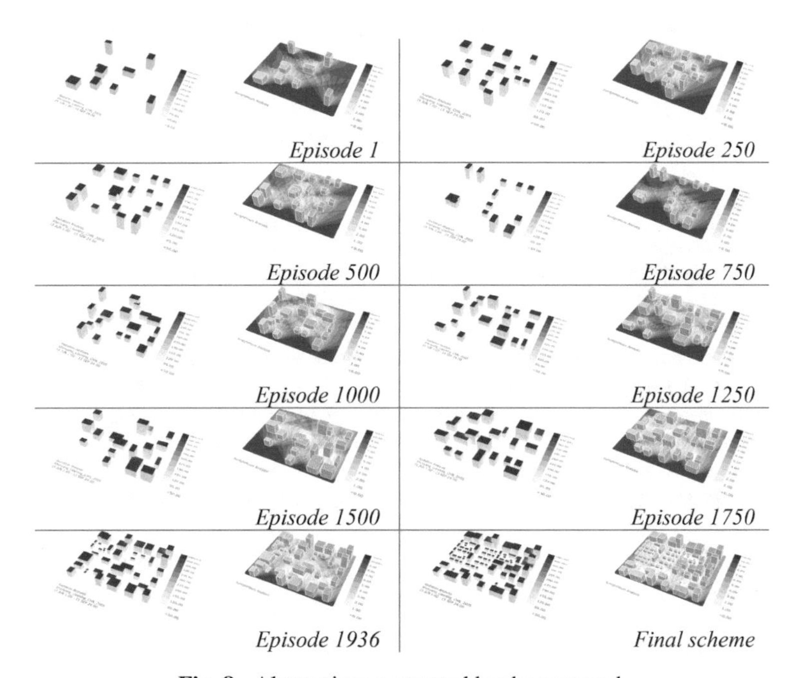

Fig. 8. Alternatives generated by the approach

According to the results, the agent performed better and better during the training process. A better agent is expected to be presented in the future by being trained to better action parameters.

5 Conclusions and Future Work

The generative design approach proposed in this research is a performance-based automatic urban design approach using DRL and computer vision. Compared with conventional approaches using optimization algorithms, this method is not limited by the number of the design variables thus can generate a scheme with any numbers of buildings in any shape. The DDPG agent was trained using co-simulation with MATLAB and Rhino/Grasshopper, and Ladybug was used to simulate direct sunlight hours and solar heat gains. Although the agent may need further training, this experiment proved the feasibility of the theory. The contribution of this research lies in the advancement and demonstration of an innovative and complete DRL model applied to performance-based generative design. This approach can be implemented into other cases by changing the observation, action and reward.

However, the agent training process is very time-consuming and it also need tough conditions (like an appropriate reward function, actor and critic network structures) to converge. Besides, the different design conditions need different reward functions and function approximators. The design of the function approximators or network structures is not a new problem, but so far is still a research problem for further study.

Acknowledgement. This research is supported by the National Natural Science Foundation of China (Grant No. 51628803).

References

Chang, S., Saha, N., Castro-Lacouture, D., Yang, P.P.J.: Generative design and performance modeling for relationships between urban built forms, sky opening, solar radiation and energy. Innov. Solut. Energy Transit. **158**, 3994–4002 (2019)

Duda, R.O., Hart, P.E.: Use of the Hough transformation to detect lines and curves in pictures. Commun. ACM **15**(1), 11–15 (1972)

Eilouti, B.: Shape grammars as a reverse engineering method for the morphogenesis of architectural facade design. Front. Architect. Res. **8**(2), 191–200 (2019)

Frazer, J.: Chapter 9 - Creative design and the generative evolutionary paradigm. In: Bentley, P.J., Corne, D.W. (eds.) Creative Evolutionary Systems, pp. 253–274. Morgan Kaufmann, San Francisco (2002)

Lambe, N.R., Dongre, A.R.: A shape grammar approach to contextual design: a case study of the Pol houses of Ahmedabad, India. Environ. Plan. B-Urban Anal. City Sci. **46**(5), 845–861 (2019)

Lee, J.H., Ostwald, M.J., Gu, N.: A Justified Plan Graph (JPG) grammar approach to identifying spatial design patterns in an architectural style. Environ. Plan. B-Urban Anal. City Sci. **45**(1), 67–89 (2018)

Li, C., Jiang, L., Sun, F.R., Zhang, K.: Generating circulation designs using shape grammars. Tsinghua Sci. Technol. **23**(6), 680–689 (2018)

Lillicrap, T.P., Hunt, J.J., Pritzel, A., Heess, N., Erez, T., Tassa, Y., Silver, D., Wierstra, D.: Continuous control with deep reinforcement learning (2015)

Ozdemir, S., Ozdemir, Y.: Prioritizing store plan alternatives produced with shape grammar using multi-criteria decision-making techniques. Environ. Plan. B-Urban Anal. City Sci. **45**(4), 751–771 (2018)
Rodrigues, E., Fernandes, M.S., Gomes, A., Gaspar, A.R., Costa, J.J.: Performance-based design of multi-story buildings for a sustainable urban environment: a case study. Renew. Sustain. Energy Rev. **113**, 109243 (2019)
Rodrigues, E., Soares, N., Fernandes, M.S., Gaspar, A.R., Gomes, A., Costa, J.J.: An integrated energy performance-driven generative design methodology to foster modular lightweight steel framed dwellings in hot climates. Energy. Sustain. Dev. **44**, 21–36 (2018)
Sun, C., Rao, J.: Study on performance-oriented generation of urban block models. Springer, Singapore (2020)
Yavuz, E., Colakoglu, B., Aktas, B.: From pattern making to acoustic panel making utilizing shape grammars. Brussels, Ecaade-Education & Research Computer Aided Architectural Design Europe (2018)
Youssef, A.M.A., Zhai, Z.Q., Reffat, R.M.: Generating proper building envelopes for photovoltaics integration with shape grammar theory. Energy Build. **158**, 326–341 (2018)

The Development of ‘Agent-Based Parametric Semiology’ as Design Research Program

Xuexin Duan(✉)

CAUP, Tongji University, Shanghai, China
duanxuexin@gmail.com

Abstract. A new framework, agenda and practice is called for to address the challenges and opportunities architecture must confront in the age of our computationally empowered Post-Fordist network society. This paper introduces the research agenda of ‘agent-based parametric semiology’, and explains the necessity of introducing a new tool, agent-based life-process modelling, as part of the design process, in order to cope with the new complexity and dynamism of architecture’s social functionality. The paper reviews the development of this design research program over the last 10 years. Finally, the paper describes current efforts to move from the illustrative use of life-process modelling to a scientifically grounded quantitative analysis and generative design optimization.

Keywords: Crowd simulation · Life-process agent modelling · Semiological project · Agent-based parametric semiology

1 Theory Background

As the founder of Parametricism, Patrik Schumacher put forwarded “Parametricism as Style – Parametricist Manifesto” at the 2008 Venice Architecture Biennale [3]. He argues that Parametricism is an epochal style after Modernism, and it is the response of architecture to the technological and socioeconomic transformation brought by the information age [4]. In 2012, Schumacher published his theoretical treatise “The Autopoiesis of Architecture 2 - A New Agenda for Architecture” and further put forwarded a new conceptual framework for contemporary architecture to address the challenges and opportunities facing architectural design in the context of the current social and technological environment [1, 2].

Schumacher emphasizes the need to distinguish architecture from art and engineering. He raised that the social function of architecture—engineering is mainly related to technical functions, and architecture is mainly related to social functions [5]. This means that architects consider the social interaction of users as socialized actors, while engineers consider the physical safety and comfort of users understood only as biological objects [9]. The unique societal role for the architecture is no longer just to provide a tangible shelter, but a communicated frame for social life.

Just like the achievements of Modernism reached in the 20th century, Schumacher hopes that in the 21st century, Parametricism can achieve an equivalent success in terms of

P. F. Yuan et al. (Eds.): CDRF 2020, *Proceedings of the 2020 DigitalFUTURES*, pp. 144–155, 2021.
https://doi.org/10.1007/978-981-33-4400-6_14

becoming the defining epochal style for the 21st century. As mentioned above, architecture has its unique social functions, every social system needs a clear spatial relationship to make the social communication stable and orderly [7, 8]. In the knowledge economy of the post-industrial age, fundamental changes have taken place in both social relations and communication methods—from hierarchical to network-based, how should architecture respond?

Parametricism was relaunched as Parametricism 2.0 in 2016. Schumacher insists that the real concern of Parametricism shifts to social functionality and how to innovate society [11]. He argues that all design is communication design [6], and each space can be regarded as a kind of communication, which gives potential social participants a message about what interaction types may occur and broadcasts an invitation to the public to participate. As Schumacher said "The elaboration of spatial complexes in accordance with a designed semiological code is thus a key to upgrading architecture's core competency. The semiological project implies that the design project systematizes all form-function correlations into a coherent system of signification, designed as a network of similitudes and contrasts, organized via a spatio-visual grammar. Each territory is a communication." [6]. Social order depends on spatial order: by reading the spatial semantic system, people can share and establish the appropriate, intended interaction situations [10]. The task of architecture can be divided into organization and articulation. In turn, articulation is differentiated into the sub-tasks of phenomenological articulation and semiological articulation. The "meaning" of the architectural space lies in the (subtle) events or social interaction types that may occur within the space.

New tools will rebuild architectural semiology into parametric semiology [12]. Like traffic lights, the various traffic lines on the road are set by artificial rules, and everyone has been accustomed to obey and follow it. This could be seen as the most straightforward symbol.

2 Why We Need Agent-Based Life-Process Crowd Simulation

Parametricism is rooted in digital technology and is based on advanced parametric design systems and scripting methods. In terms of parametric semiology, a key advance is the development of an agent-based life-process modelling tool that simulates social interaction processes that can achieve a well-adapted architectural design [15]. This would be an architectural optimization tool, in the hands of architects, analogous to the engineer's structural optimization software.

The crowd simulation usually can be classified into Macroscopic simulation (continuum-based model) and Microscopic simulation (agent-based model). Macroscopic crowd simulation is generally aimed at a wider range of crowd movement scenarios, which having people move like particles in a fluid and describe the crowd density and velocity by using different equations. This type of simulation of crowd behaviors has been widely used in traffic engineering. Microscopic simulation is focused on the study of individual behavior in the crowd, the interaction of the virtual human and the virtual environment. The crowd continuously senses the surrounding environment information and adjusts their reactions accordingly. In this paper, the author refers to the second type: agent-based crowd simulation.

There are various agent-based crowd modelling tools. However, concerning assisting the architectural design process, there are at present no tools that recognize the specific task of architecture as defined by Schumacher.

As Patrik Schumacher raised in 2016: "…models reproduce and predict collective patterns of movement, occupation, and interaction as emerging from individual, rule-based actions… It is of great importance that architectural semiology can hook its project onto a new design simulation tool that is bound to become a pervasive medium to test and anticipate architecture's social functionality" [12].

There are three levels that the agent-based life-process simulation needs to achieve in the architectural design process, the basic level is representation. During the period of CAD and 2D drawings, the overall image of the building only exists in the designers' imagination, later with the help of 3D renderings, designers can get a better understanding of their design before it is being built. Similarly, how the space will be used in the future exists only in designers' imagination, or not. In contrast to the current empty or "blank" design models, the life-process crowd simulation tool can be very helpful for the designers' understanding of the social usability of the different spaces. This leads to the second level: evaluation. Similar to structural simulation models, through a generate-and-test process, designers can use the life-process simulation tool to optimize the design space in terms of both organization and articulation, and test different design assumptions. Ultimately, this could be developed into an evolutionary loop, and thus would become a new generative design tool. Following these three stages, the social meaning of architecture can be really brought into and explored within the design model [13, 14]. In this way, architectural semiology can be finally implemented within a truly predictive design project.

The use of agent-based life-process simulation is very powerful in terms of bringing the semantic layer of the designed spaces into the design model. It is become possible to test the dynamic patterns of social interaction through computational agent-based crowd simulation techniques, and to use this as constructive and generative design feedback As Patrik Schumacher summarized, the three key innovations for the agent-based life-process crowd simulation are [12]:

1. *the generalisation of crowd modelling from circulation flow simulations to a generalized life process modelling.*
2. *the shift from physically conceived agents and crowds to communicatively conceived agents and crowds with the crucial augmentation of sign-or frame-dependent behaviours.*
3. *the differentiation of agents according to different social roles and social valences.*

Based on this, Schumacher has proposed an original innovative methodology for architectural design processes "using evolutionary algorithms that use agent-based life-process simulations with social interaction frequencies as success measures to optimize social functionality" [15]. This research agenda 'Agent-based Parametric Semiology' (ABPS) is currently shared and developed by three main research groups under Patrik Schumacher's leadership: a small team of Ph.D. candidates at the University of Applied Arts: Robert Neumayr (team leader), Daniel Bolojan, Josip Bajcer, Bogdan Zaha, and Michael Fuchs, a research team at Zaha Hadid Architects: Tyson Hosmer (team leader, leading the development of the ABPS Research Group and software development), Soungmin Yu, Sobitha Ravichandran, and Ziming He, as well as multiple student teams from AA DRL.

3 The Intelligence Upgrading of Agent-Based Crowd Simulation

To develop a reliable agent model beyond mere flow modelling in engineering is the key for the operationalization of the semiology project. The first step is to make the virtual humans more realistic; besides the graphical and animation level, this implies work on the level of intelligent behaviour. The goal is to let the agents interact with their environment as much as possible like real humans, creating heterogeneous crowds that behave in a life-like manner and inhabit differently designated spaces differently [16, 17].

3.1 Crowd Behaviour Pattern Analysis

One of the most popular and simplest ways to simulate a crowd is with Reynolds flocking algorithm. In 1987, Reynolds [2] proposed Boid, a simulation of the flocking behaviour of birds, which was one of the original simulations of collective behaviour. The model is based on three simple rules: Separation, i.e. avoid getting too close to other particles; alignment, i.e. turn towards the average direction of a local cluster; and cohesion, i.e. steer towards the centre of the flock. This modelling approach is being used in the film industry to simulate animal flocks. With respect to the complexities of the human crowd behaviours, this simple modelling approach is rather too crude and superseded by other approaches. Based on the algorithm of Boids, Processing has many external libraries of mature flocking intelligent algorithms, such as iGeo. Plethora etc. The agent-based crowd simulation modelling using processing is mainly developed based on this group of algorithms.

By using processing to do the agent-based crowd simulation, the social activities here can be expressed as different types of attraction points, such as music performances, sculptures, etc. The attraction points can be imported into the model and recognized by agents. The imprint of human behavior in the activity will be seen as the "output" of the model and can be translated by scripts into useful data such as duration, density, etc. Take the student project IArch: Design Me (project team: Xuexin Duan, Vahid Eshraghi, Jie Shen, Wei You) from AA DRL Studio Patrik Schumacher & Pierandrea Angius in 2011–2013 as an example, they classified the public activities into three main types: single - attraction event, multiple - attractions event, and no-attraction event. No - attraction event is always happening on the square which has a large area but no specific event on it, so people will wonder or just pass through it. Single - attraction event is like concert or solo performance in public. multiple - attractions event is like visiting a gallery (Fig. 1).

The agent-based crowd simulation in Processing basically abstracts a person as a point, this point will have basic information such as position, speed, direction, etc. It is more suitable for a fast analysis of crowd behavior patterns or visualization of crowd trajectory in space.

Fig. 1. IArch: Design Me project's analysis of human behavior. Besides the physical properties, agents also have social properties, such as patience, interests, etc. and based on the different types of events, the crowd will have different behaviors response, and the different values and properties will be generated as a behaviors pattern map. (credit: Xuexin Duan, Vahid Eshraghi, Jie Shen, Wei You from AA DRL Studio Patrik Schumacher & Pierandrea Angius in 2011–2013.)

3.2 Intelligent Agents

Crowd simulation in Processing has a big limitation in simulating complex behaviors, and how to visualize the human behaviors, as well as analyze different interaction between human and architectural space. Because we need to "see" the agents' actions and behaviors in the space, not as a 2D pattern representing different data, also not as characters in crowds for navigation and steering, which tend to be rather lifeless, the life-process crowd modelling should appear more social presence. Since 2011, under the parametric semiology topic, in AADRL Patrik Schumacher's studio, the research already has started on trying to make the interaction rules dependent not on physical constrains but on visually expressed information about the spaces' designation on defining the expected social situations. At an early stage, the attempt on building the link between virtual human and virtual environment is by using Miarmy or Softimage. In the IArch: Design Me project, the team introduced a furniture system as a kind of environment clue to trigger people's different behaviors in order to mutate different events. It starts from very simple interaction rules: by changing the furniture's basic parameter, such as height, to change the meaning of the furniture which can change people's different behaviours, then to achieve ordering people. The agent behavior is based on these simple rules (Fig. 2).

In the IArch: Design Me project, the scenario is fashion week which contains three sub-scenarios: opening, catwalk, and after-party. During the openings, the two curved partitions as the trigger elements give a sign to the two types of agents and guide them to enter from different entrances. When the opening ended, some agents choose to leave, some agents stay and start to chat, when the two curved partitions start to move and together with the changing on the lighting condition, agents "know" the space is mutating to another sub-scenarios: catwalk will start soon. And when the show is over, people start to leave, and the furniture is changing to a more fragmented configuration, also the changing on lighting condition helps to differentiate territories. All these give a "hint" to the agent that the after party will begin soon. As the function of spaces is conceived as dynamic patterns of social communications, by mutating the 3 sub-scenarios, we can change the communication mode. Although people are in the same physical space, because the social meaning the designer gives to the space is different, we can see how people respond to the space in different ways (Fig. 3).

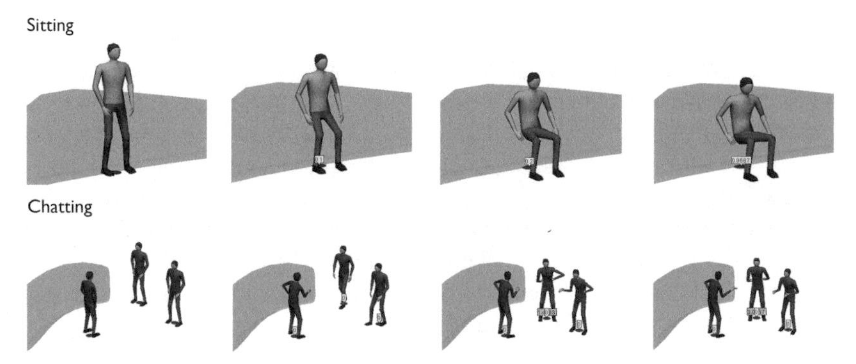

Fig. 2. The interaction rules based on the height of furniture: people will sit or start to chat beside a counter, also the agent has internal properties, like patience value, which will determine how long the agent will stay in one place or having a conversation with other agents. (credit: Xuexin Duan, Vahid Eshraghi, Jie Shen, Wei You from AA DRL Studio Patrik Schumacher & Pierandrea Angius in 2011–2013.)

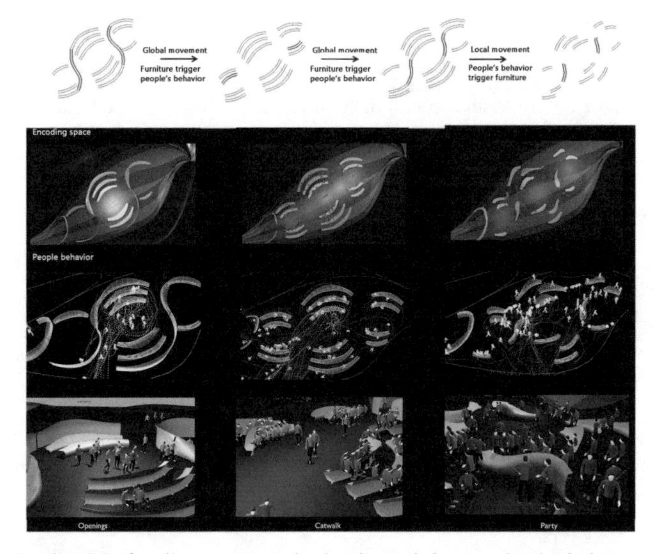

Fig. 3. IArch: Design Me furniture system: by having minimum physical changes, this furniture system can mutate the three sub scenarios. (credit: Xuexin Duan, Vahid Eshraghi, Jie Shen, Wei You from AA DRL Studio Patrik Schumacher & Pierandrea Angius in 2011–2013.)

3.3 Semantic Virtual Environment

In order to make the virtual humans have meaningful interactions in the virtual environment, the designer needs to create a semantically rich environment, as well as building up all the relevant links between agents and the environment. Recently Schumacher's AADRL studio is achieving a more complete Spatio-visual Semiological System. All semiological design must be carried out by establishing two interrelated and differentiated systems: the system of signifiers or symbols and the system of the signified or meanings [13]. For example, in one of the student projects from Schumacher's studio (Yihui Wu, Lei Wang, and Yanling Xu 2016–2018), the team set up a Manual including an explicit vocabulary and grammar for the space. Start with simple rules, like concave configuration links to individual work, convex configuration links to group work like meeting, and by introducing other distinctions, the grammar's combinatoric power can produce many different meaningful expressions. Another aspect of the parametric semiological design method is to consider that a distinction might not be introduced as a strict dichotomy, but as a gradual gradient spectrum generated by defining the two poles. However, this method is only meaningful when we can define meaningfully corresponding gradients in the field of social meaning (Figs. 4, 5 and 6).

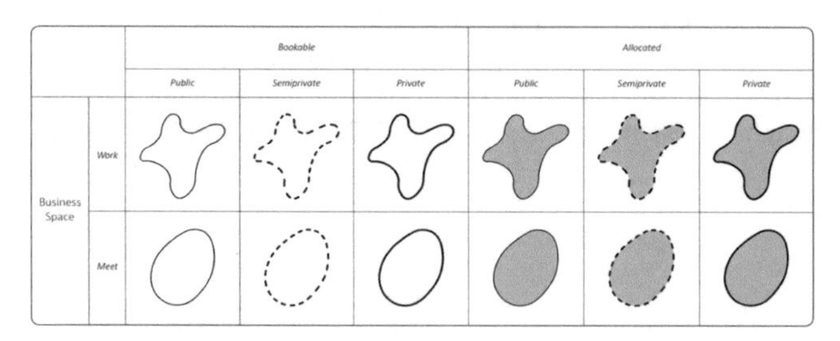

Fig. 4. Dictionary of spatio-visual vocabulary. (credit: Yihui Wu, Lei Wang, and Yanling Xu, from AA DRL Studio Patrik Schumacher & Pierandrea Angius in 2016–2018.).

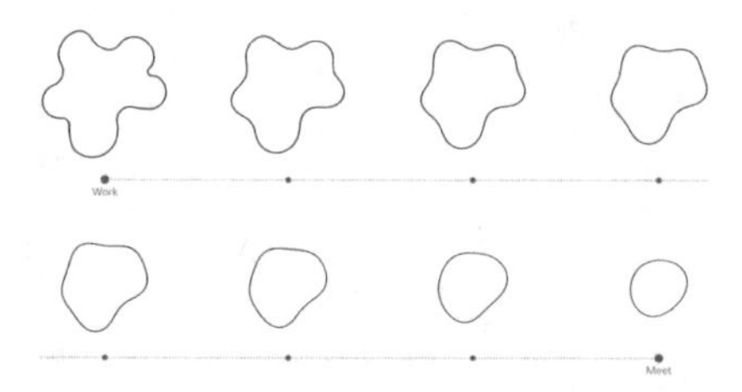

Fig. 5. The continuous spectrum of shapes can be translated to a gradient design field. (credit: Yihui Wu, Lei Wang, and Yanling Xu, from AA DRL Studio Patrik Schumacher & Pierandrea Angius in 2016–2018.).

This example of a semiological research project is intended to illustrate how a relatively complex and subtle semiological project is gradually elaborated based on a series of simple and intuitive systems of form-function association. Here, this student project, as a preliminary attempt, gives us a glimpse of the ambitions of the parametric semiological project, which is trying to create an all-encompassing and increasingly informative design language for the built environment.

4 Quantitative Analysis, Evaluation, and Optimization

The agent-based life-process crowd simulation methodology developed under the research agenda can be seen as an upgrade of the current engineer's crowd simulations. This new type of simulation is more focused on social functionality, as designers

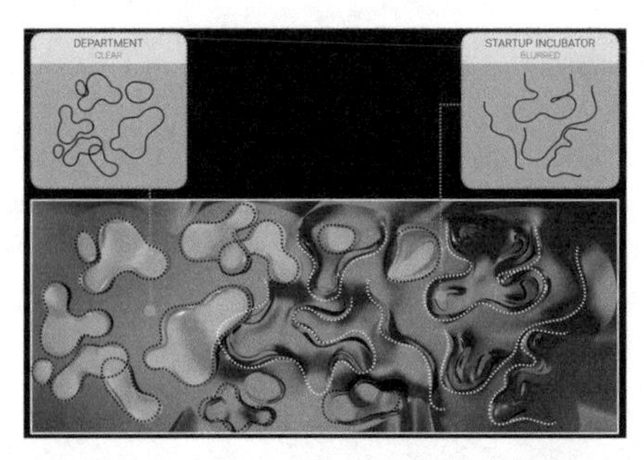

Fig. 6. The project shows one of the situations gradient spectrum, one end represents the absolute meeting situation, the other end represents the absolute individual concentrated work situation. (credit: Yihui Wu, Lei Wang, and Yanling Xu, from AA DRL Studio Patrik Schumacher & Pierandrea Angius in 2016–2018.).

treat the designed spaces as a semantically coded environment which will order people's interaction and communications. And in order to deliver a convincing semiologically informed simulation, Patrik Schumacher listed the innovations that the current research groups are working on [15]:

1. *expansion of action/behaviour repertoire*
2. *differentiation of agent population*
3. *designation dependency of behaviours*
4. *information empowered, semiology competent agents*
5. *agent decisions via dynamic utility functions*
6. *focus on social interactions and event scenarios*
7. *domain tailoring and client customization*

4.1 Methodology and Toolset

Zaha Hadid Architects ABPS group's recent research work focusses on a scientific research approach which is aiming for quantitative analysis, evaluation, and optimization in accordance with relevant social functionality criteria. The team attempts to build a methodology and toolset for a marketable, professional design service of a new quality. Their work focusses on three parts. First is the simulation: the agent-based life-process modeling. Second is the empirical validation of the model. This requires the collection and analysis of real-world data to calibrate the model. The third is the generative design which is to develop a generative design tool based on the first two parts. This paper will elaborate mainly on the first part. In order to integrate the social interaction into a heterogenous crowd simulation, the ABPS group is developing their own AI decision-making system within a software framework based on the Unity game engine.

The simulation approach contains three main parts: Agent modeling, environment modeling, and data analysis. In the agent modeling, besides the basic mobility and pathfinding using A* graph algorithm, it mainly focusses on three aspects, first is the customized action types based on different scenarios, like a base action library for the intelligent agent behavior modeling. For example, in the working scenario, it should include basic actions like having a meeting, working at a desk, etc. Second is the agent types and internal state. All these parameters will decay based on the events and time and will influence the agent's decisions. The third aspect: the AI decision-making framework. Each agent contains an AI brain that holds a set of different actions on one hand and internal state on the other hand, based on the internal state and evaluation of the environment as well as other agents' behaviors. The whole process is dynamic and autonomous (Fig. 7).

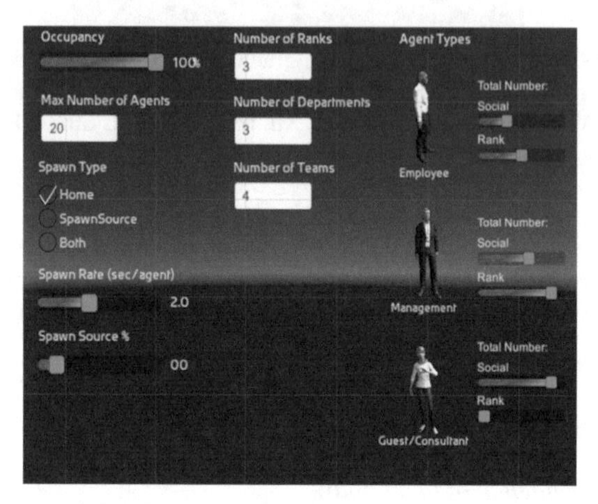

Fig. 7. ZHA ABPS's example interface of agent types, in the working scenario, the agents can be classified into different working teams, and agents can be differentiated as team leaders who will have more meetings, and team members who will work at the desk more. And each agent contains a set of different needs parameters, like thirst, hunger, concentration, motivation to work, etc. (credit: Tyson Hosmer (team leader), Soungmin Yu, Sobitha Ravichandran, and Ziming He)

The environment modeling includes three aspects: destinations, zones, and events. The destinations are places like meeting rooms which might contain sub-goals like chairs, desks, etc. Each goal has a specific script, so the agent is aware of how to interact with it. Zones are areas that are marked by special features, e.g. the floor material or ceiling colour, which will influence the behavior of the agent, such as walking speed, etc. An event is a dynamic interaction system. There can be scheduled events or unscheduled events. In the working scenario, a scheduled event could be a team meeting which will trigger the particular team to go to the defined meeting room. Unscheduled events could be spontaneous conversations triggered by an encounter between agents based on their internal state and the location of the encounter.

The data analysis includes input data and output data. Input data include agent types, internal state parameters, global influence parameters, etc. Output data include event durations, locations, different agents' encounter counts, conversation duration, etc. All the data is correlated with the spatial features.

4.2 Scenario and Example

The ABPS team is focusing its research on office work environments, especially corporate spaces and incubator spaces, as these are the best scenarios for complex dynamic interactions in information-rich environments. In such environments, many diverse interactions happen simultaneously, and it would be very difficult to predict the occupancy and communication patterns by merely looking at the drawing and imagining scenarios. The new tool of agent-based life-process simulation assists the designers to understand the social effects of the space arrangements better, to decide, for instance, where meeting zones or social zones would be best located. The tool is not only helpful on the organizational level, but also on the articulation level. The designers can test different social codes and protocols attached to different spaces. The goal is that through this design process, designers can really create a productive, socially effective architectural order (Fig. 8).

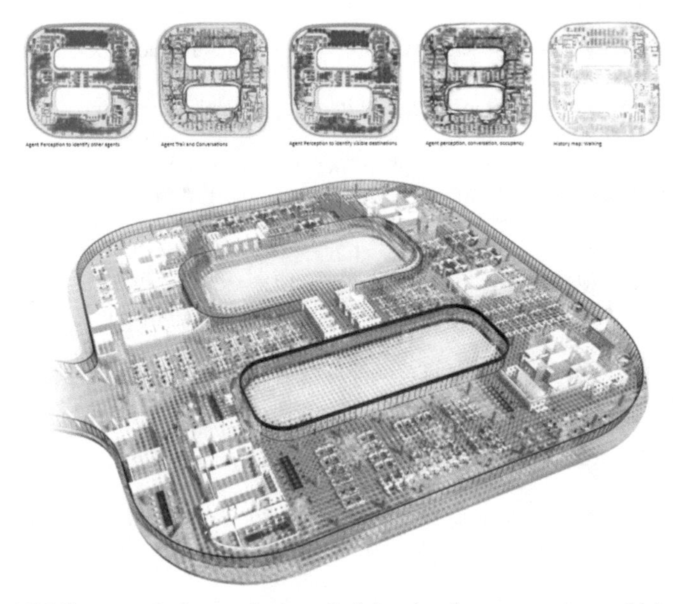

Fig. 8. ZHA ABPS's example data analysis on Infinitus headquarter project, which shows the different social interaction patterns and analysis, such as agent perception, conversation, occupancy, history map, etc. (credit: Tyson Hosmer (team leader), Soungmin Yu, Sobitha Ravichandran, and Ziming He)

5 Discussion

The development of agent-based occupancy modelling for architectural design has started on a new promising trajectory. Patrik Schumacher's original, theoretically grounded design research programme of Agent-based Parametric Semiology outlines a compelling new methodology that aims to upgrade the capacity of architecture to fulfil its societal function as posed anew in the context of the Post-Fordist network society. Over the last ten years, this research programme has been clearly theorized and well-illustrated in many speculative design projects at the AADRL. During the last three years, serious research and implementation work are underway at ZHA's ABPS research team, pushing this project closer to the point where it delivers a compelling new evidence-based, marketable design service.

With the development of Game AI, more and more researchers have begun to pay attention to this possibility. A productive technology transfer is underway that could massively upgrade our discipline. I believe we are witnessing a rare opportunity for architectural design research. This research is still only in its infancy and remains full of exciting challenges.

References

1. Schumacher, P.: The Autopoiesis of Architecture. A New Framework for Architecture, vol. 1. Wiley, London (2010)
2. Schumacher, P.: The Autopoiesis of Architecture. A New Agenda for Architecture, vol. 2. Wiley, London (2012)
3. Schumacher, P.: Parametricist manifesto. In: Out There – Architecture Beyond Building, Volume 5: Experimental Architecture, Catalogue of the 11th Architecture Biennale, Venice (2008)
4. Schumacher, P.: Parametricism: a new global style for architecture and urban design. In: Leach, N. (ed.), AD Digital Cities, Architectural Design, vol. 79, Issue no. 4, July–August 2009 (2009). Reprinted in: Carpo, Mario, The Digital Turn in Architecture 1992–2012 (AD Reader)
5. Schumacher, P.: Parametricism and the autopoiesis of architecture. In: Log 21, Winter 2011. Anyone Corporation, New York (2011)
6. Schumacher, P.: Design is communication. In: Hiesinger, K.B., Hadid, Z. (ed.) Form in Motion, Philadelphia Museum of Art in Association with Yale University Press, New Haven (2011)
7. Schumacher, P.: Parametric order – architectural order via an agent-based parametric semiology. In: Adaptive Ecologies – Correlated Systems of Living by Theodore Spyropoulos. AA Publications, London (2013)
8. Schumacher, P.: Parametric semiology – the design of information rich environments. In: Lorenzo-Eiroa, P., Sprecher, A. (eds.) Architecture in Formation – On the Nature of Information in Digital Architecture. Routledge, Taylor and Francis, New York (2013)
9. Schumacher, P.: The societal function of architecture. In: Prix, W.D., Balliet, K. (eds.), Massive Attack – IOA Sliver Lecture Series: Selected Friends and Enemies, Angewandte (edn.). Birkhaeuser Verlag, Basel (2015)
10. Schumacher, P.: Parametricism with social parameters. In: Lazovski, I., Kahlon, Y. (eds.) Exhibition Book 'The Human (Parameter) - Parametric Approach in Israeli Architecture'. ZEZEZE Architecture Gallery, Tel-Aviv
11. Schumacher, P.: Introduction: parametricism 2.0 - gearing up to impact the global built environment. In: Castle, H. (ed.) AD Parametricism 2.0 – Rethinking Architecture's Agenda for the 21st Century, AD Profile #240, March–April 2016 (2016). Guest-Edited by Schumacher., P.

12. Schumacher, P.: Advancing social functionality via agent-based parametric semiology. In: Castle, H. (ed.) AD Parametricism 2.0 – Rethinking Architecture's Agenda for the 21st Century, AD Profile #240, March–April 2016 (2016). Guest-Edited by Schumacher., P.
13. Schumacher, P.: DIGITAL - the 'Digital' in architecture and design. In: AA Files No. 76. Architectural Association, London (2019)
14. Schumacher, P.: Social performativity: architecture's contribution to societal progress. In: Kanaani, M. (ed.) The Routledge Companion to Paradigms of Performativity in Design and Architecture: Using Time to Craft an Enduring, Resilient and Relevant Architecture. Routledge, Taylor & Francis Group, New York & London (2020)
15. Schumacher, P.: From intuition to simulation. In: 'Posistions: Unfolding Architectural Endeauvors', Angewandte (edn.). Birkhaeuser, Basel (2020)
16. Allbeck, J.M., Functional crowds. In: Workshop on Crowd Simulation Co-Located with the 23rd Annual Conference on Computer Animation and Social Agents, Saint Malo, France (2010)
17. Pelechano, N., Allbeck, J., Kapadia, M., Badler, N.I.: Simulating Heterogeneous Crowds with Interactive Behaviors. CRC Press, Taylor and Franics Group, Boca Raton, London (2016)

Machine Learning

Machine Learning Aided 2D-3D Architectural Form Finding at High Resolution

Hang Zhang(✉) and Ye Huang

University of Pennsylvania, Philadelphia, USA
kv333q@gmail.com, 392057135@qq.com

Abstract. In the past few years, more architects and engineers start thinking about the application of machine learning algorithms in the architectural design field such as building facades generation or floor plans generation, etc. However, due to the relatively slow development of 3D machine learning algorithms, 3D architecture form exploration through machine learning is still a difficult issue for architects. As a result, most of these applications are confined to the level of 2D. Based on the state-of-the-art 2D image generation algorithm, also the method of spatial sequence rules, this article proposes a brand-new strategy of encoding, decoding, and form generation between 2D drawings and 3D models, which we name 2D-3D Form Encoding WorkFlow. This method could provide some innovative design possibilities that generate the latent 3D forms between several different architectural styles. Benefited from the 2D network advantages and the image amplification network nested outside the benchmark network, we have significantly expanded the resolution of training results when compared with the existing form-finding algorithm and related achievements in recent years.

Keywords: Machine learning · Architectural design · Form finding · 3D · GAN

1 Introduction

The amazing development in machine learning neural network algorithms in recent years brings us brand-new tools in design with the help of high-performance graphic cards. Many design issues can be solved by those new machine learning algorithms. Some of them are working pretty well such as the Deep Learning and GAN system.

However, most of the relative works about machine learning applications in the architecture field are working in 2D. One possible reason for the lack of a 3D architecture machine learning algorithm is the comparatively lagging development of 3D machine learning algorithms. Compared with 2D images, the complexity of 3D form issues increases dramatically, not only because of the new z-dimension but also because of different methods of 3D form representation such as point-cloud, voxel, and mesh.

When it comes to the architecture field, architects always face the issue of 2D and 3D. Traditionally, architects have used standardized 2D architectural drawings, such as floor plans, elevations, and sections, to represent 3D architectural forms. These 2D drawings, however, are limited to describe general information of the 3D building. Therefore, it is

P. F. Yuan et al. (Eds.): CDRF 2020, *Proceedings of the 2020 DigitalFUTURES*, pp. 159–168, 2021.
https://doi.org/10.1007/978-981-33-4400-6_15

obvious that the simulation and reconstruction of the 3D architectural model require 3D spatial data as the basis. In terms of the 3D model's generation, instead of using emerging 3D machine learning algorithms which has poor performance at present, we are thinking of using the more sophisticated 2D algorithm which has astonishing performance in the 2D image generation and style transfer, to help us generate 2D architectural drawings and then, use them to construct corresponding 3D architectural models.

2 Relative Work

More recently, there have been several attempts to explore three dimensional architectural forms based on machine learning. Kyle et al. presented their results of generative architectural design by transforming the 3D model data into 2D multi-view data through the neural network which is trained to perform image classification [1]. Through the concept of a 3D-canvas with voxelized wireframes, Sousa et al. introduced a methodology for generation, manipulation and form finding of structural typologies using variational autoencoders, a machine learning model based on neural networks [2]. Zheng proposed an remarkable method regarded to 3D Graphic Statics (3DGS) that quantifing the design preference of forms using machine learning and finding the form with the highest score based on the result of the preference test from the architect [3]. Zhang applied StyleGAN to train 2D architectural plan or section drawings, exploring the intermediate state between different input styles then generating serialized transformation images accordingly to build a 3D model [4].

Nevertheless, it is still very hard to directly apply 3D Machine Learning on the architectural design, as most of those previous approaches are all suffered from the extreme limitation of the overall resolution of generated results.

3 Method

This article mainly uses StyleGAN as the base network of form-finding training [5]. Compared to the other GANs, StyleGAN typically talented in generating three categories of results: (a) similar fake images, (b) style-mixing images, (c) truncation trick images. The similar fake images share common features with these input original images but look like another brand new designs. The style-mixing images are the result of 2 similar fake images. One has its basic content information and the other one transfers its image style to the content one. The truncation trick images are a series of images continuing transforming style from one type to another one.

In order to translate 3D problems into 2D, as Fig. 1 shows, we start from sections of 3D models because sections contain very rich information about the target building not only its outline but also its interior space. Instead of simply slicing the model to get just a few key sections, we start with slicing the target models for 64 times to get 64 section pieces, leading to the snapshot of each piece in the resolution of 128 * 128. After getting 64 drawings, we array them into an 8*8 grid one by one and finally get a 1024 * 1024 image, which contains the former target 3D model information in the resolution 128 * 128 * 64.

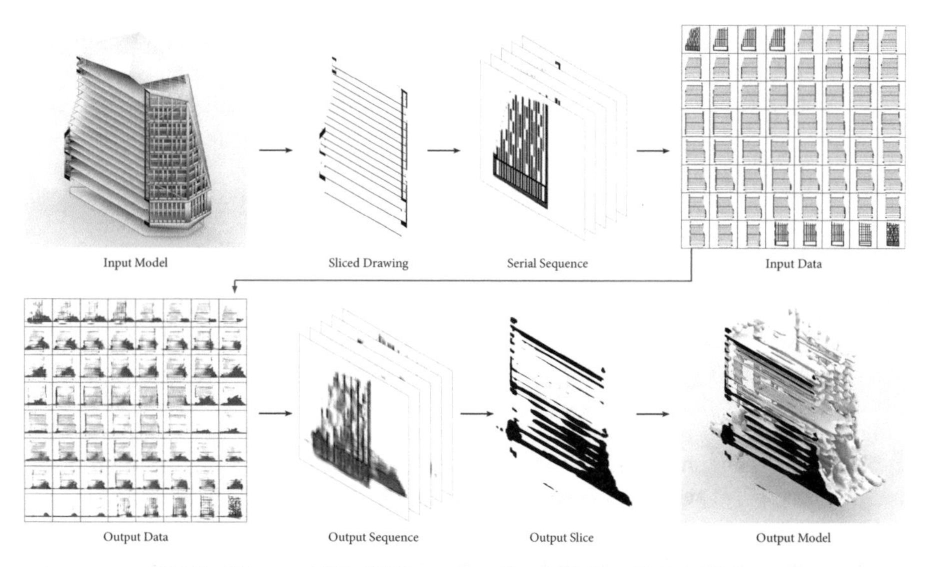

Fig. 1. These are 2D-3D form encoding and decoding work flow.

In addition, with the help of the Waifu2X for image super-resolution [6] and the Pix2Pix for image-to-image translation [7] as extra training networks, the pixel size of training model could be further expanded. Figure 2 shows the secondary network with these two layers of auxiliary model nested outside the primary network which will serve for the resolution magnification.

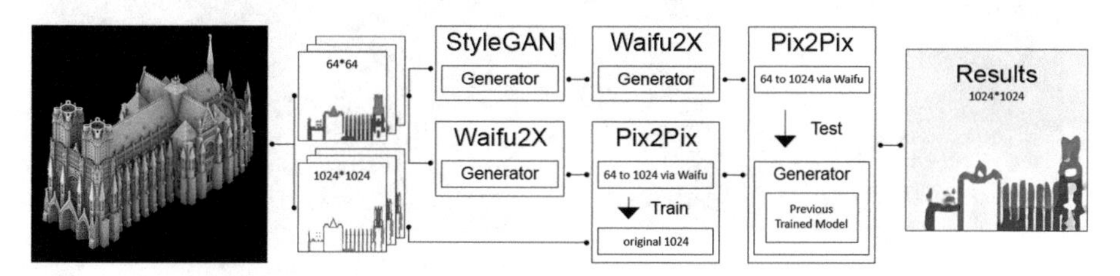

Fig. 2. These are whole process of multi-level training network.

The process on the top row is primarily the main network training via StyleGAN. Firstly, the 3D model was split into 256 slices of 2D bitmaps. While sections pieces are 4 times as large as before, the information of a single 2D layer was compressed to 64 * 64. When training finished, the pixel of generated results is processed by Waifu2X to enlarge by 16 times, which will lead to the 1024 * 1024 resolution. However, these results are distorted and blurred due to the lack of details. Here, we introduce the extra process on the bottom row. The original 3D model was sliced with up to 2000 layers in the 2D resolution of 64 * 64 and 1024 * 1024 respectively, then enlarging all 64 * 64 sequences to 1024 * 1024 via Waifu2X. The original information and enlarged information group are established on this basis, also be used as learning resources to provide image pairing information for Pix2Pix. After obtaining pairing logic from the trained model, the pixel enlarged result under the main network process is fed back to the sub-network. At last, the final output result via translation of this paring trained rules will be generated.

4 Results

4.1 Training Data Preparation

In terms of train data, we pick several styles of 3D architectural models, all of which have either historical value or form value. We divide them into several groups in which has two counterpart models, to find out the possibilities of form synthesizing within these two buildings in the context of 2D-Image Encoding. The final paring is showed in Fig. 3, from A to F they are (A) Fallingwater. (B) Sydney Opera. (C) High Rise. (D) NYC Guggenheim. (E) LV Foundation. (F) Gothic Church. We tried the following combinations: (1) E and F. (2) B and D. (3) B and F. (4) A and F. (5) B and C.

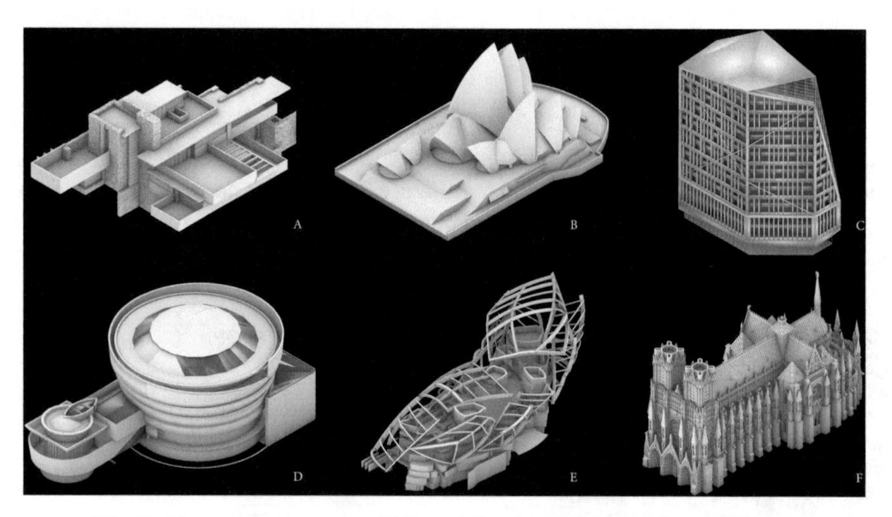

Fig. 3. These are resources of 3D models with different styles and forms.

Each 3D building model will be processed by one-way slicing, and the relative scale will be unified to fit the 2D neural network. In order to ensure that the model data mapped to 2D can be accurately decoded and restored to 3D, we fixed the position and order of each layer of 2D slices. Figure 4 shows the effect of reverse-compiling 2D mapping information into 3D.

4.2 Main Network Training

In the training of the core basic network, we used an 8 * 8 slice array, with a total of 64 layers of pixel data from the 3D model. Because the size of the whole 2D network is 1024 * 1024, every single image could have up to 128 * 128 pixels information. These

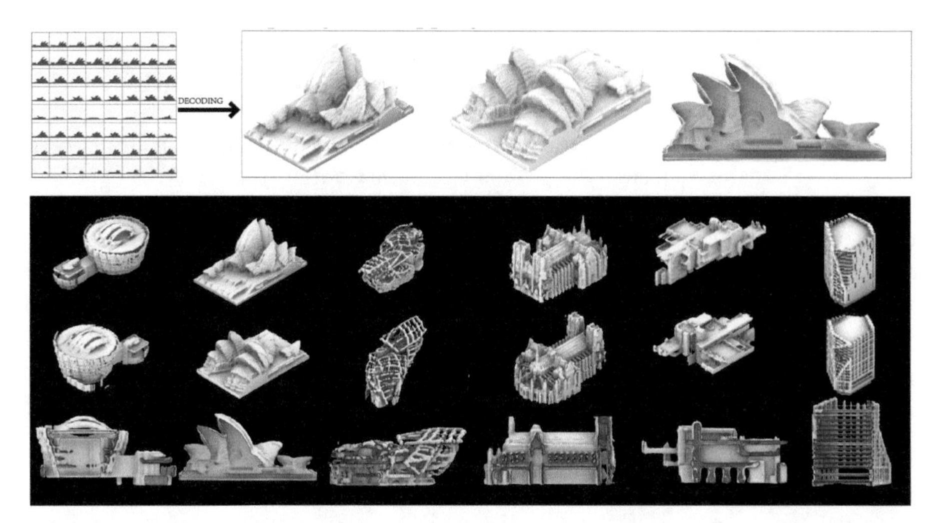

Fig. 4. These are the effect of decoding from 2D mapping information back to 3D model.

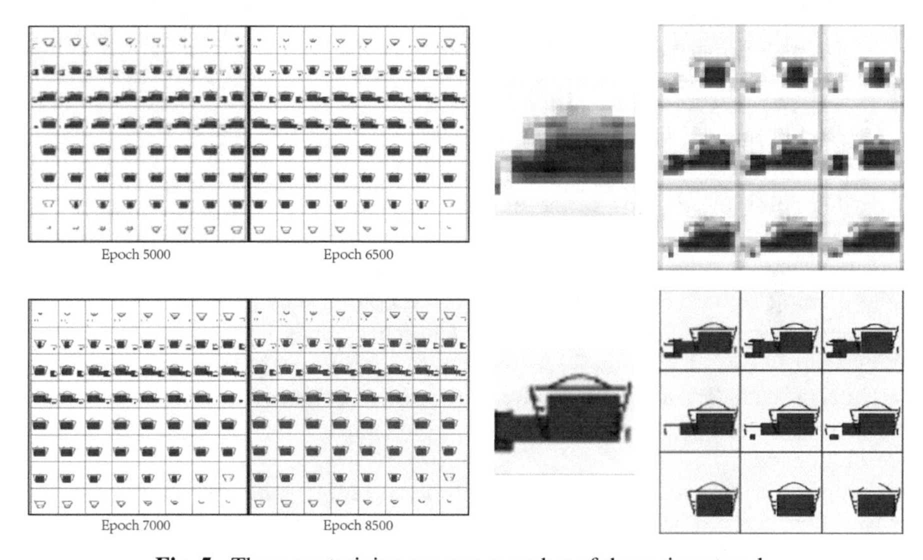

Fig. 5. These are training process snapshot of the main network.

images will function as the single-layer model information by using the pixels of the specific color gamut.

Figure 5 shows the progress snapshot during the training process. In the process of training, the whole network could keep the pixel contour information of each layer of pixel steadily, but it cannot converge to the pixel precision under the condition of insufficient training. Before 7000 kimgs, the reserved pixel area remained at least 4 * 4. As the compromise of training the 3D model through 2D is to increase the pixel limit and reduce the training time and hardware requirements, it is essential for StyleGAN to be trained at least 8500 kimgs to achieve the pixel convergence of 1 * 1. When this condition is satisfied, the training results could be effectively reconstructed into a 3D model with an pixel accuracy of 128 * 128 * 64.

Figure 6 shows one sample of trained results via StyleGAN training on 2D networks, the styles are transferred from FallingWater to Gothic Church. The row from left to right illustrates the gradual integration of one architectural style into another. By observing the different generated data on the fixed slicing position, it can be found that the original architectural form will extend or shrink from the original pixel position of the original form when it is evolving. The intermediate state in the middle of this process is where a mixture of the two architectural styles happens.

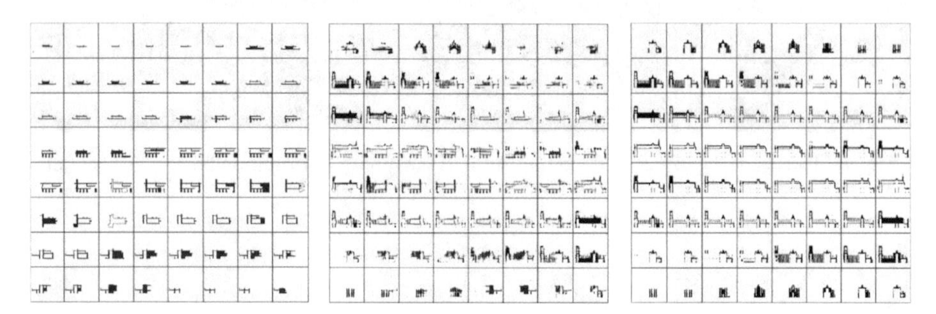

Fig. 6. These are 2D samples of trained results via StyleGAN training on 2D networks.

At this point, the training of the main network has been completed, but 2D slice images cannot directly explain the spatial sequence transformation under the influence of machine learning. Figure 7 shows the final 3D model based on the reverse compilation of the training results. From the top row to the bottom they are: NYC Guggenheim to Sydney Opera; LV Fondation to Gothic Church; Sydney Opera to Gothic Church; Fallingwater to Gothic Church; Sydney Opera to High Rise. On the whole, many pairs of architectural models of different styles shift to the other side of the style and synthesize the results of the remarkable intermediate state, and reach to the 1024 * 1024 * 64 pixels of resolution.

Compered with several previous similar projects [2, 8, 9], the generated results from our workflow are still competitive. Figure 8 shows the comparison of these 3 types of results. However, some of the results have serious noise and obvious lamination. This is because the limit of pixels makes the local details of the building lose information, and the insufficient number of vertically cut layers causes the information between layers to jump.

4.3 Multiple Network Training

As mentioned earlier, multiple nested networks produce two sets of parallel picture delivery workflows. Figure 9 presents the sample images during the whole training process: (a) 64 * 64 sliced layer from the original 3D model. (b) 1024 * 1024 sliced layer from the original 3D model. (c) 1024 * 1024 results from a via Waifu2X. (d) Final 1024 * 1024 results from c via Pix2Pix.

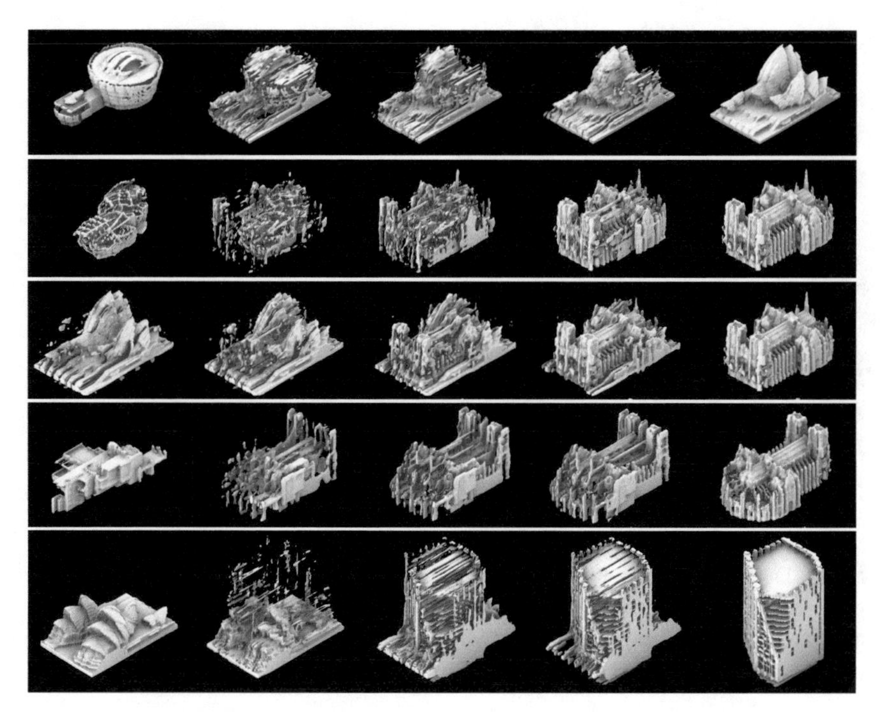

Fig. 7. These are the final 3D model based on the reverse compilation of the training results.

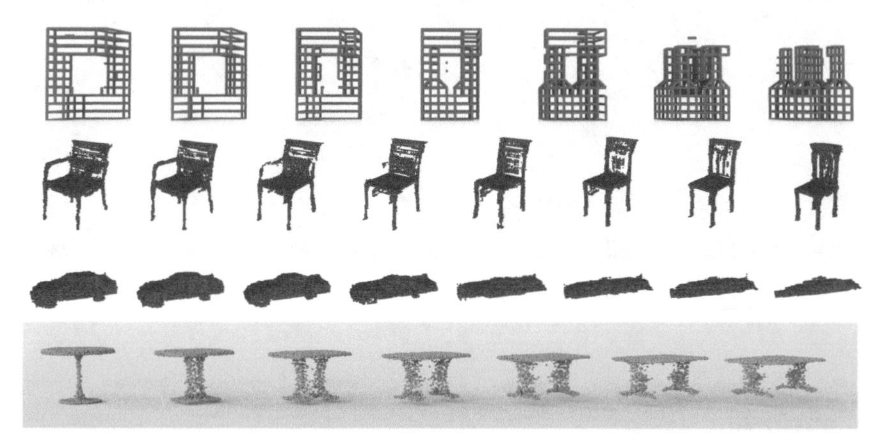

Fig. 8. These are comparison results of 3 other types of machine learning based projects. Top row: Sousa et al. with 15 *15 *12 resolution. Middle row: Jiajun et al. with 15 * 15 * 12 resolution. Bottom row: Wu et al. With total 2048 points.

Different from the previous 3d volume of 128 * 128 * 64, the result derived from multiple networks has a single-layer resolution of 1024 * 1024 and a total of 256 slices. Correspondingly, due to the disadvantages of the complex network, the generated results will produce more noise. Figure 10 shows the training results obtained based on multiple

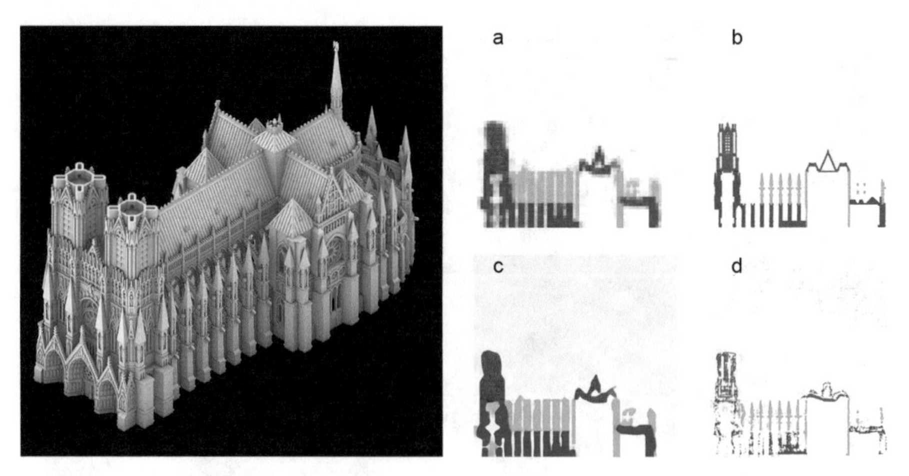

Fig. 9. These are 2D samples from Gothic Church model during the process of each networks.

networks. Due to the difficulty in controlling the noise, such nested rules are difficult to become an applicable exploration method. Of course, it is possible that training with the right parameters takes enough time to achieve more than expected results, but this goes against the principle that training time should not be too long.

Fig. 10. These are the training results obtained based on multiple networks.

5 Conclusion

In this paper, by using StyleGAN, Waifu2X, and Pix2Pix, we successfully reconstruct existing building models through the 2D-3D en-coding strategy and realize style blending and generation of new architectural forms based on existing forms at the relatively high resolution. These combinations attempt to explore how to synthesize the morphological features among various styles and forms of several buildings with distinctive features, also leading to the innovative design through these unexpected forms the AI gives us. This style blending strategy provides totally an innovative method of architecture design which expands the boundary of form finding.

On the other hand, some predictable improvements are still needed for this encoding strategy. First, the running time issue is still a difficult problem we have to deal with. As mentioned before, the training process of those input lasts for 10 days in total with 1 single Nvidia RTX Titan. Even though the reconstruct result is satisfying enough for model decoding after 8000 kimgs (5 days), the running time is unacceptable for most designers, not to mention that nested networks are more time-consuming.

Besides, the resolution of this encoding is indeed way much higher than the existing 3D machine learning algorithm most of which are about 32 * 32 * 32 or 64 * 64 * 64. But our 2D-3D encoding strategy which has the resolution of 1024 * 1024 * 64 (or 1024 * 1024 * 256 on an unreliable state) still needs improvement to get a better representation of a building system.

Last, in most situation, 64 sections in one direction is a little bit redundant for a simple building. The traditional way of representing a simple building such as a three-layer house may only have 10–20 sections and many other plans or detail. Therefore, based on the conclusion above, our first next move about our strategy is to develop a better framework for building information representation. Based on the new framework, we can extract the crucial sections or plan information from input building models in a more efficient way, so as to reduce the number of sections. In this case, the running time will be the same but the building still conveyed clearly and the detail level of the final reconstruct model and style blending model will be increased a lot.

References

1. Kyle, S., Kat, P.: Adam M. fresh eyes a framework for the application of machine learning to generative architectural design, and a report of activities at smartgeometry 2018. In: Proceedings of the 18th International Conference, CAAD Futures 2019 (2019)
2. Sousa, J.P., Xavier, J.P., Castro Henriques, G.: Deep form finding - using variational autoencoders for deep form finding of structural typologies. In: Proceedings of the 37th eCAADe and 23rd SIGraDi Conference on Architecture in the Age of the 4th Industrial Revolution (2019)
3. Zheng, H.: Form finding and evaluating through machine learning: the prediction of personal design preference in polyhedral structures. In: Yuan, P., Xie, Y., Yao, J., Yan, C. (eds.) Proceedings of the 2019 DigitalFUTURES (2019)
4. Zhang, H.: 3D model generation on architectural plan and section training through machine learning. Technologies (2019)
5. Karras, T., Laine, S., Aila, T.: A style-based generator architecture for generative adversarial networks. CoRR, abs/1812.04948 (2018)
6. Dong, C., Loy, C.C., He, K., Tang, X.: Image super-resolution using deep convolutional networks. IEEE Trans. Pattern Anal. Mach. Intell. **38**(2), 295–307 (2015)
7. Isola, P., Zhu, J.Y., Zhou. T., Efros, A.: Image-to-image translation with conditional adversarial networks. In: Proceedings of the IEEE Conference on Computer Vision and Pattern Recognition, Honolulu, HI, USA, 21–26, pp. 1125–1134 (2017)
8. Achlioptas, P., Diamanti, O., Mitliagkas, I., Guibas, L.: Learning representations and generative models for 3D point clouds. arXiv preprint arXiv:1707.02392 (2017)
9. Wu, J., Zhang, C., Xue, T., Freeman, B., Tenenbaum, J.: Learning a probabilistic latent space of object shapes via 3d generative-adversarial modeling. In: Advances in Neural Information Processing Systems, pp. 82–90 (2016)

Exploration of Campus Layout Based on Generative Adversarial Network
Discussing the Significance of Small Amount Sample Learning for Architecture

Yubo Liu, Yihua Luo, Qiaoming Deng(✉), and Xuanxing Zhou

School of Architecture, South China University of Technology, Guangzhou, China
dengqm@scut.edu.cn

Abstract. This paper aims to explore the idea and method of using deep learning with a small amount sample to realize campus layout generation. From the perspective of the architect, we construct two small amount sample campus layout data sets through artificial screening with the preference of the specific architects. These data sets are used to train the ability of Pix2Pix model to automatically generate the campus layout under the condition of the given campus boundary and surrounding roads. Through the analysis of the experimental results, this paper finds that under the premise of effective screening of the collected samples, even using a small amount sample data set for deep learning can achieve a good result.

Keywords: Deep learning · Layout design · Campus planning · Generate design · Small amount sample

1 Introduction

In recent years, with the improvement of computing power, the emergence of more data and more effective algorithms, Artificial Intelligence (AI) has once again ushered in a development boom. Among them, deep learning, which overcoming the limitations of high-dimensional data processing and feature extraction, is widely used in the field of computer vision, audio, and natural language processing, and further promotes the development of AI.

The models and methods of machine learning are mainly based on the theory of statistical learning. Currently popular machine learning models, especially the deep learning, generally need a large number of sample data to ensure that the models are fully trained to achieve the learning effect and obtain universal rules. However, in the field of architectural design, the design issues we are concerned with are not only related to the universal rules, but also related to the value orientations and aesthetic tendencies of specific architects. The particularity and difference of these design works result in the variety of architecture. This characteristic of architecture design makes it possible for the training of small amount chosen sample data sets to achieve a relatively ideal effect.

By taking the campus layout as the pointcut, this paper discusses the ability of Pix2Pix model to automatically generate the campus layout under the condition of the

P. F. Yuan et al. (Eds.): CDRF 2020, *Proceedings of the 2020 DigitalFUTURES*, pp. 169–178, 2021.
https://doi.org/10.1007/978-981-33-4400-6_16

given campus boundary and surrounding roads, which is based on deep learning and artificial chosen small amount sample data sets. With the premise of effective screening of the collected samples, we succeed in the experiment by using a small amount chosen sample data sets.

2 Related Work in the Field of Architectural Layout

Layout design has always been a research hotspot in the field of AI, such as: document layout, game map layout, cover layout, interior furniture layout, etc.

In the field of architectural layout, Merrell et al. (2010) propose a method of automatically generating residential building layout based on Bayesian networks. They select 120 residential building layouts from the 1500 best-selling residential building layouts for machine learning. Fan et al. (2014) propose a method to complete the missing or unoccluded facade layouts with a statistical model and a planning algorithm. They selected 100 complete and unoccluded building facades with uniform style as the data set of training. The author with his team also try to expand the existing 15 samples to 140 samples in the form of artificial data augmentation, and explore the generation of specific architectural forms through the neural network method (Liu et al. 2019).

In recent years, the development of deep learning has provided new ideas and methods for the field of architectural layout. Huang and Zheng (2018) use Generative Adversarial Network (GAN) for the recognition and generation of apartment floor plans. A total of 115 samples were collected, including 100 training samples and 15 testing samples. Wu et al. (2019) propose a CNN-based method of generating floor plans for residential buildings with given boundaries and they select more than 80K floor plans from more than 12K collected floor plans of the Asian real estate market to construct a database named RPLAN, of which 75K floor plans are used as training samples, half of the left is used as testing samples and the other half is used as verification samples. Chaillou (Chaillou 2019) chooses nested GANs to generate a furnished floor plan from the parcel, using about 700 floor plans as samples. Newton (2019) trains GAN to generate floor plans with the style of Le Corbusier and discusses the effectiveness of different augmentation techniques when dealing with small data sets, using 45 hand-drawn drawings of Le Corbusier as samples and expanding the samples to 135/180/540 through different augmentation techniques.

In the above work, we found two ways to collect and filter data samples as datasets: 1) One is using tens of thousands of samples for training to obtain the universal rules in a statistical manner, the specificity of specific samples is neutralized through the mass of sample data, such as Wenming Wu's research (Wu et al. 2019). 2) The other is using a small amount chosen samples which meet the training standards data set for training, only to obtain the rules of its particularity, such as David Newton's research (Newton 2019). This paper adopts the latter method, constructs specific chosen sample data sets through artificial screening according to the value orientations and aesthetic tendencies of specific architects, finally get relatively ideal results with small amount chosen sample data sets for deep learning training.

3 Methods

The main process of campus layout generation based on deep learning with small amount chosen samples data sets is as follows:

1) *Expected goal.* Automatically generate a reasonable campus layout under the condition of the given campus boundary and surrounding roads.
2) *Data screening.* According to the characteristics of the campus layout and the value orientations and treatment methods of specific architects, we have screened 85 university samples and 302 primary school samples. The specific screening rules are as follows:

The specific screening rules for the primary school:

1. The site plan is clear.
2. Decentralized campus design.
3. An independent playground.
4. Single-loaded teaching building connected by corridors.
5. A basketball court or scale is used as a scale reference.
6. The campus is located in the subtropical monsoon climate zone with balanced economic development.
7. The campus land is flat, avoiding the special terrain such as slopes or water.

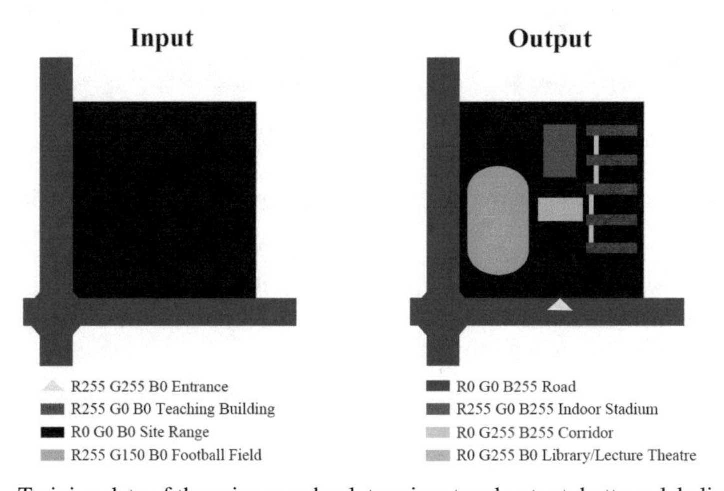

Fig. 1. Training data of the primary school, top: input and output, bottom: labeling rule

The specific screening rules for the university:

1. The site plan is clear.

2. The campus layout conforms to the design concept of a specific design company.
3. The loop traffic system and the central area.
4. A basketball court or scale is used as a scale reference.

3) *Model construction.* This paper selects the Pix2Pix model based on Generative Adversarial Networks to complete this research. The generator uses the U-net framework and the discriminator uses the Patch-GAN framework (Isola, et al. 2017). By inputting pictures of the actual campus boundary and surrounding roads and outputting corresponding pictures of the actual campus layout (Fig. 1, 2, and 3), this model is trained to learn the rules of actual campus layouts and generate campus layout automatically. It is worth noting that the color selection is related to the RGB value interval to avoid recognition errors. Appendix provides more details.

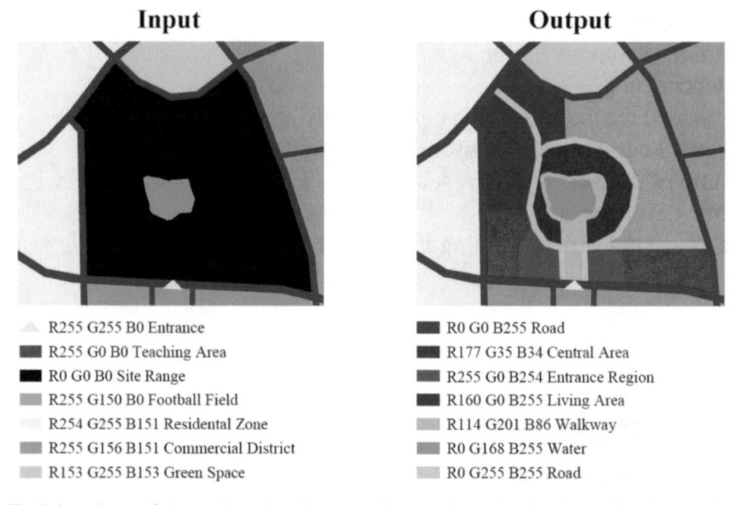

Fig. 2. Training data of the university for step 1, top: input and output, bottom: labeling rule

4) *Experimental test.* Use the testing set to test the trained Pix2Pix model, analyze and evaluate the training effect of the model by comparing the generated campus layout with the actual campus layout.

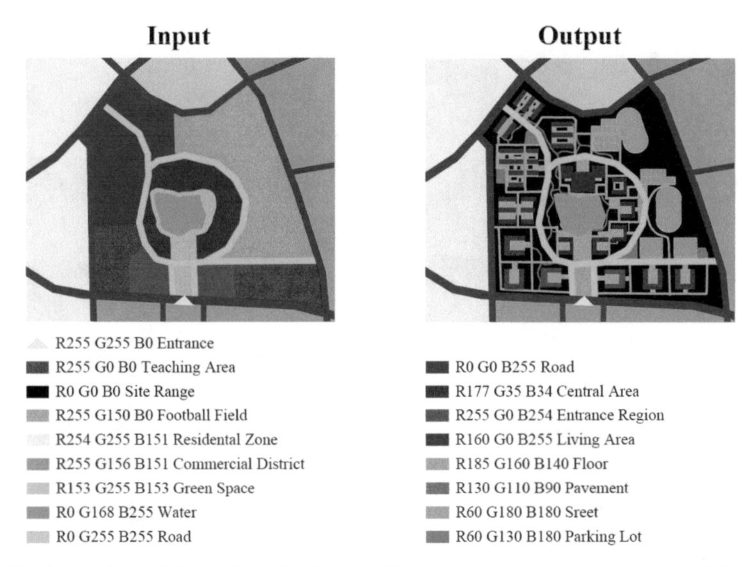

Fig. 3. Training data of the university for step 2, top: input and output, bottom: labeling rule

4 Experimental Results and Analysis

First, we tried to train a model that automatically generates the campus layout of the primary school from the campus boundary and surrounding roads. The experimental results show that the model we trained has learned the rules of the layout of the primary school to some extent. Figure 4 shows the selected results from the testing set.

Compared with the actual layout, the layout generated in No. 265 has a more reasonable relationship between the teaching building, the gymnasium and the stadium, leaving a transition road in the middle. The layout generated in No. 301 pays more attention to the north-south orientation of the teaching building than the real layout. There are also some poor layouts. For example, the layout in No. 275 fails to generate a complete teaching building group, and as for the layout in No. 294, the campus entrance is not reasonable enough at the intersection of roads, but the main building chooses to strive for north-south orientation instead of parallel to the surrounding road.

On the basis of some achievements in the campus layout of the primary school, we tried to automatically generate the campus layout of the university. Due to the complexity of the campus layout of the university, we split the training process into two steps: The first step is to train a model that generates the main functional zoning of the campus based on the campus boundary, surrounding roads and surrounding functional distribution. The second step is to train another model which is based on each functional zoning of the campus and generates the internal architectural layout of the them. In this way, a complete campus layout of the university is achieved.

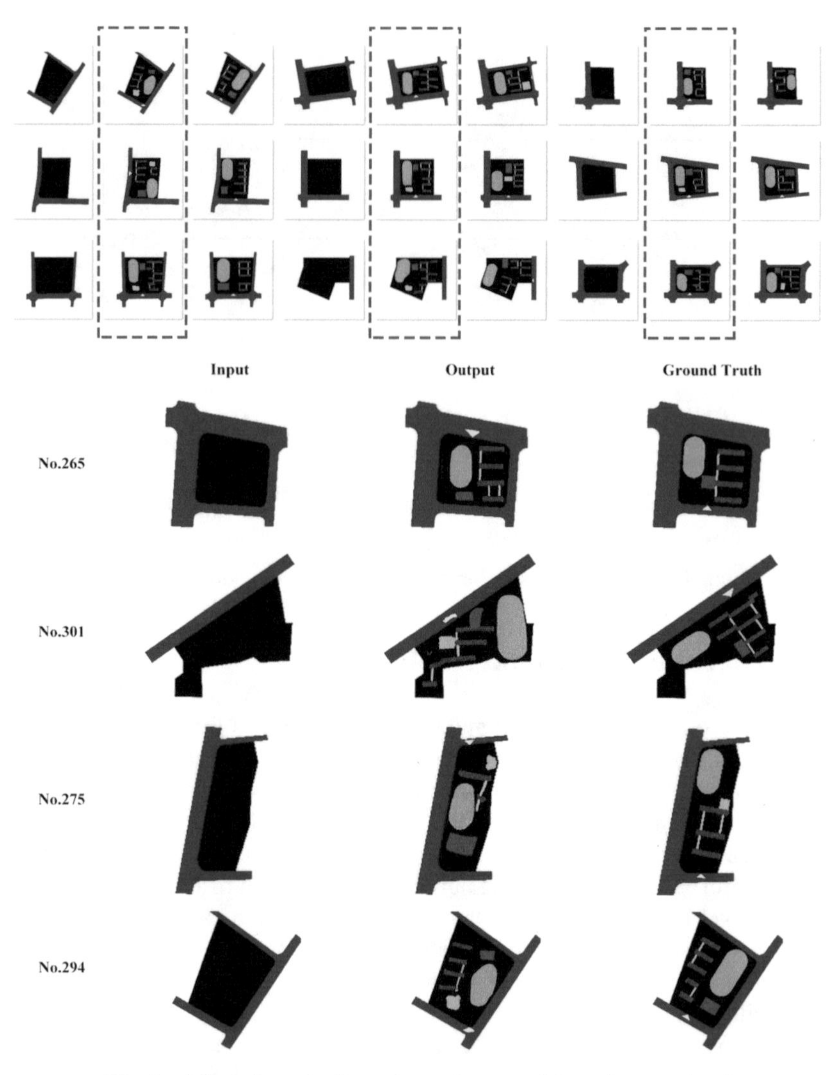

Fig. 4. Selected results from the testing set of the primary school

In the first step, we chose the campus planning scheme with a specific style designed by a design company as samples, highlighting the team's adherence to the layout concept of the loop traffic system and the central area which emphasizes that teaching buildings and landscapes jointly shape the communication space, and the final effect reached the expected. From the perspective of functional zoning, the generated layout is highly consistent with the original plan, forming an obvious layout pattern of the loop traffic system and the central area. The relationship between each functional zoning and the external environment is relatively reasonable. Almost all the testing samples successfully output the loop traffic system, forming a continuous campus walkway from the entrance area to the central area (Fig. 5).

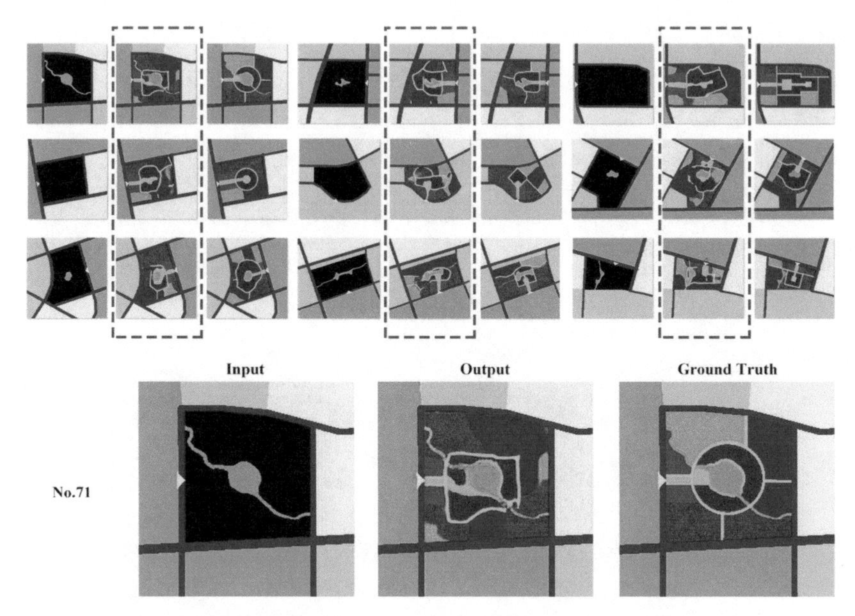

Fig. 5. Selected results from the testing set of the university in step 1

In the second step, we also continue to select samples that consistent with the design concept of the design company, which emphasized the organic combination of the architectural layout of the central area and the core open space of the campus, and emphasized that the layout of the dormitory area should be combined with the main pedestrian traffic roads to form the street space. The final effect is also relatively successful (Fig. 6). The output results not only grasped the laws of north-south orientation of building layout and stable building spacing, but also coordinated the relationship between the buildings, public space and water in the central area. The layout of the dormitory area has also initially formed the spatial relationship we want to shape. Unexpectedly, the layout of some buildings also initially reflects the overall style of lightness and flexibility, which is pursued by the design company. As discussed in the paper, in small amount sample learning, the clearer the layout rules of the samples are, the better the learning effect will be. For example, the layout of the dormitory area and sports area in the samples is simple and clear, the generation effect is good. But the central area is rich in architectural form and flexible in layout, there are fuzzy areas in the generated results. We can achieve improvement by adjusting the samples and enhancing the regularity of the layout of the central area. However, this fuzzy area is very similar to the sketch for the architect, and maybe has a stronger design inspiration. Related research attempts are still underway.

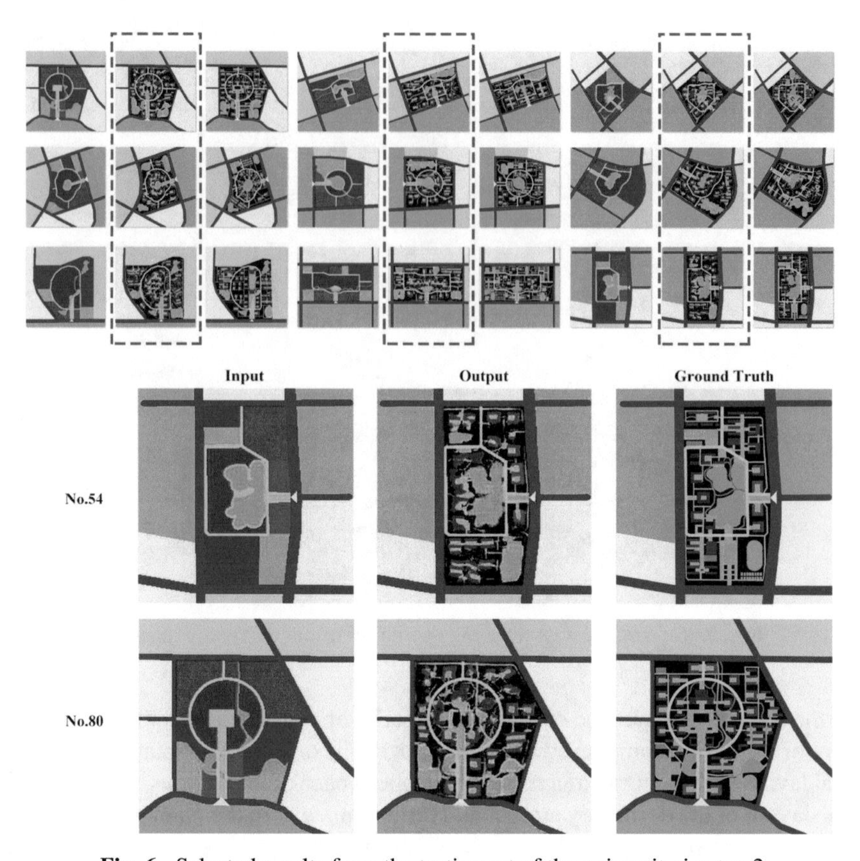

Fig. 6. Selected results from the testing set of the university in step 2

5 Discussion

This paper aims to explore the idea and method of using deep learning with a small amount sample to realize campus layout generation. The experimental results show that when the collected samples are screened by the characteristics of campus layout and the value orientations and treatment methods of specific architects, even using small amount sample data sets for deep learning can achieve a good result. The inspiration of this study is that when combining the field of architectural design and AI, we should pay attention to the particularity and difference in the architecture, found out own way, not just follow the way of computer discipline. From this point of view, we hope to find a way to deep learning based on small amount sample data sets which conforms to the characteristics of architecture discipline and points to specific design concepts and design techniques.

Acknowledgments. This research is supported by National Natural Science Foundation of China (No. 51978268 & No. 51978269) and State Key Lab of Subtropical Building Science, South China University Of Technology (No. 2019ZA01).

Appendix

In the process of deep learning, each value of RGB values is divided into five intervals, 0–50, 51–100, 101–150, 151–200 and 201–255 (Fig. 7). At least one value of the RGB values of the two colors is in a different interval, so as to distinguish the colors, for example, the color (255, 255, 0) can be understood as (5, 5, 1).

RGB Values Interval	Classification
0-50	1
51-100	2
101-150	3
151-200	4
201-255	5

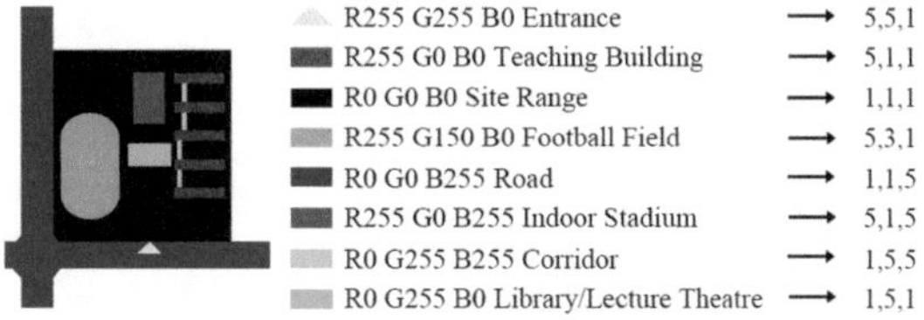

R255 G255 B0 Entrance → 5,5,1
R255 G0 B0 Teaching Building → 5,1,1
R0 G0 B0 Site Range → 1,1,1
R255 G150 B0 Football Field → 5,3,1
R0 G0 B255 Road → 1,1,5
R255 G0 B255 Indoor Stadium → 5,1,5
R0 G255 B255 Corridor → 1,5,5
R0 G255 B0 Library/Lecture Theatre → 1,5,1

Fig. 7. Further explanation of the choice of colors used to label different elements

References

Chaillou, S.: AI & Architecture—an experimental perspective (2019). https://towardsdatascience.com/ai-architecture-f9d78c6958e0/ Accessed 6 May 2019

Fan, L., Musialski, P., Liu, L., Wonka, P.: Structure completion for facade layouts. ACM Trans. Graph. **33**(6), 210–211 (2014)

Huang, W., Zheng, H.: Architectural drawings recognition and generation through machine learning. In: Proceedings of the 38th Annual Conference of the Association for Computer Aided Design in Architecture, Mexico City, Mexico, pp. 18–20 (2018)

Isola, P., Zhu, J.Y., Zhou, T., Efros, A.A.: Image-to-image translation with conditional adversarial networks. In: Proceedings of the IEEE Conference on Computer Vision and Pattern Recognition, pp. 1125–1134 (2017)

Liu, Y.B., Lin, W.Q., Deng, Q.M., Liang, L.Y.: Exploring the building form generation by neural networks—taking the Free University of Berlin as an example. In: Proceedings of the 14th National Conference on Architecture Digital Technologies in Education and Research, Chongqing, China, pp. 67–75 (2019)

Merrell, P., Schkufza, E., Koltun, V.: Computer-generated residential building layouts. In: ACM SIGGRAPH Asia 2010 Papers, pp. 1–12 (2010)

Newton, D.: Deep generative learning for the generation and analysis of architectural plans with small datasets. In: Proceedings of the 37th eCAADe and 23rd SIGraDi Conference, vol. 2, Porto, Portugal, pp. 21–28 (2019)

Wu, W., Fu, X.M., Tang, R., Wang, Y., Qi, Y.H., Liu, L.: Data-driven interior plan generation for residential buildings. ACM Trans. Graph. (TOG) **38**(6), 1–12 (2019)

A Preliminary Study on the Formation of the General Layouts on the Northern Neighborhood Community Based on GauGAN Diversity Output Generator

Yuzhe Pan, Jin Qian, and Yingdong Hu(✉)

School of Architecture and Design, Beijing Jiaotong University, Beijing, China
ydhu@bjtu.edu.cn

Abstract. Recently, the mainstream gradually has become replacing neighborhood-style communities with high-density residences. The original pleasant scale and enclosed residential spaces have been broken, and the traditional neighborhood relations are going away. This research uses machine learning to train the model to generate a new plan, which is used in today's residential design. First, in order to obtain a better generation effect, this study extracts the transcendental information of the neighborhood community in north of China, using roads, buildings etc. as morphological representations; GauGAN, compared to the pix2pix and pix2pixHD, used by predecessors, can achieve a clearer and a more diversified output and also fit irregular contours more realistically. ANN model trained by 167 general layout samples of a neighborhood community in north of China from 1950s to 1970s can generate various general layouts in different shapes and scales. The experiment proves that GauGAN is more suitable for general layout generation than pix2pix (pix2pixHD); Distributed training can improve the clarity of the generation and allow later vectorization to be more convenient.

Keywords: Neighborhood community in north of China · General layout morphology generation · GauGAN · pix2pix · Machine learning

1 Introduction

Driven by the value of urban areas, from the 1980s to the 1990s, high-density housing has gradually replaced neighborhood community as the mainstream. The original pleasant scale and enclosed residential spaces have been broken, ignoring human scale and psychological considerations. The settlement environment has become larger in scale and richer in form, but the inclusiveness for diversified functions and behaviors has become less and less, and it is more difficult to change the fact that neighborhood relations are gradually indifferent. How to create a habitable public space and reshape the neighborhood social network has become the difficulty of residential planning today.

These authors contributed equally to this work and should be considered co-first authors. Project supported by Beijing Jiaotong University Training Program of Innovation and Entrepreneurship for Undergraduates.

P. F. Yuan et al. (Eds.): CDRF 2020, *Proceedings of the 2020 DigitalFUTURES*, pp. 179–188, 2021.
https://doi.org/10.1007/978-981-33-4400-6_17

Neighborhood community originated from the former Soviet Union were widely used in cities construction in China because it met the need of rapid construction new settlements in the Early Stage of the Republic. Most buildings are 3–5 floors, with streets outside and courtyards inside. The semi-public space organization in the courtyard and the mode of the dense road network in small blocks are the main reasons for its good neighborhood relations [1].

Nowadays, if you want to reproduce the neighborhood scene, you need to explore the characteristics of the public space organization and street scale of neighborhood community first. Existing neighborhood communities are the result of a long time trial and error process by predecessors, and it is undoubtedly more efficient than groping. However, the analysis of potential mechanism problems has always been a major problem facing the architectural discipline. There are certain limitations in the related research, no matter the methods and tools used. First, relying on the subjective judgment of researchers, deducing the constituent elements of independent cases, limited by differences of knowledge and background, cognition is also different, and researches based on a small number of cases cannot accurately extract common characteristics; secondly, generation laws are complex, it is difficult to qualitatively and quantitatively analyze and convert into measurable indicators, specific design methods and executable programs. This research hopes to extract objective factors and discover hidden laws [2].

Recently, advances in deep learning technology, represented by computer vision, have made it possible for computers to collect and analyze implicit feature laws. This study uses machine learning to conduct large-scale sample training, to cover the potential laws of neighborhood community and automatically generate output to provide a more rational and comprehensive working method and ideas for the protection and innovation of northern neighborhoods.

2 Research

2.1 AI Application in Architecture

The application of computers in the architecture field has gone through four stages: Modularity, Computational Design, Parametricism and AI. In the first three stages, computer was used as design aids due to its high computing performance, such as BIM and digital construction, which couldn't design independently without architects. The development of AI in the 21st century makes it possible to replace designers in specific fields like building form generators, digital construction and so on, especially in the field of building layout in which predecessors have used computer vision to explore.

2.2 Deep Learning Architectural Plan Generator Application

Through the retrieval of existing articles, the following three used related methods of computer vision achieving substantial results in the field of building plan generation. The first two articles are mainly focused on the generation of indoor floor plans, while the last article is aimed at general layout. The results all adopt a similar method: first convert building layout into bitmaps, and then use image to image algorithm to train the deep learning model (Table 1).

Table 1. Comparison of various studies in the research

Author	Algorithms	Result	Deficiency
Hao Zheng	pix2pixHD	Indoor floor plan and Mutual mapping of functional color block diagrams	Can't match irregular boundaries (lack of corresponding samples); Unable to output diversely
Stanislas Chaillou	pix2pix	Step training, designers can intervene in the intermediate link, controllability; Realize the output from the outline of the plane to the indoor layout	Can't match irregular boundaries (lack of corresponding samples; limitation of algorithm itself); Unable to output diversely
Yubo Li	Deep learning	Generic flat contour and road generate function color map	Unable to output diversely

In 2018, Hao Zheng trained 100 samples using pix2pixHD to realize the recognition and generation of architectural layouts. First, it trains with indoor layout as label and function color map as result; at the same time, it reserves training to achieve the mutual map of the indoor layout and the function color map; The visualization of the training process proves that pix2pixHD has similarities with human cognition. The author mentioned that results couldn't match the irregular boundary well that may be due to samples which have irregular boundary are few [5].

Stanislas Chaillou produced the indoor layout generator program ArchiGan in 2019. Using GANs and step training, three models can realize from site conditions to building outline to function color map to indoor layout. When inputting the site condition, it can output the indoor layout. In the third step, it conducts separate training the furniture layout of each type of room through different colors. Step training is to control the single steps of machine learning, to realize the human intervention and ensure generation quality. It trained four architectural styles to achieve style transfer to be selected according to pros and cons when designing [6].

The schgan produced by Yubo Liu in 2019 can automatically generate a functional layout of elementary school based on road and contour conditions. According to the user evaluation mechanism set by it, the plan generated by AI is higher than the original plan. Specific experimental method was not disclosed, but the results prove that the deep learning can achieve generation of general layouts [8].

The above three articles have the following questions: (1) Due to the limits of algorithm and the number of samples, the generated results have a lot of noise, and the visual recognition is greatly affected; (2) The above articles exist that the direction of elements in the results can only be orthogonal, and cannot change according to site condition. This experiment proves that reasons include not only the above-mentioned oblique arrangement, but also the limitations of pix2pix itself; (3) The above researches can not realize diversity generated results with a single input, which is impossible to provide designers with multiple possibilities.

3 Methodology

3.1 GauGAN

Compared with pix2pix [7] used by predecessors, GauGAN [7] performs better in high-resolution image synthesis, and natively supports diverse output. The reason lies in the SPADE generator and Multiscale Discriminator that constitute the GauGAN neural network structure. The first image to image algorithm is pix2pix, then improved pix2pixHD [8]. Compared to pix2pix, pix2pixHD removes the U-Net structure in the generator, and uses Resnet and local Enhancer to increase the network depth, and improves on the basis of pix2pix's patch discriminator to multiscale discriminator, which contributes to the better high resolution Images (above 512 × 512) synthesis. And the difference between GauGAN and pix2pixHD is mainly the generator model architecture, which has the following advantages:

1. GauGAN's generator has a better ability to obtain information because it uses SPADE (Spatially-adaptive denormalization) as the norm layer. Compared to the instance/batch norm layer used in pix2pix and pix2pixhd, SPADE can better retain the information in the labelmap, and finally can make the generator output more in line with the information provided by the labelmap. This may also be the reason why the generated result of gaugan can correspond to the inclined profile. Figure 1 shows the difference between the output results of gaugan and pix2pix when facing oblique land use conditions. It can be seen that GauGAN adopts the SPADE norm layer, which makes the shape of the plane of the oblique layout of the building correspond well in the generated results. The outline of the hypotenuse of the land used in the labelmap, but pix2pix does handle with the outline of the hypotenuse in the labelmap, and still reflects the orthogonal form.

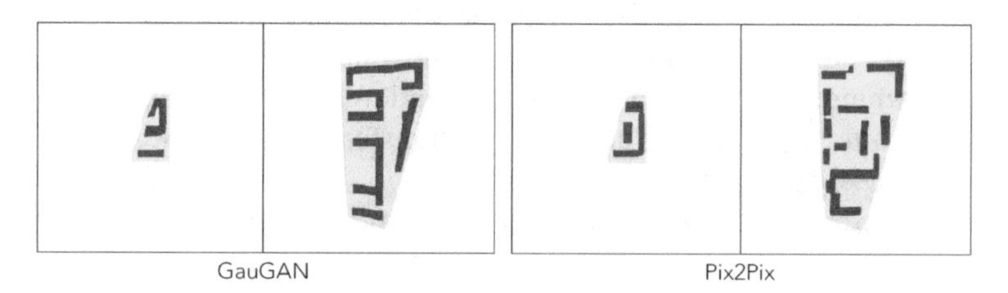

Fig. 1. Comparison between GauGAN and Pix2Pix

2. Multi-modal synthesis is natively implemented in GauGAN. Similar to ordinary GANs, the input of SPADE Generator is a normal distributed random vector (latent), and labelmap is the input of each SPADE Resnet in the Generator, so GauGAN can make generation diverse as general GANs. However, the generator input of pix2pix and pix2pixHD is directly a semantic map without latent, so the result diversity cannot be directly achieved. The results of generating diversity are conducive to designers to filter, reflecting the concept of human-computer interaction.
3. The calculation efficiency of GauGAN is higher than pix2pixHD. The resolution changing part of the pix2pix and pix2pixHD generator is executed by the deconv/conv

layer, while this part of SPADE Generator is executed by the interpolate sample layer result in less parameters and higher efficiency than pix2pixHD.

3.2 Step Training

In order to improve the clarity of the generated results, this experiment compares the overall and step training. The overall training is to train the model to directly map the contour graph to the layout morphology, while step training is to map the contour graph to the building layout graph first, and then map the building layout graph to the road network graph. When testing, matrix multiplication is applied on the obtained building layout and road network to obtain the final result.

Comparing the generated results obtained by testing two methods (Fig. 1), it can be distinguished that the generated results of step training are clearer and less noisy, so the step training method is finally adopted in this experiment (Fig. 2).

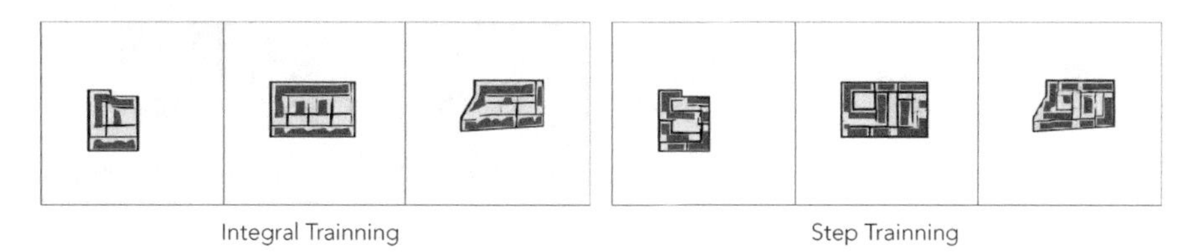

Fig. 2. Comparison between Integral training and Step training

4 Machine Learning for the General Layout Shapes of the Northern Neighborhoods in China

See Fig. 3.

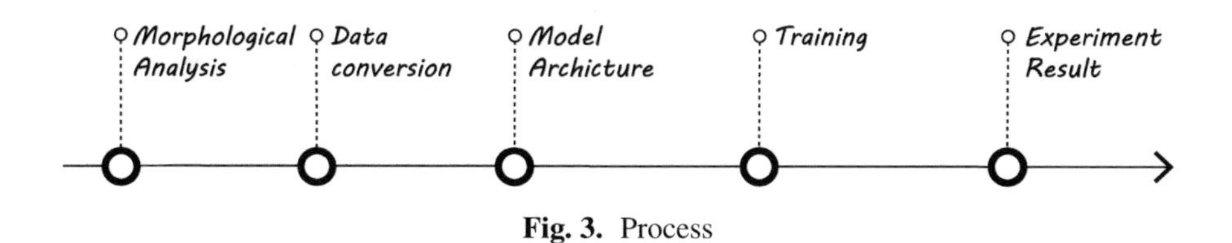

Fig. 3. Process

4.1 Morphological Analysis

First step is to analyze the composition of the general layout of neighborhood community and extract the elements. The public space formed by buildings and roads is the main place to communicate. The buildings determine the division, and the road as a special public space also affects communication. So, this study extracts roads, buildings, and courtyard spaces as morphological representations, shields redundant information that is not relevant, using machine learning to extract common features of public space organization and street scale in the samples.

4.2 Data Conversion

This step is to convert the general layout into bitmap data that the program can process. This study explored 167 neighborhood communities in Beijing and Tianjin, compiling and plotting as data sets. In order to adapt step training, each sample includes four types: outline map, non-road set, road set and complete sample set (Fig. 4). After balancing the calculation amount of neural network and the resolution, the 512 * 512 pixel bitmap is used uniformly, and the maximum range is 330 m * 330 m. All samples are entered with the same scale. The scale of the 167 sample ranges from 110 m * 110 m to 330 * 250 m, all within the maximum range. Use three colors to represent the three extracted elements, and use HSV to control the variables: building: $H = 0$, $S = 100$; courtyard space: $H = 58$, $S = 100$; road: $H = 0$, $S = 0$. Some samples have the characteristics of oblique field contours so that whether GauGAN can match irregular boundaries can be confirmed.

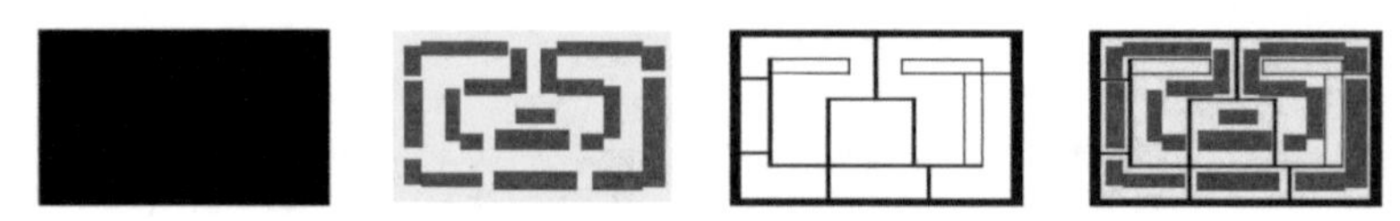

Fig. 4. Sample examples (from left to right: input, output non-road, output road, full output)

4.3 Model Architecture

This step uses the GauGAN algorithm to complete the model architecture and uses step-by-step training (outline-> buildings and buildings-> roadmap)

1. GauGAN neural network architecture

GauGAN is an image translation algorithm published by Nvidia Lab in 2019 that can achieve multi-modal synthesis. This experiment is implemented according to the paper, including VAE, which is used to achieve style guide multi-modal synthesis. The implementation of GauGAN is shown in Fig. 5.

2. Step Training

The training process of the model is not directly from contour to final result, but a step-by-step training (ArchiGAN: a Generative Stack for Apartment Building Design, 2019) to obtain a clearer bitmap result, which is divided into the following two parts: (1) Take the out-of-plane contour as the input dataset and train the building layout as the output dataset; (2) Then use the building layout as the input dataset and the road network map as the output dataset for training.

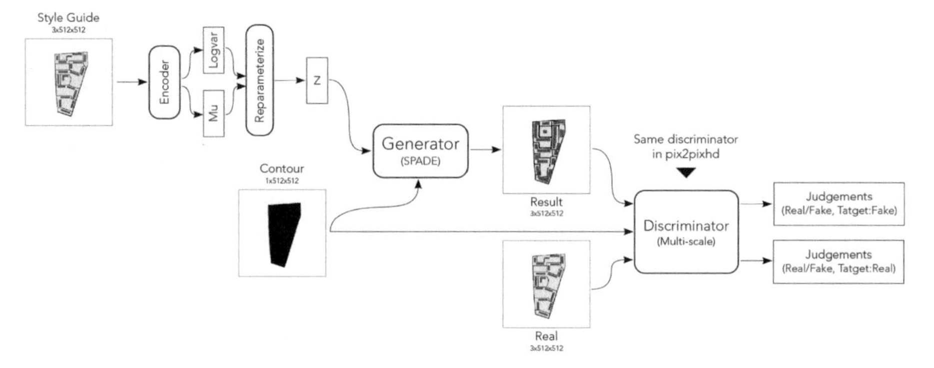

Fig. 5. Detail of GauGAN architecture

When testing, input the outline of the plane into the model trained in the first step, and the output building layout map is input into the second model to obtain the road network map, and then matrix multiplication of the building layout map and the road network map to obtain the final result. The content is shown in Fig. 6

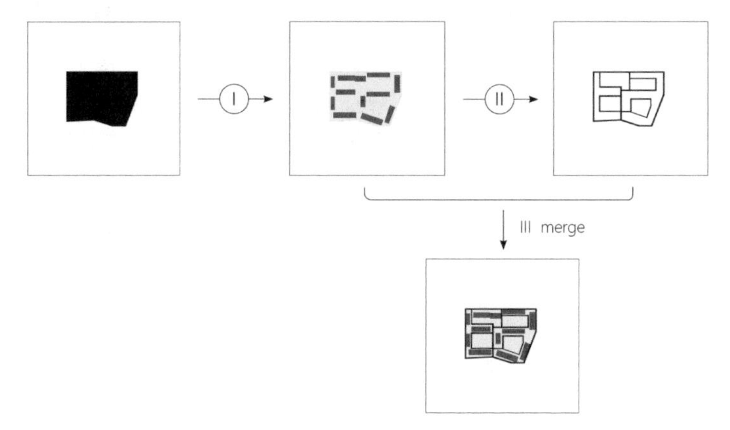

Fig. 6. Step Training

4.4 Vectorization and 3D Procedural Modeling

Models trained by the step training generates road bitmap and building layout bitmap results. The two bitmap results are vectorized and merged, and then 3d procedural modeling and visualization are performed. The black polygons represent the road net, and it can also be seen the windows and red roof, which are similar to the real world neighborhood community building style, so that the generated results are more intuitive and easier to be accepted by architects (Fig. 7).

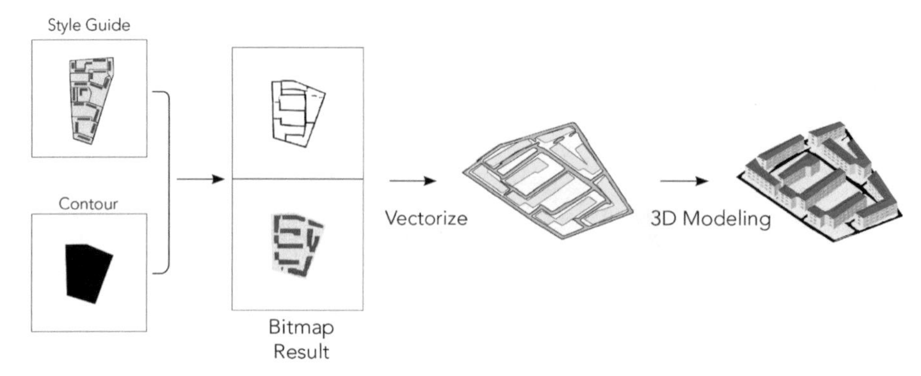

Fig. 7. Vectorization and 3d procedural modeling

4.5 Experiment Result

Using the trained model for the result generation test, compared with the real plan (Fig. 8), it can be seen that for the input of the same site contour map, the model can generate various different results, reflecting the diversity of the GauGAN algorithm. Secondly, from test results of the third sample in the figure, it can be seen that GauGAN makes good use of the information provided by the contour map to obtain generated results that conform to the input, so that the generated building shape fits the oblique outline fitted the sample status happening. However, the results also show that the style guide has less effect on the general layout generation results.

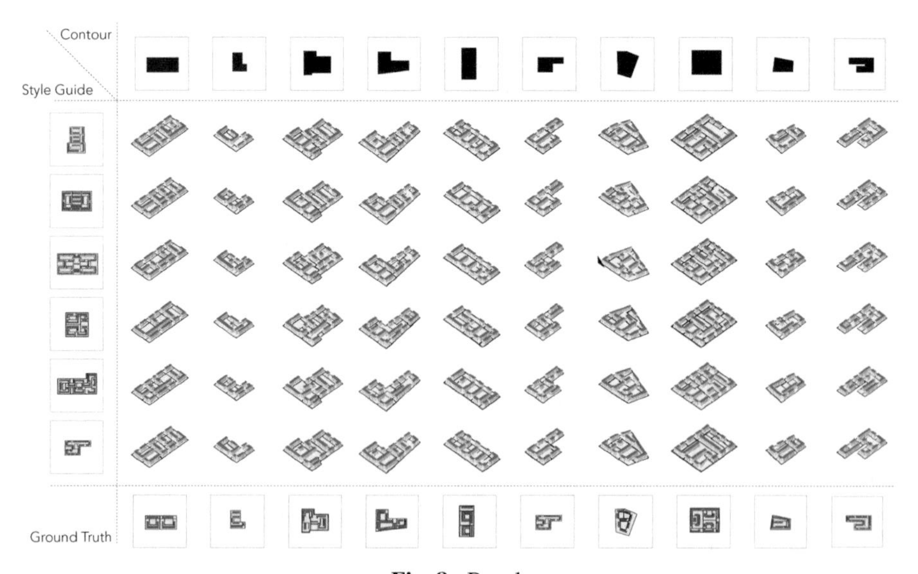

Fig. 8. Result

5 Conclusion

5.1 GauGAN Is More in Line with Architectural Design Needs Than Pix2pix (Pix2pixHD)

GauGAN is better than pix2pix in terms of diversity and contouring which is undoubtedly more in line with the needs of architectural design. Through experiments, it compared the GauGAN algorithm with the predecessor's pix2pix algorithm, and finally realized the multiple different flats including roads and buildings for a single plane outline; due to the use of GauGAN algorithm and the non-orthogonal samples included in the sample set, the results generated in this experiment are already achievable compared to the results of the predecessors Fit and change irregular contours. The above results prove that the method of this research experiment can be used to diversify and automatically generate the overall functional layout of the building, but whether Pix2pixHD can achieve contour fit is not yet accurately concluded.

5.2 The Use of Step Training Can Improve the Clarity of Generated Results and Allow the Later Vectorization to Be More Convenient

Experiments have found that the use of step training helps to improve image quality, resulting in clearer, less noisy, and more recognizable results; second, this method allows the results of building and roads to be completely separated, making it easier to separate buildings and buildings. The bitmap result of the road is vectorized to avoid the mixing of elements, which can not accurately vectorize different elements, so that the vectorized result becomes more accurate.

References

1. Xuan, Y.: The Effect of Neighborhood Mode in Modern Residential District Planning. Chinese & Oversea Architecture, pp. 94–95 (2011)
2. Jia, R., Sun, Y.: Cultural characteristics and existence value of buildings in Beijing million village. J. Beijing Inst. Civ. Eng. Arch., 76–80 (2012)
3. Huang, W., Zheng, H.: Architectural Drawings Recognition and Generation through Machine Learning. Tsinghua University, University of Pennsylvania (2018). https://www.researchgate.net/publication/328280126
4. Chaillou, S.: AI Architecture Towards a New Approach (2019). https://www.academia.edu/39599650/AI_Architecture_Towards_a_New_Approach
5. Isola, P., Zhu, J.-Y., Zhou, T., Efros, A.A.: Image-to-image translation with conditional adversarial networks. Berkeley AI Research (BAIR) Laboratory, UC Berkeley. https://arxiv.org/abs/1611.07004 (2016)
6. Wang, T.-C., Liu, M.-Y., Zhu, J.-Y., Tao, A., Kautz, J., Catanzaro, B.: High-resolution image synthesis and semantic manipulation with conditional GANs. https://arxiv.org/abs/1711.11585v1 (2017)
7. Park, T., Liu, M.-Y., Wang, T.-C., Zhu, J.-Y.: Semantic image synthesis with spatially-adaptive normalization. https://arxiv.org/abs/1903.07291 (2019)
8. Liu, Y., Deng, Q., Liang, L.: SchGAN:Primary school campus layout generation, intelligent assistant for architectural design. (2020) https://blog.csdn.net/shadowcz007/article/details/104035601

BY

Artificial Intuitions of Generative Design: An Approach Based on Reinforcement Learning

Dasong Wang and Roland Snooks(✉)

Royal Melbourne Institute of Technology (RMIT), Building 100, Melbourne 3000, Australia
{dasong.wang,roland.snooks}@rmit.edu.au

Abstract. This paper proposes a Reinforcement Learning (RL) based design approach that augments existing algorithmic generative processes through the emergence of a form of artificial design intuition. The research presented in the paper is embedded within a highly speculative research project, Artificial Agency, exploring the operation of Machine Learning (ML) in generative design and digital fabrication. After describing the inherent limitations of contemporary generative design processes, the paper compares the three fundamental types of machine learning frameworks in terms of their characteristics and potential impact on generative design. A theoretical framework is defined to demonstrate the methodology of integrating RL with existing generative design procedures, which is further explained with a Random Walk based experimental design example. The paper includes detailed RL definitions as well as critical reflections on its impact and the effects of its implementation. The proposed artificial intuition within this generative approach is currently being further developed through a series of ongoing and proposed research trajectories noted in the conclusion. The ambition of this research is to deepen the integration of intention with machine learning in generative design.

Keywords: Generative design · Machine learning · Reinforcement Learning · Intelligent formation

1 Introduction

Architectural design has been fundamentally impacted over the past three decades by the integration of emerging technologies and processual theory which have contributed to the proliferation of generative design methodologies [1]. Among these, the rapid maturity of artificial intelligence techniques, the massive increase in computational power and further development of complexity theory provides a new perspective to critically reflect on future directions of generative architectural design [2]. Framed by the limitation of contemporary generative methodologies, this paper proposes a Reinforcement Learning based approach to integrate Machine Learning with current computational design processes. This will be demonstrated through a simple design experiment based on a random walk algorithm. The broader ambition of this research is to leverage the generative potential of contemporary processes while integrating intuition through machine learning.

P. F. Yuan et al. (Eds.): CDRF 2020, *Proceedings of the 2020 DigitalFUTURES*, pp. 189–198, 2021.
https://doi.org/10.1007/978-981-33-4400-6_18

1.1 Contemporary Algorithmic Generative System

Contemporary generative design approaches can be categorized into two broad types, roughly summarized as parametric-based and behavioral-based. Parametric design processes operate through the manipulation of parameters that have an established linear relationship to a set of known geometric procedures. While, in behavioral-based systems, the control operates through encoding design intentions into a series of local behaviors to form a bottom-up, self-organizing process [3].

While both approaches are capable of compelling and sophisticated design outcomes, a series of limitations, or bottlenecks, still exist as obstacles towards its further development. Initially, the parametric-based approach relies on the linear relationship between parameters and the system/geometry, which leads to models that are limited to their predefined conditions. While, the behavioral-based system privileges micro interactions over macro awareness [4], establishing a global ignorance that limits the integration of overall design intentions. Furthermore, the integration of real-time materialization and structural performance [5] within non-linear generative design processes remains problematic due to the inherent volatility of these methodologies.

1.2 Artificial Intuitions

The paper speculates on a generative process driven by machine learning, which is capable of gradually developing typical and specific artificial "intuition" towards a series of design intentions. In natural processes of evolution, intuitions emerge from intelligent creature's inheritance of long-time accumulation of knowledge from generation to generation. The approach posited in this paper is intended to form a higher level of (machine) intelligence within generative design by undertaking a self-training, learning, and incrementally evolving process.

The research presented in the paper is part of an ongoing research project, Artificial Agency, with aims to explore the operations of machine learning with generative design and autonomous fabrication process, undertaken at the RMIT Architecture, Snooks Research lab.

2 Background

2.1 Machine Learning with Generative Design

The proposed intuitive generative approach is inspired by, and based on, the development of machine learning techniques. Contemporary machine learning consists of three fundamental types of frameworks: supervised learning, unsupervised learning and reinforcement learning [6] (Fig. 1).

Supervised Learning (SL) is essentially an algorithm that trains a predictive model with a labelled dataset (known outcomes). In recent years, enormous progress has been achieved with the rapid development of SL across a wide range of fields: data-prediction, image-synthesis, language-processing, etc. [7]. However, the impact of SL on three-directional generative design is still to be explored. Firstly, SL relies on massive labelled dataset, which is considered as a highly inefficient [8] and unrealistic process. While

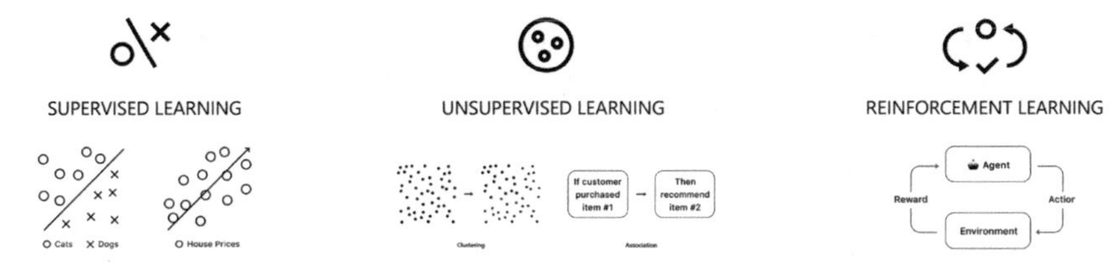

Fig. 1. Diagram of three typical types of machine learning: Supervised Learning (SL), Unsupervised Learning (UL), Reinforcement Learning (RL).

the labelling operation could be undertaken algorithmically, the feedback from the two sides of the ANN (Artificial Neural Network) is a linear procedure regardless of the data parsed during the generating process, which is opposed to the ambitions of existing generative design. Additionally, three-dimensional geometry representations are problematic with SL, and in particular with 3D GAN (Generative Adversarial Network) algorithmic frameworks [9], due to the substantial computational requirements.

Comparatively, Unsupervised Learning (USL) is based on training a clustering and association model with non-labelled datasets. Generally, USL doesn't have a clear training objective, but instead it aims to uncover invisible relationships within a massive dataset. Consequently, this approach is problematic when working with generative approaches that involve specific design intention.

2.2 Reinforcement Learning

Reinforcement Learning (RL) is closely associated with the field of optimal control, in which an agent seeks an optimal policy by interacting with its environment through a feedback between observation states and quantified rewards, modeled as a Markov Decision Process [10] with following specific elements (Fig. 2).

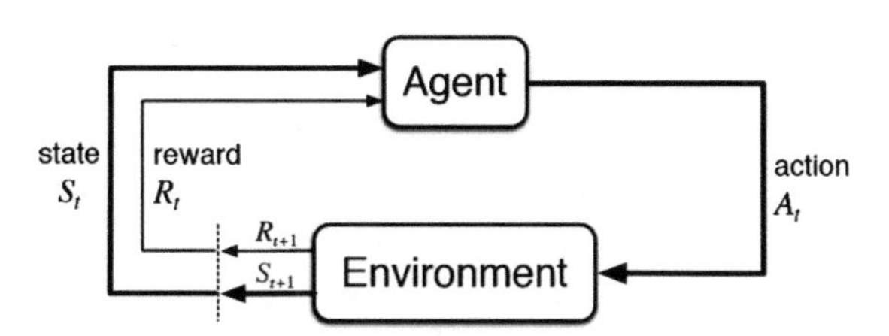

Fig. 2. Diagram of Reinforcement Learning (RL) with main elements: agent, environment, state, reward, action.

- Observation State (S): State is a concrete and immediate information summary of the agent itself and its interaction with the environment.
- Agent Action (A): Action is a set of possible moves the agent can take to interact with the environment.
- Reward (R): Reward is the feedback that measures the success or failure of an agent's actions in a given observation state.

- Policy (π): Policy is the strategy that the agent employs to determine the next action based on the current state. It maps states to actions, undertaking the actions that return the highest reward.

Under the overall structure of RL, there are diverse implemented algorithms: Q-Learning (Value-Based), Policy Gradient (Policy-Based), Actor-Critics, as well as further research fields: Hierarchical RL, Multi-Agent RL, etc. Contemporary RL has achieved significant progress with its application in Gaming AI, Self-Driving Vehicles and Robotics fields since 2017 [11].

It can be seen that RL has a clear correlation with, and enormous potential impact on existing generative design processes. Firstly, RL operates in a heuristic mode with no direct human knowledge, as opposed to the labelling process of SL. This heuristic mode is conceptually similar to the objective of generative design: to create the unpredictable and previously unimagined through logical design intentions. Secondly, RL operates on a sequential decision tree rather than the simultaneous processing of massive datasets (SL), which is suitable to be implemented with the constantly evolving generative controlling process. Thirdly, there are multiple technical approaches to implementing RL within generative design in three-dimensional environment, such as Gym toolkits [12] by OpenAI and ML-Agents toolkits [13] within Unity3D platform.

3 Methodology

The framework of the proposed design approach is to integrate RL with existing generative processes, in which RL is acting as a brain to further control the algorithmic system instead of creating an entirely new procedure. The methodology is further demonstrated with a Random Walk based design experiment from the overall training setup to detail definitions.

3.1 Intuitive Random Walk Formation

Random Walk (RW) is a long-standing algorithmic model inspired by a natural stochastic process [14], with applications in numerous scientific fields. As shown in Fig. 3, the goal of this example is to train a RW with a series of basic architectural intuitions, initially inspired by Le Corbusier's Domino System [15] and further developed with more abstract and critical design intentions of spatial and structural logic. With the implementation of RL, it is expected that the opposing characteristics of Random Walk's stochastic operation and the Domino System's formality can be integrated with a synthetic design process.

3.2 RL Actions Definition

Within the training framework, RL actions can be based on the underlying generative system or customized methods to further control the generating process, depending on the characteristics of the system and training task (Fig. 4).

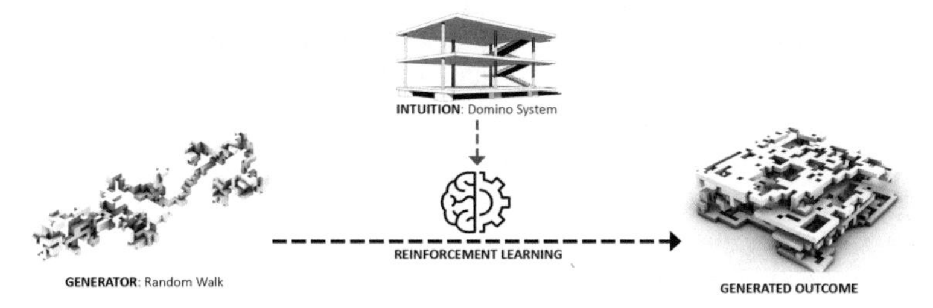

Fig. 3. Diagram demonstrating the training intention of the Domino system as a type of intuition within the RW generative formation process.

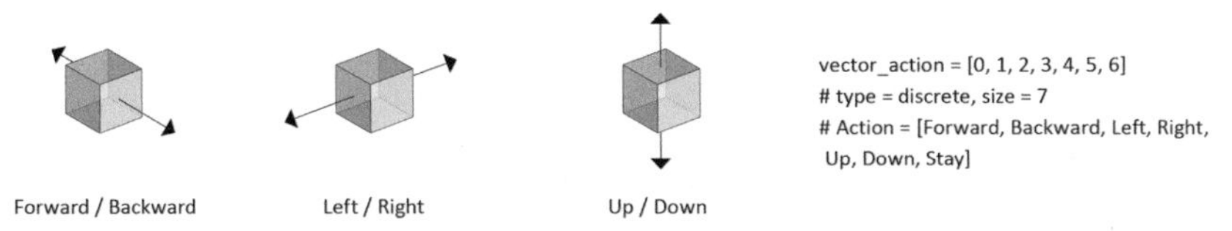

Fig. 4. Diagram of the definition of RL actions within the RW generative process.

The action definition of the RW experiment is that an agent takes random decisions to move towards six directions within a limited three-dimensional voxel gird, from which the walking trail is recorded as a generated form. In this case, the RL action is considered as a discrete [10] action, with a vector[1] size of seven.

3.3 RL Observations Definition

The definition of observation states describes the current condition within the generating process, which normally consists of two types of information: the overall matrix data type representation of the form and a series of significant reward-oriented values (Fig. 5).

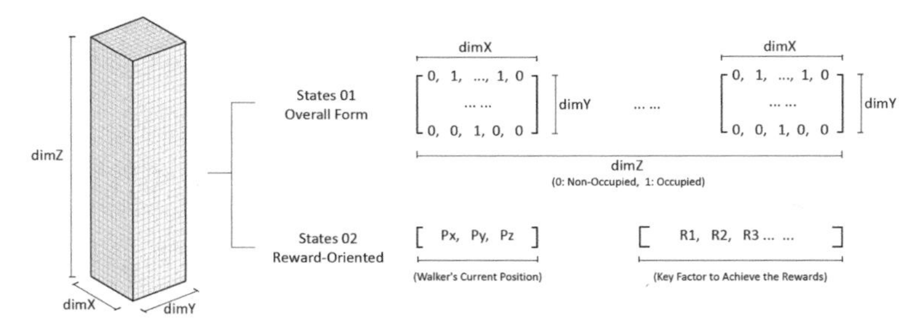

Fig. 5. Diagram of the definition of RL observations within the RW generative formation process.

In the RW example, the form is converted to a three-dimensional representation: voxel-based matrix of integers (1 or 0), representing a Boolean describing whether the

[1] Vector: refers to a one-dimensional array (being distinguished from vector definition of Computer Graphics) .

voxel is occupied or not. Additional reward-oriented information is also included in the states, such as the current position of an agent, and its real-time reward evaluation figures.

3.4 RL Reward Definition

As the most critical part of the RL training process, the reward definition is normally a quantitative evaluation structure based on design intention. In this case, the initial reward definition is simply identifying some reward locations (representing domino floors) in the voxel grid and encouraging the walker to seek and connect the floors. With further development, a more comprehensive structure is setup with more detailed design intentions, showed in Fig. 6.

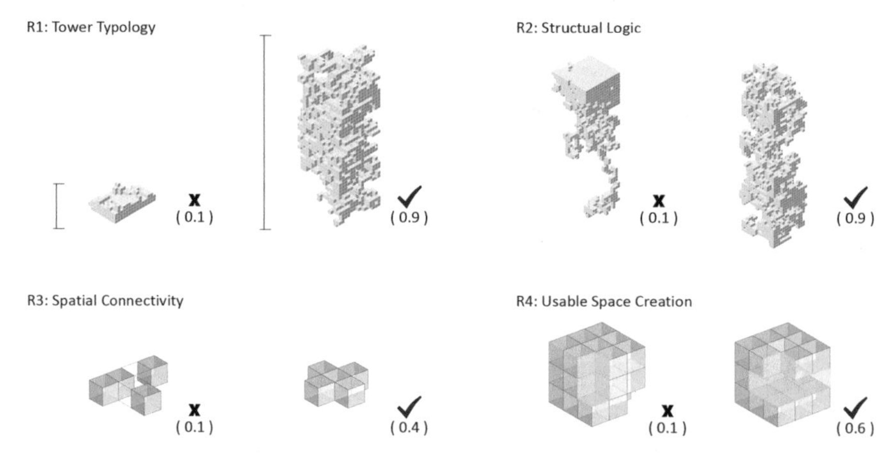

Fig. 6. Diagram of the definition of RL rewards within the RW generative formation process.

- Tower Type Reward (R1): Agent is encouraged to generate a tower-like form. The reward calculation is based on the height of the form.
- Structural Logic Reward (R2): A pyramid-like structural logic is implemented such that a reward at the bottom part should be larger than the stacked part above.
- Spatial Connectivity Reward (R3): Horizontally, if one generated voxel is connected to its four neighbouring voxels, the agent will receive a positive reward of spatial connectivity.
- Spatial Creation Reward (R4): The greater the void space generated in between two voxels in the vertical direction, the greater the positive reward the agent receives.
- Site Response (R5): Some existing voxels are setup in the grid to represent site context. When the agent collides with these voxels, a negative value will be added to the reward calculation as a form of punishment.

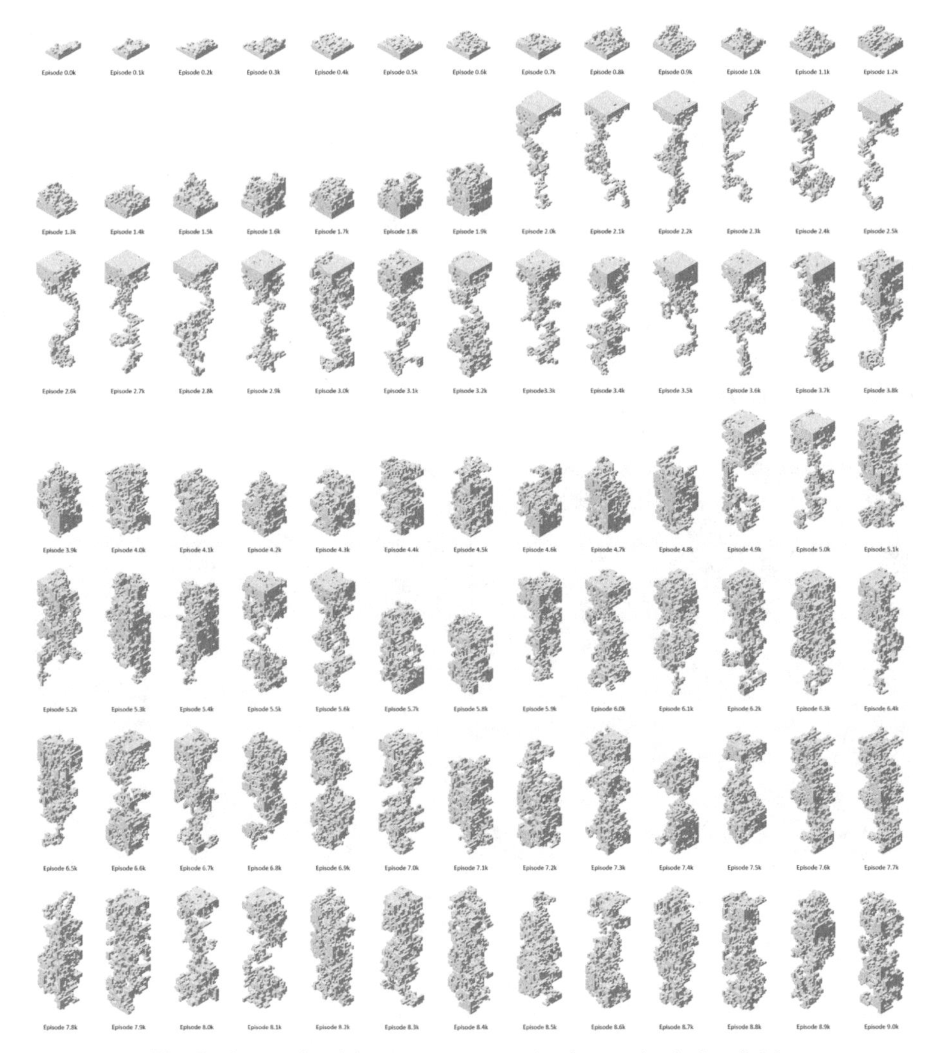

Fig. 7. Selected training outcome samples from episode 0 to 9.4 k.

4 Discussions

4.1 Training Process and Outcomes

The Random Walk design experiments operates with a customized Deep Q-Learning Algorithm in Python and Tensorflow Environment. Totally, the training process undertaken 10,000 episodes, calculating on a local computer with a time consumption of about three hours. In order to assess training outcomes, the generated form is recorded every 100 episodes, shown in Fig. 7. Overall, the training result is remarkably successful. The intense and squeezed form that resulted from the initial episodes (0.0 to 1.9 k) are significantly improved and evolved in the latter iterations (7.7 k to 9.4 k) in terms of the predefined reward.

The characteristics of forms generated through this process evolved unexpectedly over time, creating a clear sequence of design intentions. From episodes 0.0 to 1.9 k, the Random Walker System doesn't generate any effective intuitions. However, from 2.0 k to 4.0 k, it starts to understand the predefined intention as tower type (R1). Episodes 4.0 k to 5.9 k demonstrate how the form balances structural performance (R2) and tower type (R1) reward. Throughout the RL process the response to design intention of space connectivity (R3) and creation (R4) are slowly improved, becoming more obvious in the final episodes. The site response (R5) was not implemented in this case due to the resolution and complexity of the particular intention. Despite that all the rewards operate simultaneously and were defined prior to the training being launched, the process generates a clear, multi-stages characteristic that achieves one significant reward prior to addressing the others through a gradual process of improvement (Fig. 8).

Fig. 8. Visualization prospective of a preferred training result. (The generated voxels have been further developed into structures and platforms with a set of algorithmic stochastic operations.)

There are some obvious limitations in the posited RW design experiment that result from using a single walker to generate form within a low-resolution grid. However, as an early and speculative case to explore and demonstrate the potential approach of applying RL within generative design processes, it still shows concrete effects and significant flexibility to be deeply integrated with other existing generative design processes.

4.2 Further Research

A number of ongoing research trajectories have emerged from the posited application of RL in algorithmic generative design and digital fabrication, which are summarized as follows:

- Complex Generative System Training with RL: The research is focused on integrating RL with a complex self-organized generative system in response to non-programmable design intentions, such as the architectural typology logic.

- Multi-Agent Global Awareness Training with RL: The research aims to generate global intuitions for multi-agent systems that combines with their logic of local interactions. These global concerns include the control of form, topology and structural networks.
- RL with Real-time Robotics: As a collaborative direction, the research intend to apply RL with real-time robotic behavior in order to advance the concept of automated assemblies.

5 Conclusions

The proposed RL based design approach integrates heuristic design ituitions within known algorithmic generative processes, augmenting the processes to establish a greater level of sophistication and design capacity. Both a theoretical foundation and technical methodology are presented in this Intuitive Random Walk Formation case to demonstrate the concrete effects and potential flexibility of cultivating intuitions for generative systems. The subsequent reflections in the paper aim to indicate potential ways of applying this emerging tool to existing design methodologies as well as anticipating a closer correlation of designer and computational intelligence.

References

1. Frazer, J.: An Evolutionary Architecture. Architectural Association (Themes VII), London (1995)
2. Leach, N.: Design in THE age of artificial intelligence. Landsc. Archit. Front. **6**, 8–20 (2018)
3. Leach, N., Snooks, R.: Swarm Intelligence: Architectures of Multi-agent Systems. Tongji University Press, Shanghai (2017)
4. Snooks, R.: Behavioral Formation: Multi-Agent Algorithmic Design Strategies. RMIT Ph.D. thesis (2014)
5. Hensel, M., Menges, A., Weinstock, M.: Emergent Technologies and Design: Towards a Biological Paradigm for Architecture. Routledge, Abingdon (2013)
6. Mitchell, T.: Machine Learning. McGraw-Hill Science/Engineering/Math (1997)
7. Paliouras, G., Karkaletsis, V., Spyropoulos, C.: Machine Learning and Its Applications. Springer, Heidelberg (2001)
8. Zheng, H.: Form finding and evaluating through machine learning: the prediction of personal design preference in polyhedral structures. In: Proceedings of the 2019 Digital FUTURES (CDRF 2019) (2019)
9. Wu, J., Zhang, C., Xie. T,, Freeman, W., Tenenbaum, J.: Learning a Probabilistic Latent Space of Object Shapes via 3D Generative-Adversarial Modeling. arXiv:1610.07584 [cs.CV] (2016)
10. Sutton, R., Barto, A.: Reinforcement Learning: An Introduction. The MIT Press, Cambridge (1998)
11. Henderson, P., Islamm, R., Bachman, P., Pineau, J., Precup, D., Meger, D.: Deep reinforcement learning that matters. In: The Thirty-Second AAAI Conference on Artificial Intelligence (AAAI-18) (2018)
12. Brockman, G., Cheung, V,, Pettersson, L,, Schneider, J,, John, S., Jie, T., Zaremba, W.: OpenAI Gym. arXiv:1606.01540v1 [cs.LG] (2016)
13. Juliani, A., Berges, V., Teng, E., Cohen, A., Harper, J., Elion, C., Goy, C., Gao, Y., Henry, H,, Mattar, M,, Lange, D.: Unity: A General Platform for Intelligent Agents. arXiv:1809.02627 [cs.LG] (2018)

14. Kaye, B.: A Random Walk Through Fractal Dimensions. Wiley-VC, New York (1994)
15. Corbusier, L.: Vers une Architecture (Towards a new Architecture). France (1923)

Collection to Creation: Playfully Interpreting the Classics with Contemporary Tools

Ana Herruzo[1,2(✉)] and Nikita Pashenkov[1(✉)]

[1] Woodbury University, 7500 N Glenoaks Blvd, Burbank, CA 9150, USA
{Ana.herruzo,Nikita.Pashenkov}@woodbury.edu,
ae.herruzo@alumnos.upm.es

[2] Universidad Politecnica de Madrid, Pº Juan XXIII, 11, 28040 Madrid, Spain

Abstract. This paper details an experimental project developed in an academic and pedagogical environment, aiming to bring together visual arts and computer science coursework in the creation of an interactive installation for a live event at The J. Paul Getty Museum. The result incorporates interactive visuals based on the user's movements and facial expressions, accompanied by synthetic texts generated using machine learning algorithms trained on the museum's art collection. Special focus is paid to how advances in computing such as Deep Learning and Natural Language Processing can contribute to deeper engagement with users and add new layers of interactivity.

Keywords: Machine learning information visualization · Real-time generative graphics · Creative technology

1 Introduction: Generations to Generative

Last academic year, Applied Computer Science program at Woodbury University was invited by The J. Paul Getty Museum, a premier art institution in Los Angeles, to design an installation for the College Night event and exhibition scheduled to take place in April 2019. In conversations with educational specialists at the museum, interest was shown in exploring human emotions as a thematic element in a project that would merge art and technology, which successfully aligned with the undergraduate program's mission.

To approach the design and development of this project we created a collaboration between two sophomore classes and designed a new syllabus for each course to include dialogue with the museum. The syllabus also included guided visits to the museum to study some of the art works in its collection, learning about the artists' intentions, and what the works are communicating in their portrayal of emotions and facial expressions.

While studying the museum's art collection [5] that comprises Greek, Roman, and Etruscan art from the Neolithic Age to Late Antiquity; and European art from the Middle Ages to the early 20th Century; with a contemporary lens, questions arose regarding static and finished pieces of art, in contrast to interactive and responsive artworks [6]. In our attempt to find ways to connect the student project to the museum and its art collection,

P. F. Yuan et al. (Eds.): CDRF 2020, *Proceedings of the 2020 DigitalFUTURES*, pp. 199–207, 2021.
https://doi.org/10.1007/978-981-33-4400-6_19

we decided to explore solutions that would allow for creation of dynamic generative visuals in combination with content somehow derived from the existing collection.

A key observation guiding our concept was that artworks in exhibitions and museums are typically accompanied by a title and a brief description. In our case, unique visuals would be generated based on user interaction, so we proposed the creation of new synthetic titles and descriptions to accompany each user engagement. It seemed fitting that the textual output would be generated by machine learning algorithms trained on existing texts from the museum's art collection, to create a unique connection bridging carefully curated static content with dynamic generated visuals.

The resulting installation utilizes a Kinect sensor to analyze the users' movements and a separate camera to read facial expressions [3] via computer vision and Deep Learning algorithms, using their outputs in the next stage as a basis for text generation based on natural language models trained on the text descriptions of the Getty Museum's art collection.

2 Process: Beyond Codified Interaction

Our goal was to create a new media art [4] piece that has an intimate connection with the users while simultaneously generating new periodic content that is always evolving, changing; never the same.

The installation consists of a vertical video wall composed of three landscapeoriented screens, with two sensors embedded on top: Microsoft Kinect ONE and a USB web camera. We used two software platforms, PyCharm as an integrated development environment for the Python programming language; and Derivative TouchDesigner as a real-time rendering, visual programming platform. The two platforms communicated with each other via TCP/IP sockets over the network.

The title of the project, "WISIWYG", is a play on the popular acronym "What You See Is What You Get" (WYSIWYG), based on the idea that the installation incorporates computer vision processing and machine learning algorithms, in a sense allowing it to generate outputs according to what it sees from its own perspective.

The user experience was a central component of the project and was carefully crafted. After several sessions of user testing to determine the ideal flow for the installation, we designed a sequence of animated events, to successfully guide the user through the experience lasting a total of about two minutes (Fig. 1).

As a result, two types of media proposals were developed in parallel: media to be displayed during users' interaction, and media generated in-between interactions (Fig. 2).

Once the user experience had been designed, the following strategies were developed in order to connect and interact with the attendees.

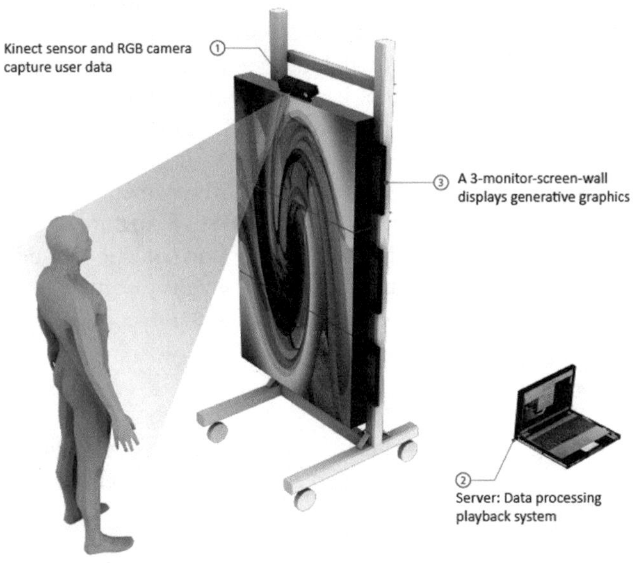

Fig. 1. Installation diagram

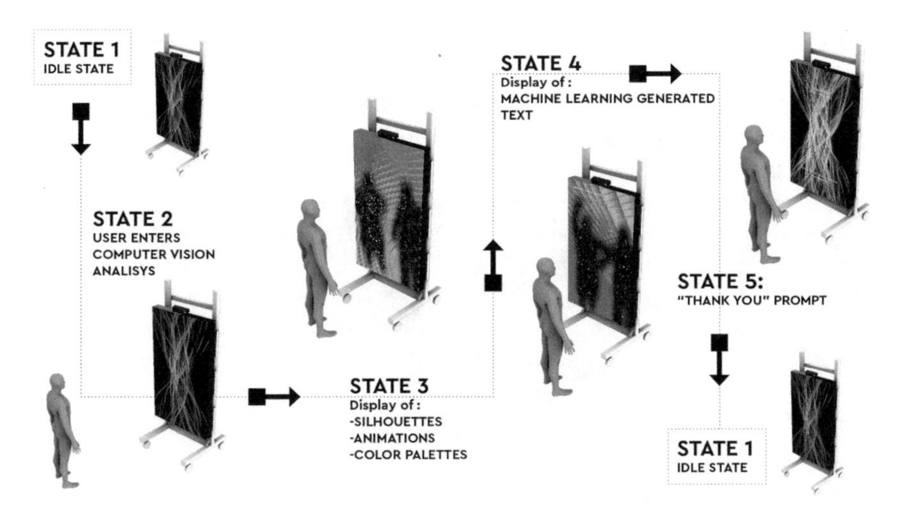

Fig. 2. User experience diagrams

3 User Analysis

The primary driver of the project is the camera input, processed through computer vision and machine learning algorithms. During the course of users' engagement, computer vision is first used to isolate faces and determine the number of people in camera's field of view. This step is accomplished via a traditional computer vision face tracking method using the popular Open Computer Vision (OpenCV) library. The second step uses a Deep

Learning model based on a Convolutional Neural Network (CNN) [7] programmed in Python with the help of Keras and TensorFlow frameworks.

We utilized a 5-layer CNN model made popular by the Kaggle [2] facial expression recognition challenge. The model has been trained on the Feb2013 dataset distributed with the challenge, which consisted of 28,000 training and 3,000 test images of faces stored as 48 by 48-pixel grayscale images. In order to provide the image data in the format that the CNN model expects, sub-windows of the camera feed with faces detected by the OpenCV library were scaled down to 48 × 48 size to be passed on in this step. The Python code to construct the facial expression detection model in Keras, as well as other Deep Learning models and associated weight parameters utilized in the project and discussed further, are available in an open GitHub code repository.

4 Media Creation

When visitors approach the installation, depending on how many people are in front of the screen; their facial expressions and estimated ages; a unique animation is generated on the video wall, comprising the following elements:

- Silhouettes: A mirror-effect reflection of users' silhouettes on the screen featuring real-time generated visuals, playing inside and outside the silhouettes.
- Scenes: Each student designed four real-time scenes with animated content, driven by hard-coded rules based on the parameters obtained through computer vision algorithms.
- Color palettes: Students generated parametrized color animations using the computer vision inputs (estimated ages, facial expressions, and number of people) (Fig. 3).

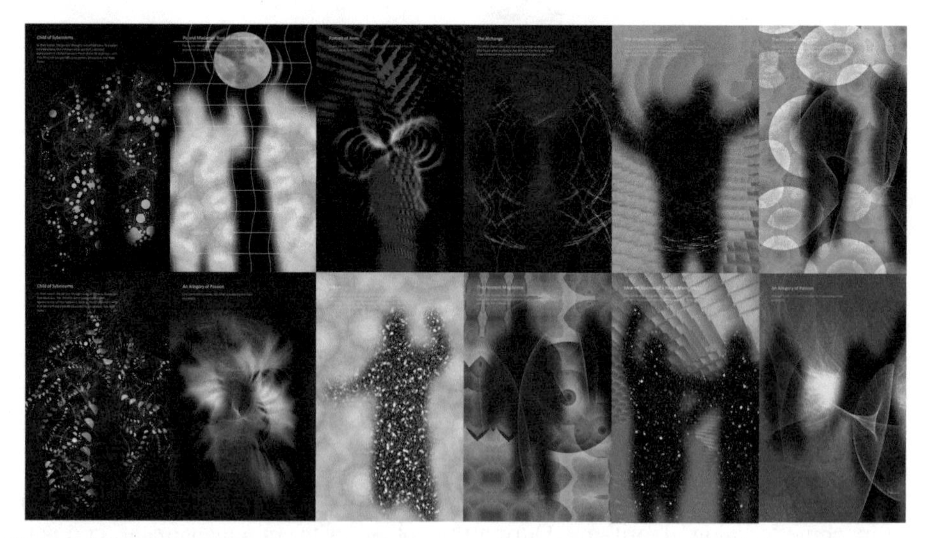

Fig. 3. Gallery showcasing several of the animations designed by the students and the different color palettes applied based on the users' facial expressions and movements.

5 Synthetic Text Descriptions

This portion of the project began by screening the Getty's art collection online, and selecting all artifacts depicting people. As the next step, a database was created by recording the artworks' titles and descriptions, as well as subjectively estimating the number and ages [10] of people featured.

In analyzing the Getty's art collection, students experimented with the Deep Learning natural language model GPT-2 [9], an acronym for Generative PreTrained Transformer, released by the non-profit foundation OpenAI in February 2019. The language model generates convincing responses to textual prompts based on set parameters such as the maximum length of response and 'temperature' of indicating the relative degree to which the output conforms to the features resembling training data.

Our project utilized the GPT-2 model with 144 million parameters, the largest made available by the OpenAI Foundation as of April 2019. The model has been "fine-tuned" by the students, a process in which an existing machine learning model was re-trained to fit new data, using descriptions of artworks on display at the Getty Center as the training dataset. To accomplish this task, we utilized the Google Colaboratory [1] notebook environment that allows Python code to be shared and executed online. A Collaboratory notebook designed for fine-tuning the GPT-2 model allowed each student to individually analyze and modify the Python code to read new text data, re-train the language model, and produce new text descriptions based on interactive text prompts.

The text prompts were pre-generated by the students based on their own analysis of artworks and consisted of short singular and plural descriptions like "sad young person" or "two happy people." The text prompts were interactively input to the GPT-2 model to generate responses that were entered into a database and associated by rows with tagged columns for age, number of people and facial expressions. The database content was then programmatically correlated with the outputs of computer vision processing and Deep Learning classification using the Pandas library in Python, by selecting a random database cell containing a response that matches detected facial expression tags. Finally, the selected response was rendered as the text description that accompanied visual output onscreen.

It was beyond the scope of the project to build students' expertise in constructing their own Deep Learning models or invest resources into training the models from scratch, especially in view of time and computing power expended in the process. Researchers at University of Massachusetts, Amherst, have shown that training the GPT-2 language model, for example, can consume anywhere between $12,902–$43,008 in cloud computing costs [11]. While training such computationally expensive models was beyond the scope of our project, each student had a chance to install the relevant Python code and libraries, analyze them, fine-tune and run the models on their personal computers.

In addition to facial expression detection, the project incorporated Deep Learning models for age and gender detection, though ultimately we decided not to pursue gender recognition due to non-technical concerns with bias and non-binary gender identification. The age detection was built on a successful CNN architecture based on Oxford Visual Group's (VGG) model [8] that gained recognition in the ImageNet Large Scale Visual Recognition Competition (ILSVRC) in 2014 [11]. The datasets that formed that basis of the VGG model, known as IMDBWIKI [13], consist of over 500,000 labeled faces.

In order to utilize the model effectively, we again made use of pre-trained weights found online [14], applying transfer learning techniques to work with the real-time video data in our project. As a result, each student was able to directly experiment with deploying machine learning models in Python code to process computers' built-in camera input and generate predictions for detected facial expressions and age estimates of each other, initially in the classroom environment and eventually in the public setting of the museum exhibition (Fig. 4).

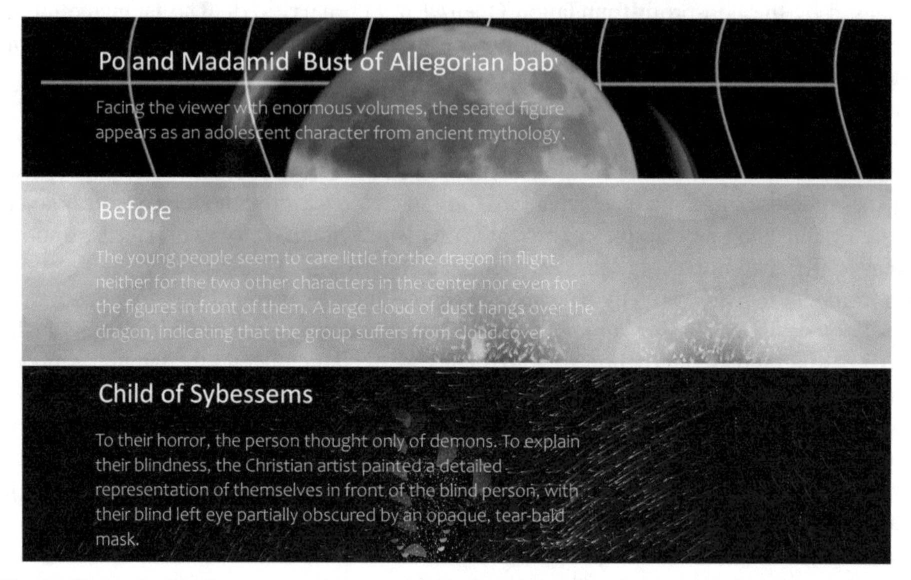

Fig. 4. Example of texts generated by machine learning algorithms overlaid on the real-time generated animations.

6 Thoughts

With the ongoing evolution in computing hardware and software, machine learning is becoming more accessible and widespread. We're now able expand the levels of interaction to more complex layers, allowing recognition of patterns, movements, facial expressions and more. Along with the rapid evolution of realtime rendering engines, programmable shaders, and new algorithms, it is now possible to effectively create real-time data-driven media at high resolutions and with great rendering quality. Advances in technology and computing exemplified by Deep Learning can contribute to deeper connections and new layers of interactivity.

Different simultaneous levels of interactivity occurred in this project, some with a direct and transparent effect, and others with more elaborate or indirect interaction. For example, we discovered that when the users recognized their own silhouettes, they would engage more actively with the installation: dancing, jumping, moving their hands or even performing backflips. The facial expression and age-based visualization parameters were not as transparent, since it is harder for the user to acknowledge this relationship during a time-limited engagement (Fig. 5).

Fig. 5. Users experiencing the installation and playing with their silhouettes on the screen.

Even though our database was quite small, the generative algorithms provided widely variable outputs. The level of uncertainty in understanding generated textual content created a level of engagement as well, attested by visitors inquiring on many occasions why a certain title or description appeared on the screen. It is of our interest that further explorations should address a more thorough analysis of the users' feelings or personality, in search of a deeper and more profound connection between the art piece and the subject.

After creating our database, at the beginning of the user testing process, some of the students raised concerns about the use of gender in camera-based analysis and generated texts. As a result, questions came up regarding the role of gender identity in today's society and opened the door for further discussions involving human-machine interactions. We attempted to address this issue in part by avoiding the use of gender, and programmatically manipulating the generated texts, as well as manually editing our database by screening for male and female pronouns, attempting to "de-gender" it by replacing those with plural "they," or neutral "person", wherever appropriate.

The possibility of using gender detection and other forms of identification, afforded by Deep Learning software algorithms, opens the door to a host of significant questions and concerns. Shoshana Zuboff, in her book *Surveillance Capitalism* [15], raises key elements to reflect on, now that a wide range of software applications and hardware devices the we use daily, monitor, log and process a wide variety of data obtained from the user. Zuboff warns that these technologies are often designed to obtain users data in disguise, in search of individual and society behavior modification. Users' privacy while being monitored and issues surrounding this topic were briefly addressed in this project, but are undoubtedly significant questions to be aware of, discussed and reflected upon when designing interactive experiences.

7 Conclusion

Features that make this project unique include the combination of real-time generative graphics with exciting new machine learning models, the nature of its development within an academic environment and the opportunity for the students to exhibit at a great art institution. Trying to find ways to connect the existing art collection to our project, while addressing the student learning outcomes, resulted in a project that successfully fulfills the mission of our University program: hybrid art and technology.

By tapping into different areas of study in one experimental project, we feel that we managed to offer the students an opportunity to understand how diverse disciplines can be intertwined and relevant to each other, while the deployment of a live interactive installation at a world-renowned art institution endowed them with valuable professional production experience. The number of constraints, such as addressing multiple curriculum requirements, a collaboration between academic and art institutions, maintaining appropriate workloads and so on, turned out to strengthen and boost the level of creativity, allowing the students to become proficient in several technical skills working with advanced programming frameworks and computational models.

Perhaps our primary contribution as faculty guiding this project has been to help students synthesize and leverage the somewhat disparate technical and creative tools, frameworks and resources as part of the learning process, as well as providing a critical view towards that state of technology and computer science in today's culture. We hope this project summary provides a useful reference point for others seeking an approach in creative applications of contemporary technologies.

References

1. Google Colaboratory. Colab.research.google.com (2020). Cited 25 May 2020. https://colab.research.google.com/notebooks/welcome.ipynb
2. Challenges in Representation Learning: Facial Expression Recognition Challenge | Kaggle. Kaggle.com (2020). Cited 25 May 2020. https://www.kaggle.com/c/challenges-in-representation-learning-facial-expressionrecognition-challenge/
3. Viola, P., Jones, M.: Robust real-time face detection. Int. J. Comput. Vis. **57**(2), 137–154 (2004)
4. Tribe, M., Grosenick, U., Jana, R.: New Media Art. Taschen, Köln (2006)
5. Collection (Getty Museum). The J. Paul Getty in Los Angeles (2020). Cited 25 May 2020. http://www.getty.edu/art/collection/. Accessed 10 Jan 2020
6. Krueger, M.: Responsive environments. In: AFIPS 1977 Proceedings, National Computer Conference, pp. 423–433 (1977)
7. Lopes, A., de Aguiar, E., De Souza, A., Oliveira-Santos, T.: Facial expression recognition with Convolutional Neural Networks: coping with few data and the training sample order. Pattern Recogn. **61**, 610–628 (2017)

8. Parkhi, O., Vedaldi, A., Zisserman, A.: Deep Face Recognition. Visual Geometry Group - University of Oxford (2015). Cited 25 May 2020. http://www.robots.ox.ac.uk/~vgg/publications/2015/Parkhi15/parkhi15.pdf
9. Radford, A., Wu, J., Child, R., Luan, D., Amodei, D., Sutskever, I.: Language Models are Unsupervised Multitask Learners. OpenAI Foundation (2019). Cited 25 May 2020. https://cdn.openai.com/better-languagemodels/language_models_are_unsupervised_multitask_learners.pdf
10. Rothe, R., Timofte, R., Van Gool, L.: Deep expectation of real and apparent age from a single image without facial landmarks. Int. J. Comput. Vis. **126**(24), 144–157 (2016)
11. Strubell, E., Ganesh, A., McCallum, A.: Energy and Policy Considerations for Deep Learning in NLP. arXiv.org. (2019). Cited 25 May 2020. https://arxiv.org/abs/1906.02243
12. ImageNet Large Scale Visual Recognition Competition 2014 (ILSVRC2014). Image-net.org (2020). Cited 25 May 2020. http://www.imagenet.org/challenges/LSVRC/2014/
13. IMDB-WIKI - 500 k + face images with age and gender labels. Data.vision.ee.ethz.ch (2020). Cited 25 May 2020. https://data.vision.ee.ethz.ch/cvl/rrothe/imdb-wiki/
14. Apparent Age and Gender Prediction in Keras - Sefik Ilkin Serengil. Sefik Ilkin Serengil (2020). Cited 25 May 2020. https://sefiks.com/2019/02/13/apparentage-and-gender-prediction-in-keras/
15. Zuboff, S.: The Age of Surveillance Capitalism, 1st edn. Public Affairs, New York (2019)

embedGAN: A Method to Embed Images in GAN Latent Space

Zhijia Chen, Weixin Huang(✉), and Ziniu Luo

School of Architecture, Tsinghua University, Haidian District, Beijing, China
huangwx@mail.tsinghua.edu.cn

Abstract. GAN is an efficient generative model. By performing a latent walk in GAN, the generation result can be adjusted. However, the latent walk cannot start from a selected image. The embedGAN is proposed to embed selected images into GAN and remain the generation effect. It contains an embedded network and a generative network. Application cases of residential interior design are given in the article. With advantages of a low computing cost and short training time, embedGAN shows its potential. The embedGAN algorithm framework can be applied to various GANs.

Keywords: embedGAN · GAN · Latent walk · Embedding data

1 Introduction

As a famous generative model in the field of machine learning, GAN has been widely used since proposed [1]. Further development of GAN has come up and one of the directions is to enable GAN to generate results according to users' control, for example, latent walks. The advantage of latent walks is that the pre-trained model is used and there is no need to retrain the model, which brings low computation cost and it is easy to check the effect. Latent walks can be applied to almost all types of GAN; and through the control of latent codes, directional generation can be achieved.

The problem of latent walks is that it must start from a randomly generated latent code, and cannot start from selected data. It is even unable to start from an image in the training data set, as GAN learns the sample features distribution, rather than the sample itself. At the same time, GAN training process also has the problem of insufficient diversity. To fundamentally solve the problem of generative diversity, it is necessary to expand the latent code dimension and make the training process invertible, like GLOW [9].

However, this approach requires large-scale computing power and a long training procession. If the goal is just to use any selected image for latent walk, is there a low-cost and rapid way to enable generating selected images in GAN?

The embedGAN method proposed in this article achieves this goal and its main functions are:

P. F. Yuan et al. (Eds.): CDRF 2020, *Proceedings of the 2020 DigitalFUTURES*, pp. 208–216, 2021.
https://doi.org/10.1007/978-981-33-4400-6_20

1. Embed selected images in GAN generation distribution.
2. It hardly affects the parameters of other parts of the generative network, so that random walks in latent space can still be carried out.

The contributions of this research are:

1. A low-cost and effective method to expand the latent walk range of GAN is proposed.
2. The effectiveness of this method is verified by experiments, and examples applied to interior design are given.

2 Related Work

2.1 Regenerating Data in GAN

In the process of testing different evaluation standards, Shmelkov et al. tried to regenerate data in the training set through GAN, but found that images generated by GANs have a large gap with the selected ones [3]. A similar attempt was made in BEGAN and experiments were conducted to reverse the latent codes of real images. In most cases, GANs can hardly regenerate images from the training set [4]. Zhu et al. alleviated this problem by using multiple initial points and CNN to train the mapping of images to latent code [5]. Lipton et al. proposed a method to jump out of the local optimal solution [6]. An attempt is to train the images together with latent code. However, the results shows that the latent code still cannot generate the selected images well [7, 8]. It can be seen that generating selected data of the training set in GAN is an unsolved problem. The GLOW model can solve it, but it does not have the lightness of GAN, and it needs to pay a very high computing cost [9].

2.2 GAN Latent Walk

In DCGAN, parts of generative features are controlled by linear operations on latent code. But this attempt is limited to standardized data such as human faces [10]. In 3D-GAN, similar operations can also be achieved [11]. By introducing an aesthetic evaluator, Goetschalckx L et al. performed a heuristic search on the GAN latent space, and guided the generative network to produce images in the direction of higher aesthetic evaluation scores [12]. Jahanian A et al. succeeded in "steering" the GAN generative network by introducing an evaluator of lens shift [13]. Abdal et al. used a method of embedding images for StyleGAN, and studied the generation effect of various embedded images comprehensively [14]. However, the method is only applicable to StyleGAN. The method provided in this article can embed data in the initialized latent space and is widely applicable to various GANs.

3 Method

3.1 Principle

The input of GAN is a latent code and the output is an image. The input latent codes are usually generated from a multivariate Gaussian distribution, of which dimensions are lower than the real-world features. In the process of neural network training, due to the existence of ReLU and other activation functions, some feature information will be lost. For the reasons mentioned above, the discriminative network can hardly learn low-probability sample features, resulting in the generative network also unable to learn this part of the features or regenerate images from the training set (as the real images contains low-probability features) (Fig. 1).

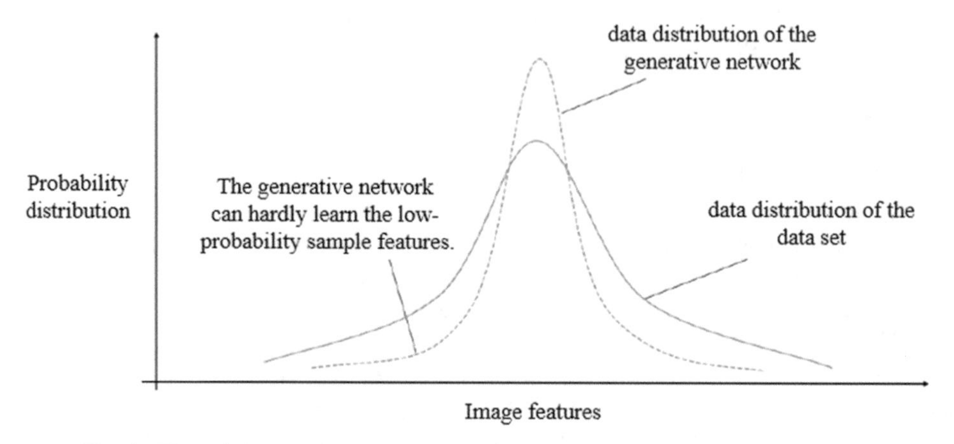

Fig. 1. The training set data distribution and the GAN generative data distribution.

In order to obtained selected images from the generative network, an idea is to fine-tune parameters of generative network and embed data. Through additional training, the generative network can be targeted to learn some low-probability features. This approach is equivalent to offsetting and expanding the data distribution, which improves the diversity of generated images (Fig. 2).

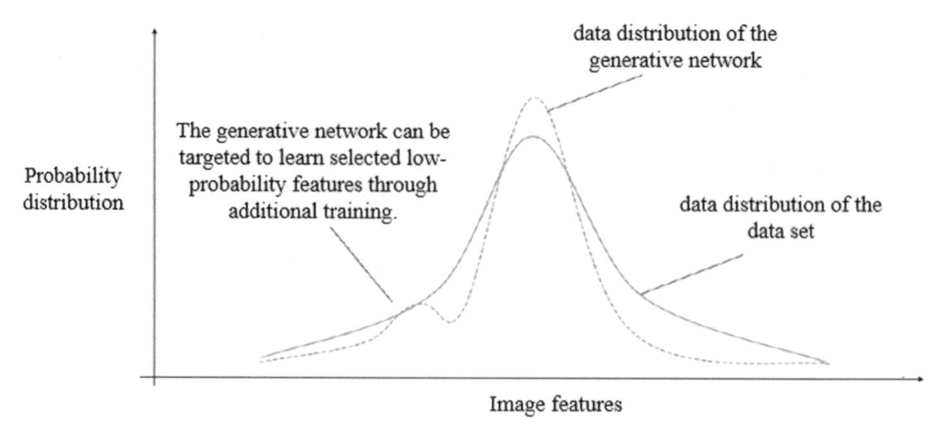

Fig. 2. The training set data distribution and the GAN generative data distribution after embedding data

3.2 Architecture

Experiments show that if random codes are assigned to the selected images, the entire data distribution of the generative network will be distorted. Therefore, the key is to find an appropriate latent code for each selected image and minimize the impact on the distribution.

To achieve the above goals, a new GAN is designed, named embedGAN. EmbedGAN consists of a generative network and an embedded network. The two networks are mutually coordinated rather than antagonistic. To guarantee the basic quality of the output images, the generative network is from a pre-trained DCGAN. The embedded network consists of convolutional layers and fully connected layers. The embedded network outputs the appropriate latent codes according to the input target images, which also provides the direction of the updating parameter for the generative network. The architecture is shown in Fig. 3.

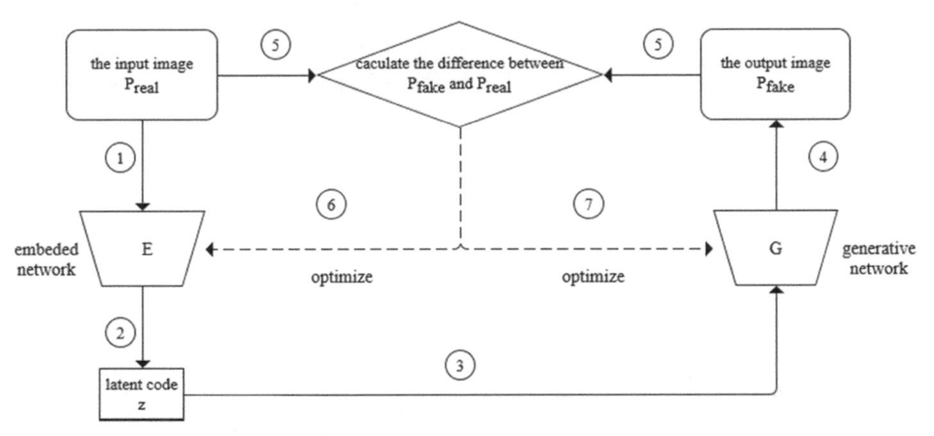

Fig. 3. The structure of embedGAN. There are two parts of network training. Firstly, the real image P_{real} is input to the embedded network and the output is the code z. After z is input to the generative network, the embedded network is optimized according to the difference of P_{real} and the generative image P_{fake}. In the second round, the difference between P_{fake} and P_{real} is measured to optimize the generative network.

The optimization goal of embed GAN is

$$\min_{G,E} V(G,E) = [G(E(p_{real})) - p_{real}]^2$$

p_{real} refers to the selected real image data. $E(p_{real})$ refers to the latent code output by the embedded network. $G(E(p_{real}))$ refers to the image generated by $E(p_{real})$. The meaning of the formula is that the generation network G and the embedded network E should minimize the difference between the generated images and the selected real images.

3.3 Training Details

In the example, the data set consists 64 randomly selected real interior images. Before formal training, the embedded network can be pre-trained for 50 epochs. After pre-training, codes generated by the embedded network are input to the generative network, and it can be seen that there is still a large gap between the generated images and the target images (selected real images) (Fig. 4(a)(b)).

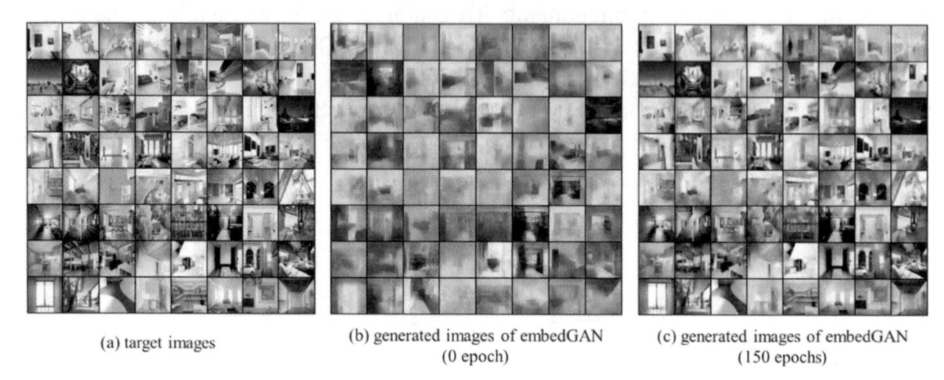

(a) target images (b) generated images of embedGAN (0 epoch) (c) generated images of embedGAN (150 epochs)

Fig. 4. Comparison of the target images and images generated by embedGAN.

In formal training, the learning rate of the generative network is 0.001 when the learning rate of the embedded network is 0.002. The loss function is MSE. The Adam optimization is used.

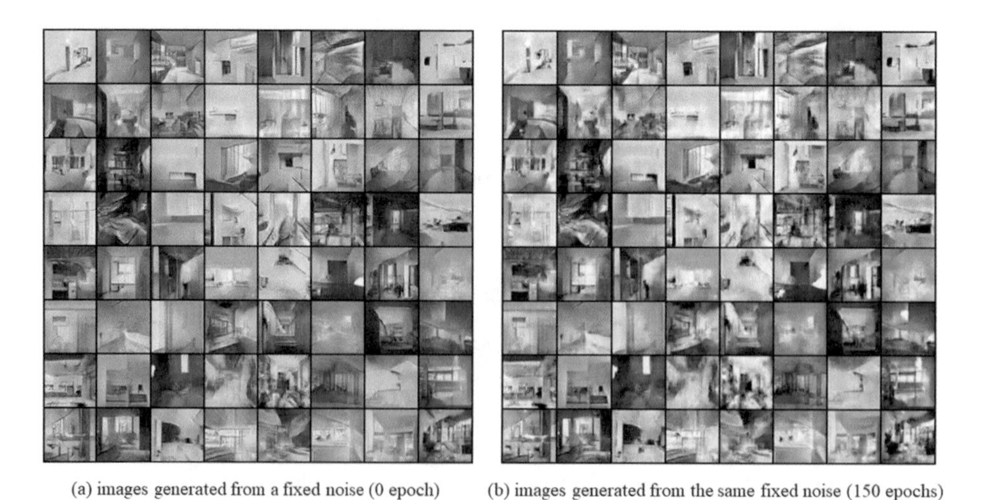

(a) images generated from a fixed noise (0 epoch) (b) images generated from the same fixed noise (150 epochs)

Fig. 5. Other images generated by a fixed random noise in embedGAN.

Images generated by embedGAN is shown as Fig. 4(c). It can be seen that after training, embedGAN does generate images closer to the target images. At the same time, images generated from other random codes have stayed high quality (Fig. 5). The training process of embed GAN only changes parts of the distribution, so that embedGAN can still generate rich interior images besides the selected images. EmbedGAN can be used to do latent walks with specific images.

4 Application

When designers need to explore various style features for existing residential projects, embedGAN can help explore the design intention. Randomly select the first image of "The Little White Apartment/Z-AXIS DESIGN" project (available on the website https://www.gooood.cn/little-white-apartment-z-axis.htm) from gooood for an application trial (Fig. 6). Limited by the training set, the size of the image is compressed to 64 * 64.

Fig. 6. Image from "The Little White Apartment/Z-AXIS DESIGN" project.

A classification network (consists of CNN layers and uses BCE as the loss function) is trained to discriminate the decorative materials used in interior spaces. Wood, concrete, metal stone, marble and veneer styles are classified and the accuracy of the testing set is 97%. Based on the classification network, we can perform a heuristic search in the latent space of embedGAN to regenerate certain features. The heuristic search is that when a certain material score of the image is increased, the result of this exploration is retained. The selected image is regenerated into different decorative material style (Fig. 7).

Similarly, complex features that are difficult to analyze intuitively can be generative. For example, by training a national discriminator of residential projects, different national styles can be regenerated (Fig. 8). As can be seen from the above applications, embedGAN achieves the ability to regenerate the selected real images into a selected feature direction.

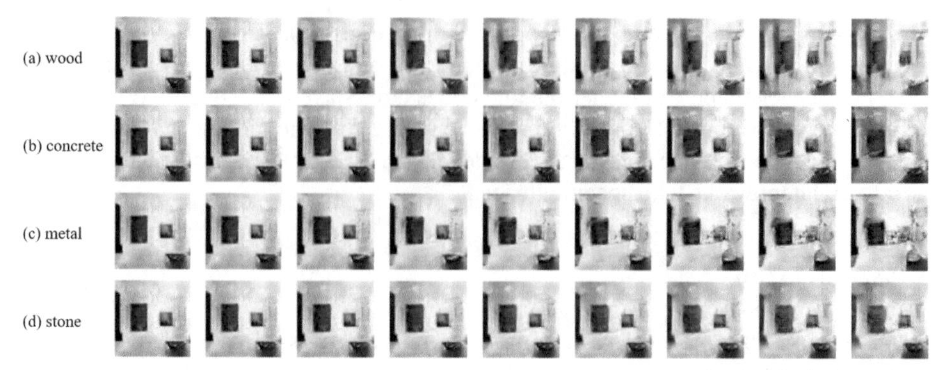

Fig. 7. Regenerated images in latent walks. From left to right, the generated image is more like the decorative style of the material

Fig. 8. Images regenerated based on the direction of nations. From left to right are generated images that are more in line with the country's style

5 Evaluation

The generation effect of embedGAN was evaluated in two aspects. On one hand, the difference between the images generated by embedGAN and the original images. On the other hand, the quality of randomly generated images.

The effects of image embedding are evaluated by MSE and SSIM. The RGB pixel mean square error (MSE) as well as structural similarity (SSIM) of the 64 embedded images and the 64 original selected images are calculated (Fig. 9). The average of MSE is 307.73 for the 64 * 64 pixel images, the mean deviation of each pixel is 0.07. The mean of SSIM is 0.72 (The closer to 1, the less different the two images are). It can be seen that embedGAN can generate images that have small differences from the selected images.

The embedGAN generative network and the original DCGAN generative network are compared to evaluate the random images generation quality. The two generative networks are tested by a same DCGAN discriminative network. It found that embedGAN has a larger loss function (Fig. 10). An explanation is that embedGAN is designed for a new problem (embedding images), which makes it works worse in the original problem

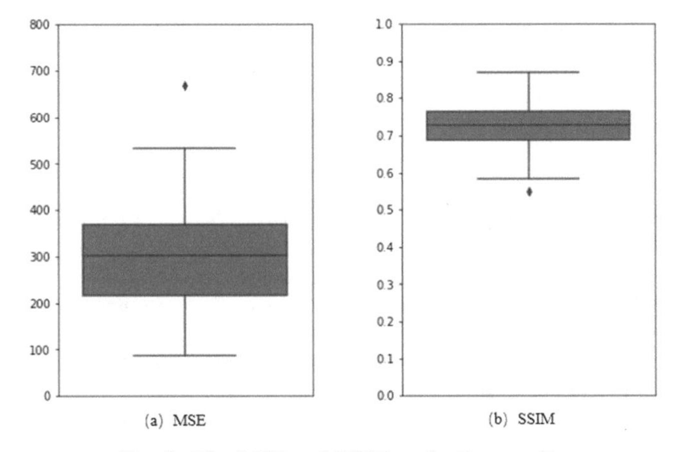

Fig. 9. The MSE and SSIM evaluation results.

(generating image from the original data set), according to the "there is no free lunch" principle. However, under the condition of generating design intent, embedGAN can still generate images closed enough to the data set. The loss of image quality is acceptable.

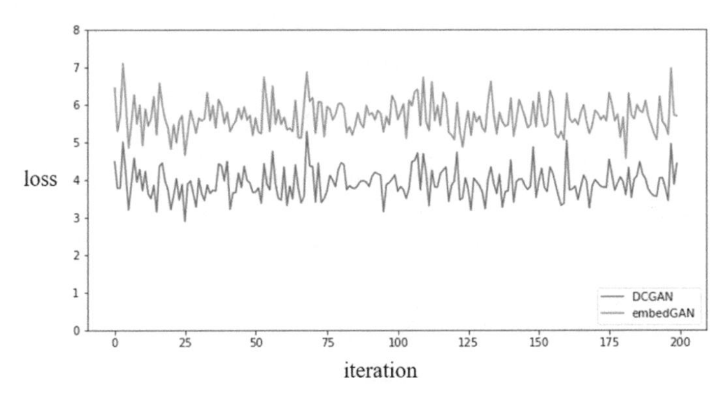

Fig. 10. Loss function of DCGAN and embedGAN generative network on a same DCGAN discriminative network.

6 Conclusion

This study is dedicated to solving the problem that when using GAN for latent walks, it is unable to start from a selected image. In the application of latent walks, embedding specific data points in parts of the generative data distribution is a method that takes both computation costs and image quality into account. This research solves this problem by proposing a framework that includes an embedded network and a generative network. This article verifies the effectiveness of this solution, and points out that after embedding the data, the quality of the generation has declined to some extents. However, under the condition of design intention, this decline is within an acceptable range. The method proposed in this paper is expected to be applied as a design tool.

References

1. Goodfellow, I.: NIPS 2016 tutorial: Generative adversarial networks. arXiv preprint arXiv: 1701.00160 (2016)
2. Karras, T., Laine, S., Aila, T.: A style-based generator architecture for generative adversarial networks. In: Proceedings of the IEEE Conference on Computer Vision and Pattern Recognition (2019)
3. Shmelkov, K., Schmid, C., Alahari, K.: How good is my GAN?. In: Proceedings of the European Conference on Computer Vision (ECCV) (2018)
4. Berthelot, D., Schumm, T., Metz, L.: Began: boundary equilibrium generative adversarial networks. arXiv preprint arXiv:1703.10717 (2017)
5. Zhu, J.-Y., et al.: Generative visual manipulation on the natural image manifold. In: European Conference on Computer Vision. Springer, Cham (2016)
6. Lipton, Z.C., Tripathi, S.: Precise recovery of latent vectors from generative adversarial networks. arXiv preprint arXiv:1702.04782 (2017)
7. Donahue, J., Krähenbühl, P., Darrell, T.: Adversarial feature learning. arXiv preprint arXiv: 1605.09782 (2016)
8. Dumoulin, V., et al.: Adversarially learned inference. arXiv preprint arXiv:1606.00704 (2016)
9. Kingma, D.P., Dhariwal, P.: Glow: generative flow with invertible 1×1 convolutions. In: Advances in Neural Information Processing Systems (2018)
10. Radford, A., Metz, L., Chintala, S.: Unsupervised representation learning with deep convolutional generative adversarial networks. arXiv preprint arXiv:1511.06434 (2015)
11. Wu, J., et al.: Learning a probabilistic latent space of object shapes via 3D generative-adversarial modeling. In: Advances in Neural Information Processing Systems (2016)
12. Goetschalckx, L., Andonian, A., Oliva, A., et al.: Ganalyze: toward visual definitions of cognitive image properties. In: Proceedings of the IEEE International Conference on Computer Vision, pp. 5744–5753 (2019)
13. Jahanian, A., Chai, L., Isola, P.: On the "steerability" of generative adversarial networks. arXiv preprint arXiv:1907.07171 (2019)
14. Abdal, R., Qin, Y., Wonka, P.: Image2StyleGAN: how to embed images into the StyleGAN latent space?. In: Proceedings of the IEEE International Conference on Computer Vision (2019)

Research on Architectural Form Optimization Method Based on Environmental Performance-Driven Design

Jinghua Song(✉) and Sirui Sun

School of Urban Design, Wuhan University, Wuhan 430072, China
113318088@qq.com, three.530@qq.com

Abstract. In the context of contemporary environment and society, the architectural form optimization based on Environmental performance-driven design is a method by using environmental performance data to optimize the architectural form. Its value lies in dealing with the interaction between architecture and environment, and developing architecture with environmental sustainability. This thesis summarizes the similarities and differences between performance-driven form design and traditional bionic form design. The traditional bionic design separates the bionic object from its complex living environment, and its simple imitation tends to fall into the local rather than the global optimum. However, performance-driven design is different from bionic design. It advocates environmental factors as a driving factor rather than a confrontational factor. It is a systematic global optimal method for studying architectural form. This paper puts forward the specific architectural form optimization simulation process based on the performance-driven thought. Taking the multilayer parking building design of the riparian zone on the south bank of Chongqing as an example, the parametric design method is used to obtain architectural optimization form adapted to the environment.

Keywords: Performance-driven design · Form optimization · Environmental adaptability

1 Introduction

With the continuous development of environment, society, economy and technology, the relationship between architecture and environment has been increasingly discussed. The study of the relationship between architecture and environment will inevitably involve environmental adaptability, which comes from the theory of environmental adaptability. This theory was originally derived from the field of biological research at the end of the 19th century, marked by Darwin's theory of natural selection, and was applied to the field of architecture and urban research by the 1850s [1]. The theory is the ideological and theoretical basis of the research in this paper. Buildings should have relative adjustment ability in a specific environment to adapt to the complex changes of the environment. Buildings can be used as a medium to respond to the environment, and it can be presented as a dynamic intelligent collection through interaction with the environment.

P. F. Yuan et al. (Eds.): CDRF 2020, *Proceedings of the 2020 DigitalFUTURES*, pp. 217–228, 2021.
https://doi.org/10.1007/978-981-33-4400-6_21

The external environmental data of buildings generally have the characteristics of complexity, periodicity, immediacy and combination [2]. Through the design, the environment interacts with the building, combining the environmental performance data with the building. A new mode of thinking is integrated into the Multivariate complex system of architecture, environment and people, making to interact and respond from among the architecture, environment and people (Fig. 1).

2 Performance-Driven Design and Its Advantages

2.1 Performance-Driven Design Theory

Along with the construction, engineering and other industries into the sustainable low-carbon era, the building performance has attracted more and more attention. Simulation technology has quantified the building performance, so architects can incorporate performance analysis into the design workflow. Performance-driven design involves computer-aided optimization techniques that make performance the standard for driving design [3]. Performance simulation technology has been widely used in architectural design for a long time, but in the early stage, it is only used as a design evaluation condition rather than a driving factor of form generation and optimization. In the field of aviation, performance simulation data is used as the design driving force to improve the aerodynamic performance of aircraft, so the performance-driven design method is derived.

Under the action of driving factors and in the process of architectural form shaping, performance-driven design can stimulate the new possibility of building in organization, space, form and performance. Make full use of the natural environment energy to drive architectural form generation and optimization, which can not only respond to the changes of wind environment, light environment, thermal environment, water environment and so on, but also intelligently respond to the external climate conditions of the building based on the energy flow and transfer [4]. The original intention of this study is to start with rational analysis and capture of environmental parameters, and to optimize the overall form of the building under the influence of environmental parameters such as water flow, wind, light and landscape [5].

2.2 Performance-Driven Design Advantages Compared with Bionic Form Design

It is generally believed that the bionic architecture is obtained by using the bionic form design method. Bionic architecture is a kind of architecture which imitates the effective system specialty of some organisms in the aspects of architectural environment, function, form and organizational structure, and it is more in line with the laws of nature and requirements of human nature [6]. In the category of bionic concept, the bionic form design thought such as life-form characteristic bionic, bionic configuration and bionic structure is close to the performance-driven design thought, which have both certain similarities and differences.

From the perspective of similarity, the goals of performance-driven design and bionic form design are to adapt to the laws of nature. Architecture must adapt to the environment to achieve the symbiotic relationship between human and nature. The buildings are integrated into the circulation system connected with the environment, so as to make more rational use of resources, maximize the use efficiency of energy and materials, reduce the energy consumption of the buildings, make the buildings become a part of the local ecosystem, and make the nature become part of the buildings [6]. Both performance-driven design and bionic form design are architectural form design theory based on the development of environmental adaptability theory.

From the perspective of difference, in the first place, performance-driven design takes performance goals as the driving force, and with the help of computer analysis ability, it can create more diverse form and function solutions under the premise of meeting the comprehensive performance requirements. Select the best scheme through simulation optimization and achieve the optimal solution of multi-objective problem [7]. In the second place, performance-driven design process maximizes the driving effect of performance indicators on the design scheme, it avoids the rework of the design scheme due to the non-compliance of performance indicators and improves the design efficiency. These two advantages promote the application of performance-driven design theory in architectural creation. The bionic form design is more to pursue the goal of the harmony between the architecture and the image of biology. It regards the unity of functional image as the objective basis of harmony and lacks the essential connection [8]. However, performance-driven form design breaks through a single perspective, and it is mainly based on the performance of the system rather than the expression of the function or form, which focuses on performance, that is, function and efficiency [9]. Therefore, the performance-driven form design is a more in-depth development of the bionic form design in dealing with the relationship between architecture and environment.

3 Performance-Driven Architectural Form Optimization Method

3.1 Combined with Parametric Design

The performance-driven parametric design is a design method that combines performance-driven design with the design of "parametric model". Parametric model is a computer design model, which is based on the geometry. This geometry, itself, contains two fixed features, known as constrained and variable attributes. Parametric design is the form of development based on a set of relations and variables (parameters) [10]. In the parametric model, when a new alternative solution is sought, the parameters will change accordingly. Therefore, it is necessary to adjust the new values of the parameters to respond to such changes, and to define different architectural forms [11].

Under the environment relation, the performance-driven parametric design method combines the parameters of the environmental performance data with the architectural form. This method enables the computer to generate architectural forms based on the building space, structure, materials, and physical environmental parameters such as wind, light, heat and sound, so as to make architectural forms respond to environmental performance [4].

4 Form Optimization Simulation Process Establishment

There is certain discreteness in architectural design so that the design process itself can be simulated by computer. To explore the generation and optimization of architectural form, we need to use performance-driven design thought to find an effective simulation method based on complex models to help complete the design process. The steps of simulation optimization design combined with performance-driven thought are as follows: firstly, traditional design method is adopted for conceptual design; secondly, model is established; thirdly, simulation program is used to analyse one or more related performances; and fourthly, simulation results are analysed and evaluated. On this basis, the design and model are modified repeatedly to find certain rules and the optimal form interval, so as to further drive the detailed design and obtain the target (Fig. 2).

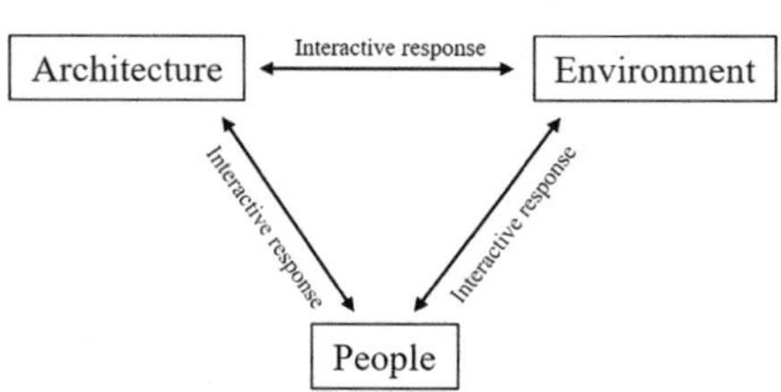

Fig. 1. Diagram of interaction response among architecture, environment and people

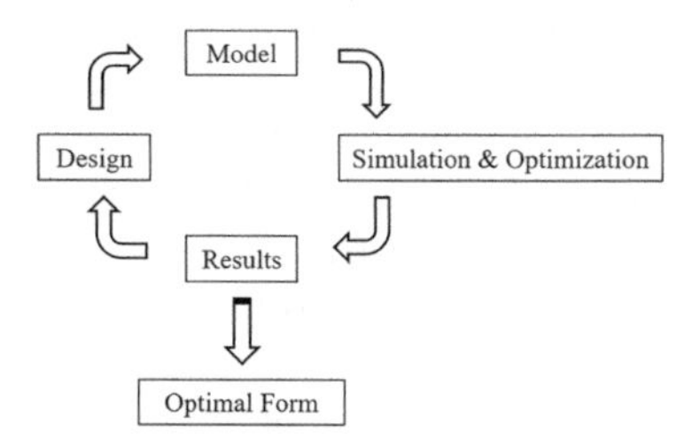

Fig. 2. The thought process diagram for performance-driven simulation optimization

Figure 3 shows a block diagram of the performance-driven form optimization simulation process. This paper will only discuss the use of water flow factor as a driving factor in relation to case studies. According to the overall analysis, the preliminary architectural form is designed, and then the parameters of form optimization and the reasonable numerical constraint range are determined. Architects use rhino and grasshopper to build the model, and require that the relevant data of the model must be in the range of parameter numerical constraints, so as to reduce the differences of the optimization results. The simulation process is optimized by the performance simulation platform of Phoenics or RhinoCFD, and the results of stage optimization are obtained. Analyse the data of the results of each stage, and summarize the numerical range of the optimal form that meets the requirements. Combined with other constraints, the optimal form is selected for detailed design, and the final architectural form is obtained.

5 Design Practice

5.1 Project Background

The project is located in the riparian zone of "Egongyan Bridge-Shijiayan" on the south bank of Chongqing (Fig. 4). According to the overall goal of comprehensive control planning of the riparian zone and the natural topographic conditions of the riparian zone, the urban multilayer parking building is designed within the project area of Height

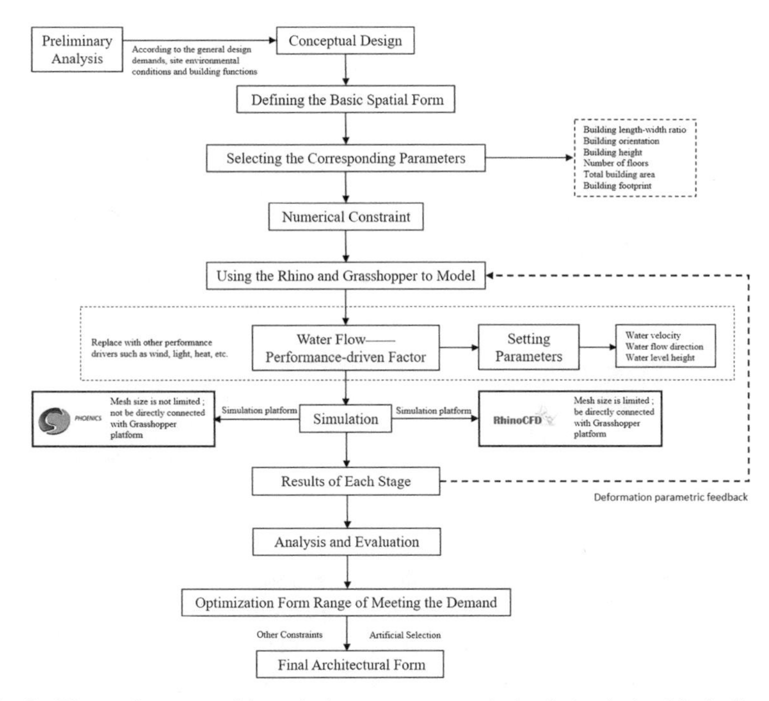

Fig. 3. The performance-driven design generates optimized simulation block diagram

range 186–199 m (Fig. 5). The building site is located in the height range of 180–193 m, which is the urban construction control area of the water level against 5–50-year recurrent floods. The purpose of this study is to demonstrate that the overall form of the design scheme adapts to the site environment. The architectural form should have the flood protection ability in special period. In addition, the building can provide a leisure places for citizens and make it a vibrant riverside area in the city.

5.2 Design Parameters Selection and Numerical Constraint

According to the general design requirements of the project, the architect first needs to do a lot of analysis on the site conditions and functionality. Next, the architect needs to define the basic spatial form of the building, select the corresponding parameters and numerical constraints, so as to ensure that the optimization goal of architectural form meets the design requirements and the rationality of space use.

Considering the influence of boundary line of building and the overall terrain environment on the architectural form, we have chosen the building length-width ratio, building orientation (Fig. 6), building height and number of floors as design parameters. According to the analysis, the numerical constraint of the building length-width ratio is limited to 1.6–3.3, and the architectural forms lower or higher than this range all show certain irrationality. For example, the building space is not suitable for the function or the poor fit with the terrain. The building orientation is limited to 25–30° north-northwest, so as to adapt to the terrain conditions and get a good riverside viewing effect. At the same

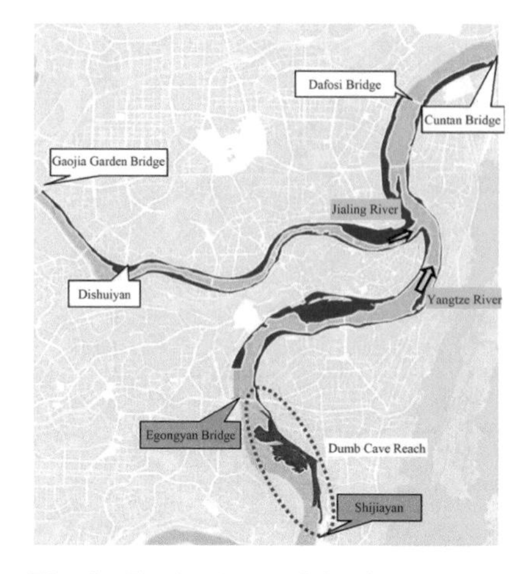

Fig. 4. The land area of the riparian zone

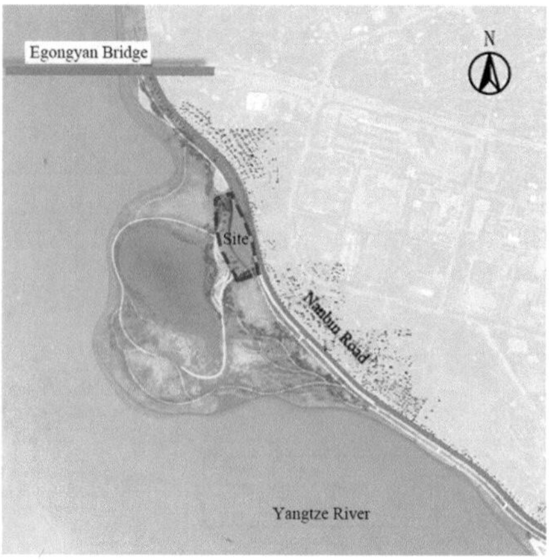

Fig. 5. The Scope of project site

time, the building can follow the water flow direction of the Yangtze River, reducing the problems caused by special circumstances. In addition, in order to avoid breaking through the requirements of the optimal design scheme, we also limit the total building area and the building footprint as parameters. The specific parameter numerical range are shown in Table 1.

Table 1. The constraint conditions on parameter range of architectural form optimization design

	Parameter name	Numerical constraints	Unit
1	Building length-width ratio	1.6–3.3	–
2	Building orientation	North-northwest 25–30	°
3	Building height	13.5–16.5	m
4	Number of floors	3	–
5	Total building area	13500–16500	m^2
6	Building footprint	4500–6000	m^2

5.3 Setting Simulation Parameters

5.3.1 Water Velocity

The engineering reach is located in the upper reaches of the confluence area of the Yangtze River and Jialing River, and is called the Dumb Cave reach. The flood fluctuation of the two rivers will affect the water level in this reach. By analysing the historical data of Zhutuo hydrological station, Cuntan hydrograph station, Beibei hydrograph station and Egongyan stage gauging station, the water velocity of simulation is 2.37 m/s.

5.3.2 Water Flow Direction

According to the water flow direction of the whole Yangtze River, in the engineering reach and from the upstream to the downstream, the water flow direction of simulation is north-northwest direction (considering a single direction temporarily to facilitate calculation and simulation).

5.3.3 Water Level Height

The numerical values of water level height related to the building are shown in Table 2. The water level height of simulation is the water level against 5-year recurrent floods—185.6 m.

Table 2. The characteristic water level height value associated with the building

	Water level name	Water level height/m	Relative building height/m
1	Water level against 50-year recurrent floods	191.7	11.7
2	Water level against 5-year recurrent floods	185.6	5.6
3	General highest water level	173.0	−7
4	General lowest water level	164.5	−15.5

Note: Yellow Sea Elevation

5.4 Form Optimization Process Diagram

The whole simulation process of form optimization is adjusted in multiple stages. The initial conceptual design was the traditional block shape. But through the Phoenics water flow simulation analysis, the concave space of the building will be affected by the severe impact of the water flow, which is not conducive to the overall structural performance. Therefore, the architectural form is pushed outwards to gradually weaken the boundary of the rectangular block, so as to adopt a soft curve form. Finally, we get the spindle shape with better performance. And on this basis, we find the optimal performance range by changing the building length-width ratio and shape for many times. Figure 7 shows the evolution process of architectural form in simulation and optimization. Figure 8 and Fig. 9 respectively show the pressure value diagram and the velocity value diagram of the water level against 5-year recurrent floods in 18 phases schemes.

5.5 Result Analysis

The shape coefficient of building (the ratio of the external surface area of the building in contact with the outdoor atmosphere to its volume) is chosen to describe the architectural form characteristics. According to research, the smaller the external surface area

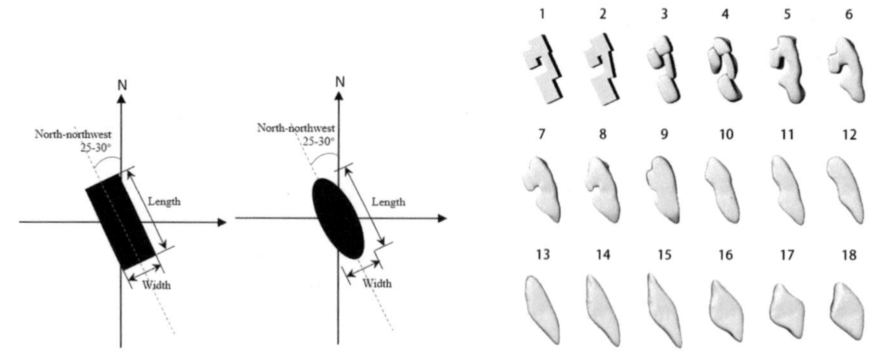

Fig. 6. Diagram for determination of building aspect ratio and building orientation

Fig. 7. Optimization of architectural form evolution

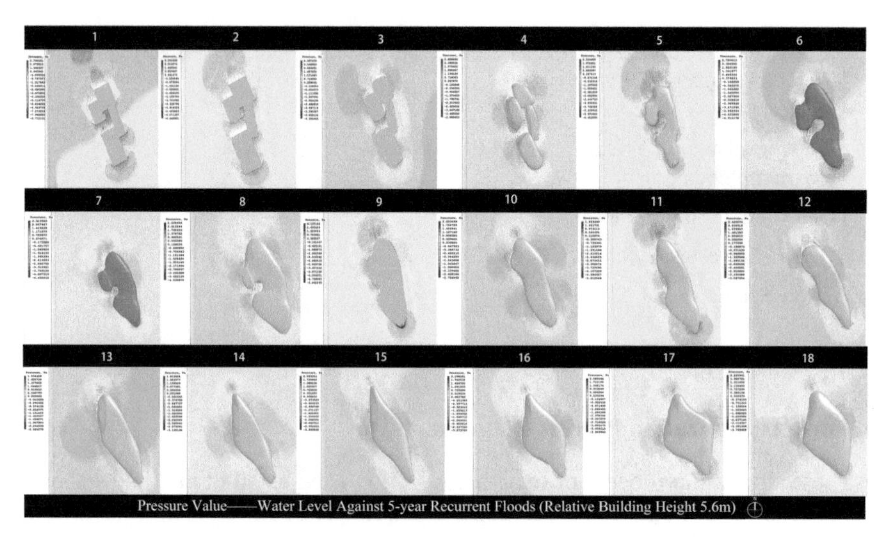

Fig. 8. The pressure value diagram of simulation process

allocated on the unit building volume, the smaller the impact of water flow on architectural form in the special period, that is, the more conducive to the stability of the overall structural performance of the building. We extract the corresponding maximum and minimum pressure values (velocity values) from each stage of simulation and average them. And then, combined with the shape coefficient of building, the average pressure value and the average velocity value, we find out the relationship among them and get the optimal form numerical range. Table 3 shows the relevant data values of the above 18 simulation phases. The data values of every phase are within the effective range of the design parameters.

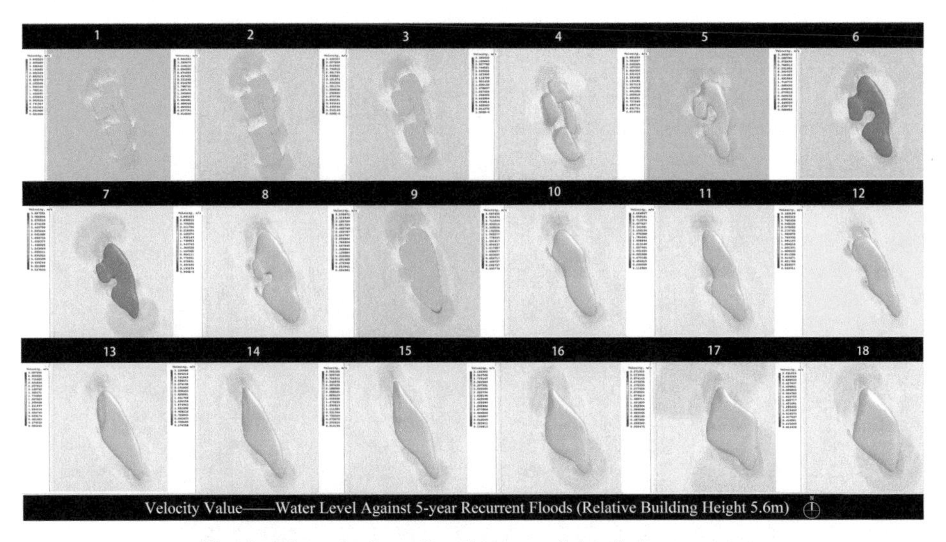

Fig. 9. The velocity value diagram of simulation process

Figure 10 shows the line chart of the relationship among the average pressure value, average velocity value and the shape coefficient of every simulation phase (the numerical variation range of the shape coefficient is too small, so we expand the numerical value by 10 times in order to clearly present the curve variation rule). In the simulation of the architectural form No. 1 to No. 13, the average pressure value, average velocity value and the shape coefficient have a general trend of gradually decreasing, which indicates that the effect of water pressure on them decreases with the change of form; the shape coefficient of the architectural form No. 14 to No. 18 tends to a relatively stable value, but the corresponding average pressure value and velocity value show an upward trend. To sum up, the architectural form No. 13 and No. 14 are relatively better. By designing them in detail and again using Phoenics simulation, the building is least affected by the water level against 5-year and 50-year recurrent floods, which is more stable, safe, beautiful and sustainable than other design forms.

Table 3. The relevant data values of every phase simulation

	Total building area /m^2	Building footprint /m^2	Building length /m	Building width /m	Building length-width ratio	Building height /m	Maximum pressure /pa	Average pressure /pa	Maximum velocity m/s	Average velocity m/s	External surface area /m^2	Volume /m^3	Shape coefficient
1	15364.21	5121.41	138.48	55.29	2.50	13.50	2.80	5.76	3.86	1.94	16752.72	69138.96	0.2423
2	15638.91	5212.97	140.05	66.65	2.10	13.50	2.59	4.62	3.56	1.79	17066.57	70375.05	0.2425
3	14785.31	5026.07	140.05	62.04	2.26	15.80	3.00	3.65	3.44	1.72	14760.49	61889.81	0.2385
4	14584.16	5614.31	145.11	58.95	2.46	16.50	2.81	3.35	3.38	1.69	15286.39	60671.71	0.2519
5	16146.12	5939.09	150.31	65.85	2.28	15.20	2.53	4.49	3.82	1.92	16350.24	68579.62	0.2384
6	16010.76	5003.09	144.73	67.21	2.15	15.20	2.79	3.85	3.39	1.71	15234.88	70464.66	0.2162
7	16253.28	5006.14	150.09	63.44	2.37	15.20	2.52	3.59	3.29	1.65	15403.99	73286.18	0.2102
8	15424.21	4784.01	149.24	60.54	2.47	15.20	2.63	3.33	3.09	1.55	15464.31	73301.31	0.2109
9	16336.52	5446.21	150.04	60.34	2.47	16.50	2.16	3.71	3.54	1.79	15452.59	71568.99	0.2087
10	15416.87	4632.56	150.21	53.13	2.83	15.20	2.02	2.85	3.09	1.59	15747.32	75449.96	0.2005
11	15270.15	4574.28	157.79	48.61	3.25	15.20	1.83	3.42	3.08	1.60	13930.61	68302.12	0.2039
12	15124.29	4528.12	154.69	48.37	3.19	15.20	2.43	3.00	3.15	1.59	13688.67	67853.43	0.2017
13	15675.26	4757.42	155.57	52.79	2.95	15.30	1.93	2.23	3.09	1.59	13973.08	69873.07	0.2001
14	15326.79	4643.76	164.39	52.77	3.12	15.30	1.82	2.50	3.11	1.64	13872.24	68051.48	0.2038
15	15348.49	4670.62	169.73	54.21	3.13	15.30	2.05	2.66	3.09	1.65	13997.35	67943.66	0.2060
16	15744.41	4665.43	142.27	67.46	2.11	15.30	2.10	2.69	3.15	1.64	13950.96	71148.68	0.1961
17	14954.44	4531.41	124.44	74.63	1.67	15.30	2.08	2.95	3.27	1.68	13050.84	66743.06	0.1955
18	15429.91	4515.17	128.60	69.83	1.84	15.30	2.27	3.02	3.23	1.62	13455.92	69264.66	0.1943

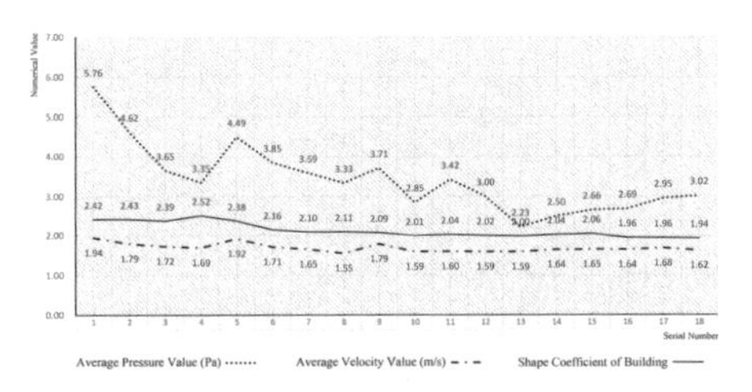

Fig. 10. The broken line diagram of relationship among mean pressure value, mean velocity value and shape coefficient

Fig. 11. A rendering of the final building scheme

6 Conclusion

In the context of contemporary environment and society, to develop sustainable energy and to protect environmental ecosystem urgently need to develop the new architectural design strategy.

Only passively seeking design inspiration from nature, and the bionic design of existing mode can no longer meet the demand of sustainable building development. Under the influence of various complex environmental factors, it is necessary to apply the performance-driven design thought to find the optimal value of the architectural form optimization design that responds to the environment. Performance-driven design gives full play to the advantages of digital technology, making the selection of its results more proactive and avoiding being limited to the design of local optimization. In short, the performance-driven design method can truly combine design with nature to drive the development of energy-efficient buildings and sustainable buildings [12].

In addition, the architectural design of riverside area is a research subject to be further developed and utilized. There are many uncontrollable environmental factors in the riverside area, which will affect the architectural form, the construction process, Use process and maintenance process of building. In the design process, it advocates environmental factors as the driving factors rather than the confrontational factors, so as to improve the energy efficiency of the building environment. The design practice explores the optimization strategy of architectural form in riverside area influenced by the performance-driven design thought. The results of simulation and optimization show that the architectural form is not only satisfied with other design conditions, but also affected least by the specific water flow factors of this area. The final optimized architectural form combines the urban space, embankment features and surrounding landscape to form a building of the modern design feature (Fig. 11). More importantly, this provides strategic guidance for further development of programs, that can be modelled, simulated, evaluated, optimized and generated simultaneously.

At present, we are gradually realizing the transformation from computer-aided design to computer-decided design. The latter will focus more on the global optimal study of architectural design problems at the level of the self-organization generation and adaptive optimization. This will be a new exploration of architectural design thinking, methods and technical tools in the context of artificial intelligence technology. It is bound to integrate more vigorous vitality into the sustainable design concept of environmental performance.

References

1. Lynch, K.: Environmental Adaptability. J. Am. Inst. Plan. **24**(1), 16–24 (1958)
2. Li, L., Ye, X., Wang, Y.: Environmental intelligent architecture. Time Archit. (2018)
3. Shi, X.: Performance-based and performance-driven architectural design and optimization. Front. Archit. Civ. Eng. China **004**(004), 512–518 (2010)
4. Li, L., Tao, S.: Evolution of green building and energy agenda of architecture. South Archit. **03**, 27–31 (2016)
5. Wu, J., Li, L., Wang, J.: Environmental responsive construction; Qingdao Linghai Hotel. Time Archit. (04), 142–147 (2016)
6. Wan, K.: New discussion about architectural bionic. Huazhong Archit. **03**, 28–31 (2005)
7. Sun, C., Han, Y., Zhuang, D.: A study of dynamic building information modelling techniques based on performance-driven thought. Archit. J. **08**, 68–71 (2017)
8. Lv, F.: Toward the 21st century – the past and future of architectural bionics. Archit. J. **06**, 14–17 (1995)
9. Yang, T.: The preliminary discussion on the generative and performative design. Archit. Pract. (2013)
10. Lotfabadi, P., Alibaba, H.Z., Arfaei, A.: Sustainability; as a combination of parametric patterns and bionic strategies. Renew. Sustain. Energy Rev. **57**(May), 1337–1346 (2016)
11. Hernandez, C.: Thinking parametric design: introducing parametric Gaudi. Des. Stud. **27**, 309–324 (2006)
12. Lin, Y., Yao, J., Zheng, J., Yuan, F.: Research on the building morphology generation method based on the wind tunnel visualization of environmental performance. South Archit. **02**, 24–29 (2018)

Optimization and Prediction of Design Variables Driven by Building Energy Performance—A Case Study of Office Building in Wuhan

Jingyi Li(✉) and Hong Chen(✉)

Huazhong University of Science and Technology, Luoyu Road, Wuhan 1037, China
{jimmylee,chhwh}@hust.edu.cn

Abstract. This research focuses on the energy performance of office building in Wuhan. The research explored and predicted the optimal solution of design variables by Multi-Island Genetic Algorithm (MIGA) and RBF Artificial neural networks (RBF-ANNs). Research analyzed the cluster centers of design variable by K-means cluster method. In the study, the RBF-ANNs model was established by 1,000 simulation cases. The RMSE (root mean square error) of the RBF-ANNs model in different energy aspects does not exceed 15%. Comparing to the reference case (the largest energy consumption case in the optimization), the 214 elite cases in RBF-ANNs model save at least 37.5% energy. By the cluster centers of the design variables in the elite cases, the study summarized the benchmark of 14 design variables and also suggested a building energy guidance for Wuhan office building design.

Keywords: Building energy performance optimization · Parametric analysis · MIGA optimization · RBF artificial neural network model · K-means cluster analysis · EnergyPlus simulation

1 Introduction

As a new design method, the parametric design method has achieved considerable development in recent years [1, 2]. More and more researcher pay attention on the parametric optimization of buildings based on building energy performance in recent years. Tuhus-Dubrow [3] studied the building energy consumption through different climatic conditions and different types of variables. He also studied the genetic algorithm for different building types and variables. By genetic algorithm research, it is found that the rectangular building plane and the trapezoidal building plane have the lowest cycle energy consumption under the five different climatic conditions. Gou [4] and Bre [5] did the sensitivity analysis of the variables in building energy optimization respectively. Jin [6] optimized the energy consumption level of a free-form building with few constraints. Zhang [7] optimized the energy efficiency of the specific classroom type of the school classroom, and different types of classrooms were set as discrete variables.

P. F. Yuan et al. (Eds.): CDRF 2020, *Proceedings of the 2020 DigitalFUTURES*, pp. 229–242, 2021.
https://doi.org/10.1007/978-981-33-4400-6_22

In 1943, McCulloch and Pitts established the first artificial neural networks model (ANNs-model). In the 1980s, Hopfield successfully applied ANNs-model to the optimization problems. Today, the ANNs technology have been widely used in function approximation, expert systems, artificial intelligence, and optimization problems [8].

Recent studies have shown that research on building energy performance based on ANNs technology can achieve outstanding results [9–11]. Among them, in the field of building energy performance, Zemella [9] used ANNs-model to optimize building façades. Ayata [11] used the ANNs technology to predict the natural ventilation of buildings.

2 Research Method

2.1 Research Objectives

The research objective is to improve the overall energy performance level of office buildings in Wuhan during the operation phase, so the research objective is to minimize E_A. The unit is $\text{kw} \cdot \text{h/m}^2$.

The total annual energy consumption (E_A) of office buildings is included: 1. annual energy consumption of building cooling system (E_C), 2. annual energy consumption of building heating system (E_H), 3. annual energy consumption of building lighting system (E_L), 4. annual energy consumption of building electronic equipment (E_E). This study will calculate E_A through 8760 simulation iteration. Because the electronic equipment in office buildings is mainly computers, the turning on and off is only related to the schedule of users. Therefore, it is hardly to change E_E by changing the design variables. E_A of the office building in this research is obtained by Eq. 1.

$$E_A = E_C + E_H + E_L \tag{1}$$

The integral electric drive chiller has been chosen as the cooling source equipment of the air conditioning system in this study. Its performance coefficient ($SCOP_T$) follow the "Design Standard for energy efficiency of public buildings GB50189-2015" and is set as 2.50. A is the total floor area of the building. The annual cooling energy consumption E_C is obtained by Eq. 2, and the annual cumulative cooling power consumption (Q_C) is the simulation result by EnergyPlus.

$$E_C = \frac{Q_C}{A * SCOP_T} \tag{2}$$

The gas boiler has been chosen as the heat source equipment for the heating system in this study. According to the "Design Standard for energy efficiency of public buildings GB50189-2015", the overall efficiency of the heating system (η_2) was set to 0.75 and the coal consumption for power generation (q_2) is 0.36 kgce/kWh, the standard natural gas value (q_3) is set to 9.87 kWh/m^3, and the conversion coefficient (ϕ) of natural gas and standard coal was set to 1.21 kgce/m^3. A is the total floor area of the building. The annual heating energy consumption E_H can be obtained by Eq. 3, and the annual cumulative heating power consumption (Q_H) is the simulation result by EnergyPlus.

$$E_H = \frac{Q_H}{A * \eta_2 q_3 q_2}\phi \tag{3}$$

The open-type luminaires have been chosen as the indoor lighting system in this study. According to the "Standard for lighting design of buildings GB 50034-2013", its efficiency (ηP) was set to 0.75. A is the total floor area of the building. The annual lighting energy consumption E_L can be obtained by Eq. 4, and the annual cumulative lighting power (Q_L) is the simulation result by EnergyPlus.

$$E_L = \frac{Q_L}{A * \eta_P} \tag{4}$$

2.2 Research Method

This study focuses on the effects of different design variables on E_A of office building in Wuhan. The study first established a parametric platform between design variables and building energy simulation software. An approximate model is established by the RBF-ANNs technology. The approximate model is optimized by the MIGA algorithm to search the elite cases of design variables. The research analyzed the elite cases of the design variables in the optimization. By K-means cluster analysis of the elite cases, the study could locate the optimal benchmark of each variable.

The energy simulation software used in this study is EnergyPlus 8.1.2. Because this study is aimed at the annual energy consumption of office buildings, the study not only needs to set up a simulation model, but also needs to set the schedule of office buildings. According to "Design Standard for Energy Efficiency of Public Buildings (GB50189-2015)", the research set the annual cooling and heating system schedule, the annual building occupancy schedule and annual ventilation system schedule. The annual interior lighting system schedule needs to be adjusted based on the natural daylight conditions. DaySim software is used to simulate the natural daylight conditions and the annual interior lighting system schedule has been set followed the DaySim simulation result [2, 13].

2.3 Multi-Island Genetic Algorithm (MIGA) and Radial Basis Functions Artificial Neural Networks (RBF-ANNs)

The variable combination optimization method used in this research is Multi-Island Genetic Algorithm (MIGA). Multi-island Genetic Algorithm (MIGA) is an improved standard genetic algorithm (SGA) [14]. Details of the MIGA algorithm settings in this study are shown in Table 1.

As an approach to reduce the complexity of the model and improve the computational efficiency, the approximate modeling method is used by more and more researchers [10, 11]. ANNs as a popular approximate modeling method, has advantages in solving highly complex function problems. ANNs is similar to biological nervous system and consist of layers of parallel basic units (called neurons) [8]. Neurons are connected to each other by a large number of weighted links through which information can enter the entire system. Basically, a neuron receives input information into ANNs' input layer, merges the input information, performs non-linear operations in the Radial basis functions layer, and then output layer gives the final result (Fig. 1). RBF (Radial Basis Functions) ANNs is used in this research, which is one of the most commonly used artificial neural networks [10]. Details of the RBF-ANNs model settings in this study are shown in the Table 2.

The research found that the shortcomings of the RBF-ANNs are obvious: 1. When the sampling data is insufficient, the RBF-ANNs will not be able to run; 2. Because of the nonlinear mapping ability of the RBF-ANNs is determined by its basis functions and the characteristics of the basis function are mainly determined by the center of the basis function, it is difficult to reach high precision by randomly selecting points as the center points to build the RBF-ANNs model.

This study will use the random sampling (RS) and Latin hypercube sampling (LHS) methods [8] to set the variables. Then use RBF-ANNs technology to establish different approximation models based on the simulation results. Finally, the MIGA is used to optimize the design variables with the best RBF-ANNs approximate model.

At the same time, the study will use testing group to analyze the prediction error of different RBF-ANNs models [15]. The study will use the Root Mean Square Error (RMSE) method to compare the RBF-ANNs' prediction results and the simulation results of the testing group. In this study, the accuracy threshold of the RMSE value is 15% .

$$RMSE = \sqrt{\frac{1}{N}\sum\nolimits_{t=1}^{N}\left(D_t^O - D_t^P\right)^2} \tag{5}$$

Table 1. MIGA algorithm setting

Related parameters	Value	Related parameters	Value
Sub-population size	10	Interval of migration	5
Number of Island	10	Elite size	1
Number of generations	50	Rel tournament size	0.5
Rate of crossover	0.9	Penalty base	0.0
Rate of mutation	0.02	Penalty multiplier	1000.0
Rate of migration	0.2	Penalty exponent	2
		Default variable bound (abs val)	1000.0

3 Research Setting

Because of the complexity of building system, researcher needs to search the design variables which have highly correlation with building energy performance. From the reference [3, 15–17], the study found that the four aspects of the design variables have a high correlation with building energy performance: 1. building plan (BP), 2. window shape (WS), 3. building orientation (BO), 4. building insulation thickness (BIT), 5. horizontal visor (HV) (Fig. 2).

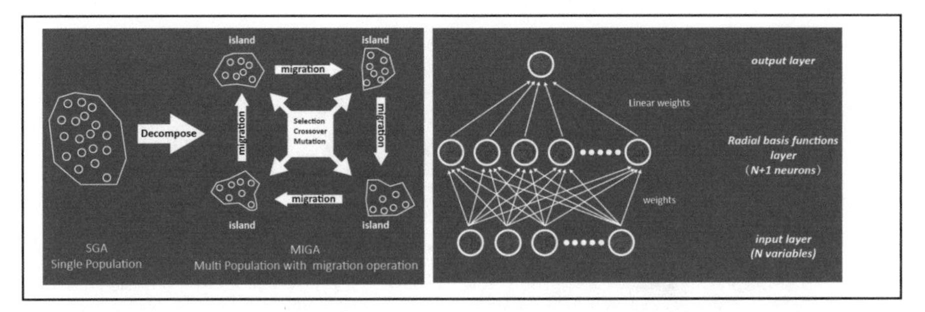

Fig. 1. Schematic diagram for the MIGA (left) RBF-ANNs workflow chart (right)

Table 2. RBF-ANNs setting

Related parameters	Model A	Model B	Model C	Model D
Smoothing filter	0.00	0.00	0.00	0.00
Basis function	Radial	Radial	Radial	Radial
Iterations to fit	50	50	50	50
Sampling method	Random (250 points)	Random (500 points)	Latin Hypercube (500 points)	Latin Hypercube (1000 points)
Sampling optimize time (minute)	0	0	5	10
Error analysis sampling method	Random 150 points	Random 150 points	Random 150 points	Random 150 points
Approximation improvement method	Sequential sampling method	Sequential sampling method	Sequential sampling method	Sequential sampling method

In terms of BP, this research focused on the values of the building plane aspect ratio (A) and floor height (H). Among them, A controls the planar shape of the building, and H controls the height of the building.

In terms of WS, this study focused on the size of windows in different orientations. The design variables are window to wall ratio (WWR) on north direction (WWR-N), WWR on east direction (WWR-E), WWR on south direction (WWR-S), WWR on west direction (WWR-W) and the window height (WH).

In terms of BO, this study focused on the angle (O) between the X axis of the building plane and the East direction of the simulation software (X axis of the 3D model software).

In terms of BIT, this study focused on the roof insulation thickness (RT) and the wall insulation thickness (WT).

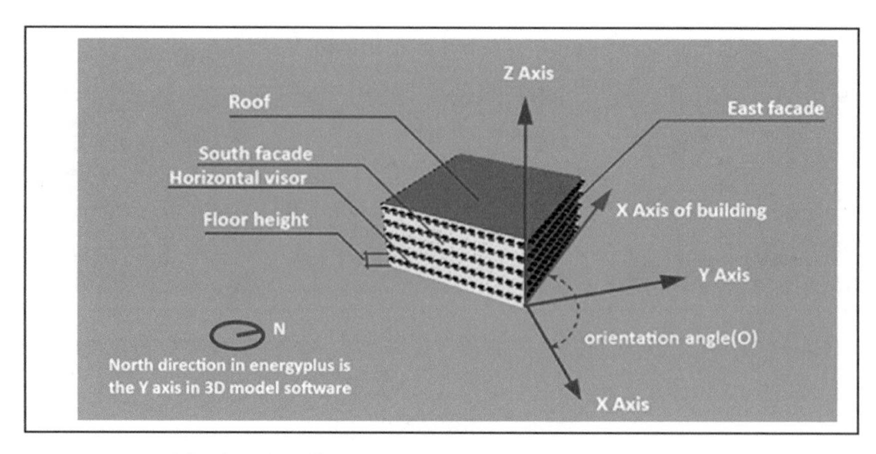

Fig. 2. The office building model in 3D model software

In terms of HV, this study set four variables for the external visors in four directions. The variables are the length of the external horizontal visors in north direction (HV-N), the length of the external horizontal visors in west direction (HV-W), the length of the external horizontal visors in south direction (HV-S) and the length of the external horizontal visors in east direction (HV-E).

The detailed of the design variables of the office building is shown in the Table 3.

Table 3. The design variables

The name of design variable		Nomenclature	The range	Unit
Building plane aspect ratio		A	1.2 to 5.0	None
Floor height		H	3.00 to 5.00	Meter
Window height		WH	0.50 to 2.00	Meter
Building orientation		O	−90 to 90	Degree
Building insulation	Roof insulation thickness	RT	0.05 to 0.3	Meter
	Wall insulation thickness	WT	0.05 to 0.3	
Window-Wall ratio (WWR)	WWR on north direction	WWR-N	0.20 to 0.80	None
	WWR on west direction	WWR-W	0.20 to 0.80	
	WWR on south direction	WWR-S	0.20 to 0.80	
	WWR on east direction	WWR-E	0.20 to 0.80	
Horizontal visor (HV)	HV on north direction	HV-N	0.1 to 2.0	Meter
	HV on west direction	HV-W	0.1 to 2.0	
	HV on south direction	HV-S	0.1 to 2.0	
	HV on east direction	HV-E	0.1 to 2.0	

This study mainly focused on the design variables could be used by the architects, the setting of other variables which have an impact on building energy performance are based on "Design Standard for Energy Efficiency of Public Buildings GB50189-2015", "Standard for lighting design of buildings GB50034—2013" and "Design standard for residential buildings of low energy consumption DB42T_559-2013". The information of the office building is listed as boundary condition in the Table 4. The construction materials are listed in the Appendix Table 6. The annual cooling and heating system schedule, annual occupancy rate schedule and annual ventilation system schedule are followed the design standard. The annual interior lighting system schedule has been set by DaySim simulation result [12, 13].

Table 4. The boundary setting of the energy simulation

The setting name	Value	Unit
The number of zone in every floor	5	None
Num of people per area	0.1	per/m^2
Total floor area	5000	m^2
Building floors	5	None
Equipment load per area	15.00	W/m^2
Lighting density per area	9.00	W/m^2
Infiltration rate per area	0.0003	m^3/s per m^2
Ventilation per area	0.001	m^3/s per m^2
Ventilation per person	0.008	m^3/s per person
Recirculated air per area	0	m^3/s per m^2
Lighting power	250	W
Lighting set point	300	Lux

4 The RBF-ANNs Optimization Process

4.1 The Comparison of RBF-ANNs Models

Before using RBF-ANNs model for the optimization process, it is necessary to check the RBF-ANNs models' accuracy. The accuracy of the RBF-ANNs model is related to the number of sampling points and the sampling method. The research set up 4 different RBF-ANNs models through different sampling numbers (250, 500, 1000) and different sampling methods (RS and LHS method). Through a RMSE method analysis for the prediction results, the study found that:

- The accuracy of the RBF-ANNs model is very sensitive to the number of input values. Because only 250 variable combination values are randomly sampled at Model-A, The RMSE values in all aspects exceed the threshold (15%).

- The accuracy of the RBF-ANNs model is very sensitive to the sampling method. Because the sampling method in Model-B is random sampling and in Model-C is Latin Hyper-cube method, the RMSE values of Model-C is better than the value of Model-B. RMSE values of E_A, E_H, and E_L in Model-B is 20.264%, 16.135%, and 18.135%, respectively. RMSE values of E_A, E_C, E_H, and E_L in Model-C is 13.319%, 9.208%, 9.28%, and 12.799%, respectively.
- By comparing the RMSE values between Model-C and Model-D, due to more sampling values in model-D (1000 values) than values in model-C (500 values), the RMSE value of model-D (E_A, E_H, E_L is 12. 132%, 7.503%, 11.829%) is better than RMSE value of model-C (E_A, E_H, E_L is 13.319%, 9.28%, 12.799%) (Fig. 3).

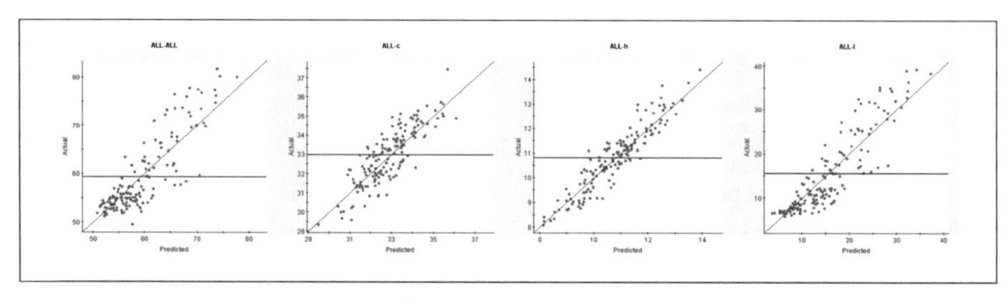

Fig. 3. Residual plots of prediction results and test values for E_A, E_C, E_H, and E_L in model-D

4.2 RBF-ANNs Optimization Process

The research uses the selected RBF-ANNs model (model-D) as the basic model for optimization. The MIGA algorithm is used to optimize the design variables in the basic model. The MIGA algorithm settings are listed in Appendix Table 1. The optimization process of RBF-ANNs model took less than 10 min. After 60 generations of optimization, the optimization results of the RBF-ANNs model have been stabilized (Fig. 4). The study found that the lowest value of E_A can reach 46.44 kW · h/m^2.

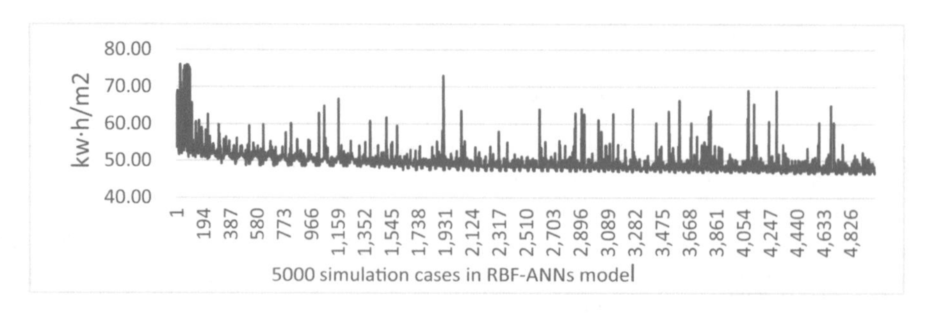

Fig. 4. The optimization history of E_A by MIGA algorithm and the RBF-ANNs model

Since the single energy simulation takes nearly 10 min, and the 100 generations MIGA optimization will perform 5000 energy simulation calculations, it will take approximately 833.33 h for the MIGA optimization. On the other hand, the optimization of the RBF-ANNs model is divided into three parts: 1. Establishing an approximate model through 1000 energy simulations (166.8 h); 2. Testing the model by 150 energy simulations (25 h) 3. Optimization using the RBF-ANNs approximate model and MIGA algorithm (0.08 h). These three steps took approximately 191.88 h. In terms of time consumption, compared with using MIGA algorithm alone, RBF-ANNs model optimization will save 76.97% of time. Therefore, in terms of efficiency, the RBF-ANNs model has great advantages [8, 9, 11].

Before analyzing the variables benchmarks, the predicted value of the RBF-ANNs model needs to be tested. In this study, the 30 lowest E_A variable combinations in RBF-ANNs model were extracted for testing.

After running the energy simulation for the 30 lowest E_A variable combinations and comparing the predicted value with the actual simulation value, the study found that:

- The E_A prediction value of RBF-ANNs model is lower than the simulation value, which indicates that there is a certain error between the prediction values and the simulation values. Despite the errors, the E_A simulation values of these 30 variable combinations are all less than 52.0 $kw \cdot h/m^2$.
- Compared with the E_A value of the Reference Case (83.19 $kw \cdot h/m^2$), the energy decline has reached at least 37.5%. This shows that it is feasible to optimize building energy performance by the RBF-ANNs model [11]
- The E_C prediction value of the RBF-ANNs model is consistent with the simulation value, which indicates that the prediction accuracy of the RBF-ANNs model in E_C is high.
- The E_H prediction value of the RBF-ANNs model is consistent with the simulation value, which indicates that the prediction accuracy of the RBF-ANNs model in E_H is high.
- The E_L prediction value of the RBF-ANNs model is significantly different from the simulation value. Due to the simulation of E_C and E_H does not take into account the changing schedule, the RBF-ANNs model can well establish the nonlinear equations between the variables and the simulation results. On the other hand, since the E_L simulation needs to consider the effect of natural daylight on the annual interior lighting system schedule, there is a more complicated nonlinear relationship between the variables and the E_L simulation results. This complex relationship is the main reason for the low accuracy of the RBF-ANNs model E_L prediction (Fig. 5).

4.3 Analysis the Optimization Results

Prior to analysis the benchmark of variable values, the study screened the optimization results. The study selected 47.0 $kw \cdot h/m^2$ as the E_A threshold value for the screening process. the study found that 214 cases in the RBF-ANNs model optimization process met the screening requirements, and these cases will become the elite cases for k-means cluster analysis. The research found that the cases that met the requirements were mainly concentrated after the 80th generation. This proved that the optimization combined MIGA with the RBF-ANNs model for the office building energy performance was successful.

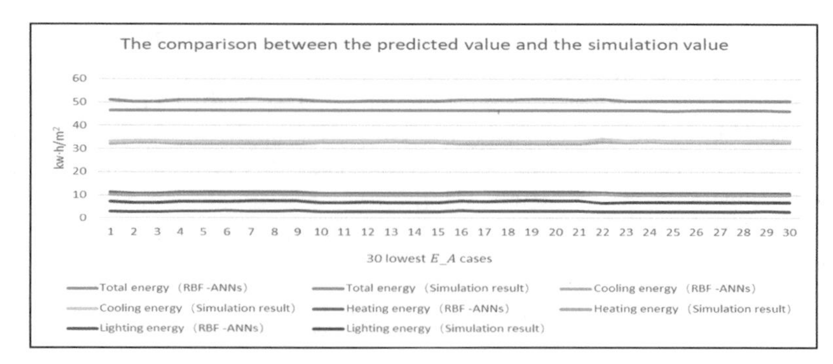

Fig. 5. Comparison of 30 RBF-ANNs model prediction values and simulation values

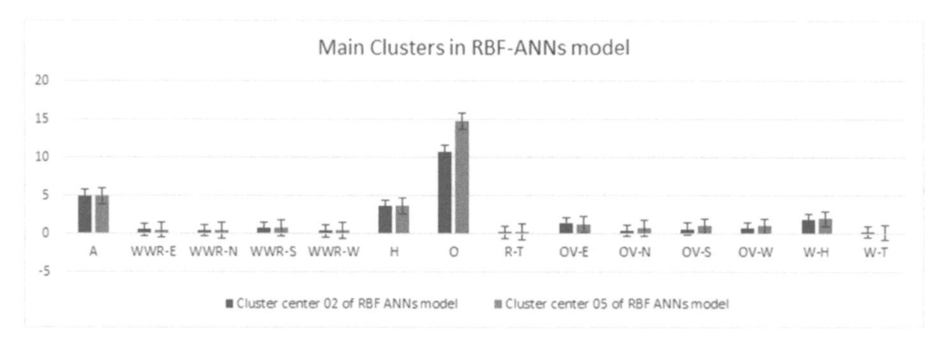

Fig. 6. Comparison chart of each design variable cluster centers

The study compared the different energy aspects of the elite cases in the MIGA optimization process and found that the proportions of different energy aspects in the elite cases are roughly same. The proportion is consistent with the high temperature and high humidity climate in Wuhan summer [18]. Due to the largest proportion in E_A is E_C, it is found that the most effective way to improve office building energy performance in Wuhan is to reduce E_C.

4.4 K-means Cluster Analysis

Clustering is a machine learning technique that is considered to be one of the most effective data mining processes and the most commonly used algorithm for building energy analysis is K-means clustering [19]. The K-means clustering would be used in this research for finding the benchmarks of variables (Fig. 6).

The study sets the number of clusters in K-means clustering analysis to 5 separately for the optimization. From Table 5, the study found that cluster 2 in RBF-ANNs cluster analysis has 118 elite cases, and cluster 5 has 70. These two clusters account for 87.8% of all elite cases in RBF-ANNs model optimization. It means that the value of each design variable in these two clusters represents the benchmarks of each design variable in RBF-ANNs model optimization.

Table 5. The final cluster centers for the elite cases

K-means cluster analysis					
Variables	Cluster 01	Cluster 02	Cluster 03	Cluster 04	Cluster 05
A	4.745	4.981	4.985	4.725	4.925
WWR-E	.569	.526	.524	.579	.499
WWR-N	.449	.434	.435	.451	.436
WWR-S	.775	.742	.733	.788	.796
WWR-W	.407	.375	.376	.373	.370
H	3.718	3.667	3.649	3.723	3.685
O	8.266	10.804	34.208	5.958	14.790
RT	.140	.201	.200	.132	.218
HV-E	.879	1.365	1.353	.844	1.182
HV-N	.346	.429	.417	.327	.694
HV-S	.961	.659	.683	.947	1.006
HV-W	.529	.729	.742	.633	.999
WH	1.889	1.866	1.909	1.825	1.974
WT	.246	.246	.194	.239	.287
Number of observations in each cluster	11.000	118.000	4.000	11.000	70.000
Effective	214000				
Omission	.000				

5 The Conclusion

The research chooses the cluster centers in the RBF-ANNs optimization as the benchmarks. The study found that:

In terms of building aspect ratio, due to the large proportion of E_L in E_A and great optimization potential for E_L, the study found that the north-south facade of the building should be as long as possible,, which will help the sun enter into the building. Since the value range of A in this study is 1.2 to 5.0, the study believes that within this range, the north-south facade can be as long as possible.

In terms of building height, as the increase of H will increase the building volume, which will increase E_C and E_H, the building floor height should be as small as possible in an acceptable range.

In terms of building orientation, due to the large proportion of E_L in E_A and great optimization potential for E_L, the value of O should be considered how to increase the length of the south-facing facade as much as possible. When the value of A is about 4.9 and O is between 10° and 14°, the diagonal of the building rectangle will be perpendicular to the south direction, so that the length of the south facade will be the largest. This would let more natural light enter into the building, and E_L would be the smallest.

In terms of the thickness of the building insulation, the research found that the thickness of the roof insulation is between 0.20 and 0.22, and the thickness of the wall insulation is between 0.25 and 0.28. Although the maximum value in the variable range has not been taken, the two thickness have high values.

In terms of WWR, the study found that the value of WWR-W is the smallest and should not exceed 0.4, the value of WWR-N and WWR-E should not be greater than 0.5, and the value of WWR-S is the largest and should not below 0.7. In general, the WWR of

Table 6. Building material setting

Construction name	Layer number	Material name	Roughness	Thickness {m}	Conductivity {W/m-K}	Density {kg/m3}	Specific Heat {J/kg-K}	Thermal Absorptance	Solar Absorptance	Visible Absorptance
Roof	1	Roof insulation	Medium Rough	The value of RT	0.049	265	836.8	0.9	0.7	0.7
	2	Brick	Medium Rough	0.12	0.89	1920	790	0.9	0.7	0.7
Wall	1	Wall insulation	Medium Rough	The value of WT	0.045	265	836.8	0.9	0.7	0.7
	2	Concrete	Medium Rough	0.2	1.7296	2243	837	0.9	0.7	0.7
Floors	1	Concrete	Medium Rough	0.15	0.53	1280	840	0.9	0.5	0.7
Horizontal visor	1	Metal	Smooth	0.001	45.28	7823.9	500	0.9	0.7	0.7
Construction name	Layer number		Material name					U-value (W/m2·K-1)	SHGC	VT
Window glass	1		Low-E glass					2.50	0.60	0.50

office buildings in Wuhan should consider façade direction. The value of WWR should be small in all directions except south façade.

In terms of building window height, the study found that the value is about 1.8 m. This shows that the building needs a high window height. Because the WWR value in west and east is generally low, a higher W-H value will make the building's east-west window opening style be high and narrow, which will help east-west shading.

In terms of HV, the research found that the value of HV-N is the smallest, and should not exceed 0.7 m. This is because there is no natural light from the north direction, so it is not necessary to block sunlight by a long visor. Due to the east-west window opening style is high and narrow, there is no need too much horizontal visor for shading, so the value of HV-W is between 0.7 m to 1.0 m. Due to the value of WWR-E is about 0.5, larger than the value of WWR-W, the value of HV-E is larger than the value of HV-W, is between 1.2 m to 1.4 m. Although the WWR-S value is larger than 0.7, due to the large proportion of E_L in E_A and great optimization potential for E_L, the value of HV-S is not too large, is between 0.7 m to 1.0 m.

References

1. Turrin, M., Von Buelow, P., Stouffs, R.: Design explorations of performance driven geometry in architectural design using parametric modeling and genetic algorithms. Adv. Eng. Inform. **25**, 656–675 (2011)
2. Wang, L., Wong Nyuk, H., Li, S.: Facade design optimization for naturally ventilated residential buildings in Singapore. Energy Build. **39**(8), 954–961 (2007)
3. Tuhus-Dubrow, D., Krarti, M.: Genetic-algorithm based approach to optimize building envelope design for residential buildings. Build. Environ. **45**, 1574–1581 (2010)
4. Gou, S., Nik, V.M., Scartezzini, J.-L., Zhao, Q., Li, Z.: Passive design optimization of newly-built residential buildings in Shanghai for improving indoor thermal comfort while reducing building energy demand. Energy Build. **169**, 484–506 (2018)
5. Bre, F., Silva, A.S., Ghisi, E., Fachinotti, V.D.: Residential building design optimisation using sensitivity analysis and genetic algorithm. Energy Build. **133**, 853–866 (2016)
6. Jin, J.-T., Jeong, J.-W.: Optimization of a free-form building shape to minimize external thermal load using genetic algorithm. Energy Build. **85**, 473–482 (2014)
7. Zhang, A., Bokel, R., Dobbelsteen, A., Sun, Y., Huang, Q., Zhang, Q.: Optimization of thermal and daylight performance of school buildings based on a multi-objective genetic algorithm in the cold climate of China. Energy Build. 139 (2017)
8. Magnier, L., Haghighat, F.: Multiobjective optimization of building design using TRNSYS simulations, genetic algorithm, and Artificial Neural Network. Build. Environ. **45**, 739–746 (2010)
9. Zemella, G., De March, D., Borrotti, M., Poli, I.: Optimised design of energy efficient building façades via Evolutionary Neural Networks. Energy Build. **43**, 3297–3302 (2011)
10. Cheng, M.-Y., Cao, M.-T.: Accurately predicting building energy performance using evolutionary multivariate adaptive regression splines. Appl. Soft Comput. **22**, 178–188 (2014)
11. Ayata, T., Arcaklioglu, E., Yildiz, O.: Application of ANN to explore the potential use of natural ventilation in buildings in Turkey. Appl. Therm. Eng. **27**, 12–20 (2007)
12. Yun, G., Kim, K.: An empirical validation of lighting energy consumption using the integrated simulation method. Energy Build. **57**, 144–154 (2013)
13. Manzan, M., Pinto, F.: Genetic optimization of external shading devices. Energy Build. 72 (2009)

14. Wang, F.P., Xu, Y., Zhang, G.Q., Zhang, K.: Aerodynamic optimal design for a glider with the supersonic airfoil based on the hybrid MIGA-SA method. Aerosp. Sci. Technol. **92**, 224–231 (2019)
15. Krarti, M., Ouarghi, R.: Building shape optimization using neural network and genetic algorithm approach. ASHRAE Trans. 112 (2006)
16. Wang, W., Rivard, H., Zmeureanu, R.: Floor shape optimization for green building design. Adv. Eng. Inform. **20**(4), 363–378 (2006)
17. Wright, J., Mourshed, M.: Geometric optimization of fenestration (2009)
18. Ichinose, T., Lei, L., Lin, Y.: Impacts of shading effect from nearby buildings on heating and cooling energy consumption in hot summer and cold winter zone of China. Energy Build. **136**, 199–210 (2017)
19. Borgstein, E.H., Lamberts, R., Hensen, J.L.M.: Evaluating energy performance in non-domestic buildings: a review. Energy Build. **128**, 734–755 (2016)

Machine Making

A Generative Material System of Clay Components-The Porosity Language

Dan Liang(✉)

The Bartlett School of Architecture, UCL, Room 603, Building 2, Residential Community, Lane 700, Guoding Road, Yangpu District, Shanghai, China
15000181281@163.com

Abstract. Compared with the pre-determined architecture design based on standard elements, the underlying structure of nature is more like a complex system. Porosity language, for example, which is inspired by nature, has been widely applied in the architecture context. Through the analysis of the underlying methodologies of topology in each case, the strategy is to illustrate how clay components can achieve this natural porosity language. With the help of parametric topology, the report will clearly show how the innovative language of clay components is inspired, optimized and applied. As the background of the literature, natural porosity and examples of existing cavity wall made by clay components will be compared and analyzed in Sect. 1. In Sect. 2, Steven Hall's porous methodology will be considered as the primary topological reference. The parametric iteration topology will be stated explicitly in Sect. 3, which will direct the randomness of porosity form to the balance between structural stability and the aesthetic value. In the last chapter, different architecture applications will be studied through the supporting of micro-climate simulation.

Keywords: Porosity · Generative design · Modularity · Structure behavior · Machine thinking

1 The Interaction Between Part (Components) and Overall Geometry

The traditional way we deal with the standard clay brick is called masonry. Although masonry is believed impenetrable,which blocks sunshine, rain, and wind, it is not the sole meaning. Cavity walls, for example, are designed to let the wind and light penetrate through the architecture surface. It takes many forms between closure and openness. This interesting inherent porosity language not only creates an appealing pattern but also has the potential to be a functional system and generate the micro-climate for the internal space. A typical example is Peter Zumthor's Kolumba Diocesan Museum in Cologne.

While Zumthor's porous wall is made by the simple combination of standardized elements (Fig. 1), porosity form in nature is more like an intricate system (Fig. 2). Needham [7] indicates the underlying common features of all the biological organization

P. F. Yuan et al. (Eds.): CDRF 2020, *Proceedings of the 2020 DigitalFUTURES*, pp. 245–254, 2021.
https://doi.org/10.1007/978-981-33-4400-6_23

in nature: Like the body consists of organs, the organs contain cells. Inevitably, all the scientific exploration of microscopic life pursues to figure out how different levels are connected and organized. Similarly, the porosity form in nature is built up with different levels of porosity which consist of different size of molecules.

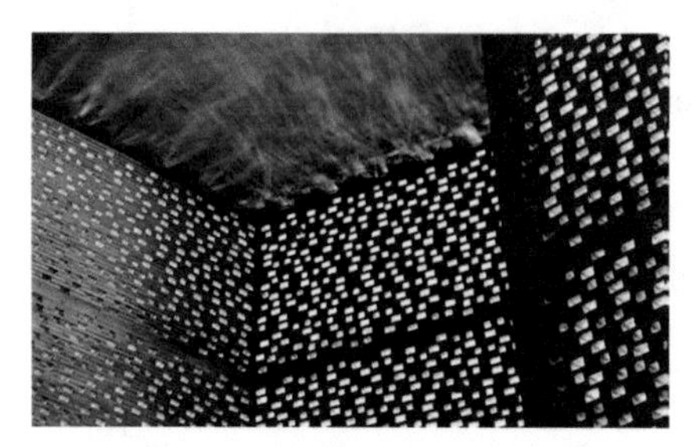

Fig. 1. Peter Zumthor's porous wall (n.d. [image on line] Available at: < https://www.designboom.com/art/helene-binet-photographs-of-the-work-of-peter-zumthor/> [Accessed by 20 January 2007])

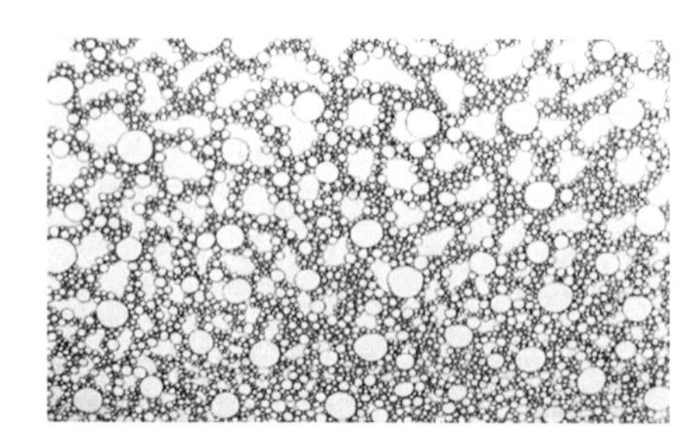

Fig. 2. Porosity form in nature (n.d. [image on line] Available at: <https://gilkalai.wordpress.com/2010/01/20/randomness-in-nature-ii/l> [Accessed by 20 Janurary 2010])

Based on the above views, the strategy is to illustrate how clay components can achieve this natural porosity language. The main topic is to explore a generative language where geometry complexity, material property and the beauty from the perspective of morphology in nature can co-exist.

2 Porous Methodology: Designing from Scratch

An exciting paradigm of how porosity was transformed into architecture context by machine thinking is the Simmons Hall at MIT (Fig. 3). The core of Hall's porous morphology is "designing from scratch" [5], which means not to frame a pre-determined porosity pattern but to grasp some grammatical transformation rules. Those rules were transformed into a formal-generative language to generate architecture form. The language is available for digital tools to generate a family of result for designers to scratch from (Table 1).

D. Kotsopolous defines this methodology as an "open-ended conceptual frame". Based on this method, the proposal is to build our own methodology- a non-predetermined generative porosity language with the clay components. With the grading rules, this language will show the potential of different architectural applications.

3 Optimization of Porosity Language- Modularity and Structure Stability

The methodology of Hall's porous morphology will be referred. The topology optimization procedures will indicate how we control random porosity in this chapter.

Leach N. [6] defines the theoretical foundation of the generative design methods as two elements: the logic basis and natural analogy. As pointed out by him, to create and

Fig. 3. The Simmons Hall at MIT (n.d. [image on line] Available at: <https://divisare.com/projects/260919-gramazio-kohler-ralph-feiner-gantenbeinvineyard-facade> [Accessed by 19 April 2016])

Table 1. The family of generated result [5]

$[C - t(x)] + t(y)$	$t(x)$	$\ni [t(x)]$
⇓ A, A		
⇓ A, A		
⇓ A, A		

optimize generative design language, nature is always an important source of inspiration. Starting from the analysis of porosity pattern in nature, we find that it is graded into different levels.

According to the above view, the modular system has the potential to build a diverse range of complex structure from a few necessary modules, and the module can be used as the seed to create a graded porosity. For this reason, we orient our research controlling the porosity both in the single module and the overall structure. Concerning poly cube geometry which is a solid figure form by joining one or more 3-dimensional components face to face, we can create the first level porosity within one single module (Fig. 4). Those modules are used as the expanding cell to align with their replicas. By removing the components in the corner of each module, we can create abundant opportunities for the modules to join and form another level of porosity pattern. Then the second level porosity is formed (Fig. 5).

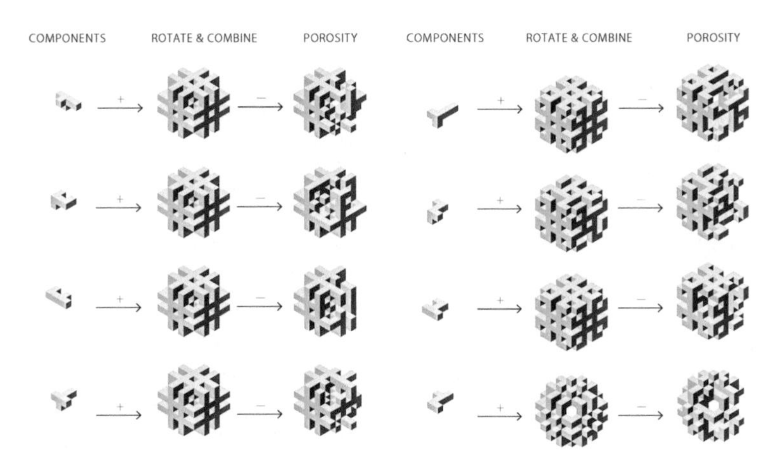

Fig. 4. The first level porosity in the component

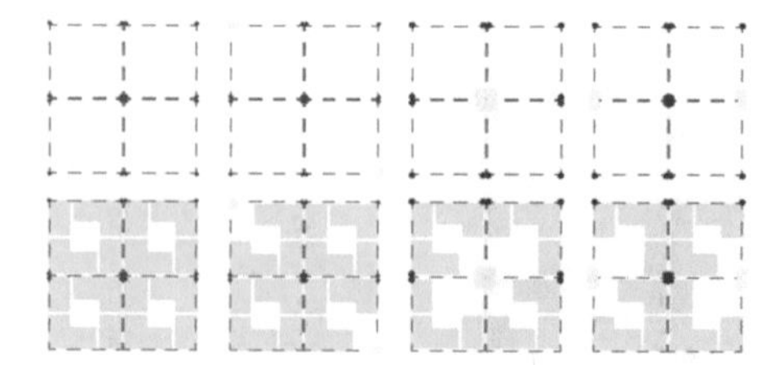

Fig. 5. The second level porosity in-between components

In order to permit a more complex porosity and fully control it, we studied the marching cube algorithm and made it another rule to control the porosity. By testing whether the vertex of the grid was inside or outside of the isosurface, the script would automatically choose the specific module cells and place them appropriately to the right place. For this reason, the control of the isosurface was to control the third level of the porosity in the overall structure (Fig. 6).

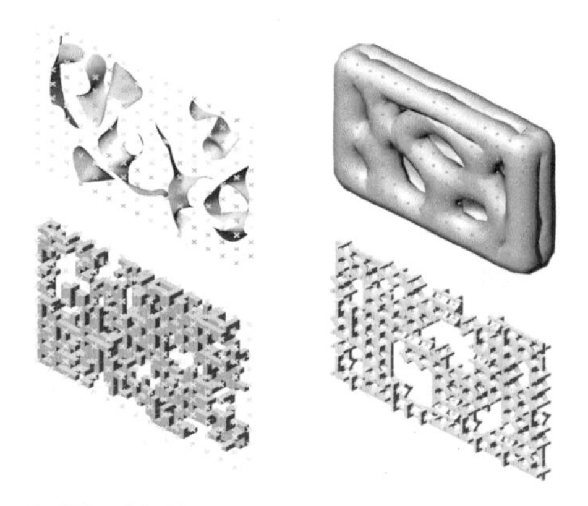

Fig. 6. The third level porosity in the overall geometry

These different levels of relations and rules were added as criteria to generate the graded porosity grammar, which was inspired by the biomorphic form (Fig. 6). This grammar defined an infinite language of porosity pattern in the overall design. By continuously running the script with random input values, we can get unique porosity pattern every single time (Fig. 7).

Since the logic is a bottom-up rule which started from the modular component. So by changing the geometry of modules, overall geometry will be influenced. Then we conducted a series of topology research. By removing volume from the initial component, we can get new modules. The more volume we remove in the single component, the wall becomes more porous. Then we decided the final geometry, which could achieve a considerable porosity. During the process, computing serves as a media rather than a tool. The outcome is not pre-determined, but it is predictable- Most of them seem entirely unrealistic in terms of construction (Fig. 8).

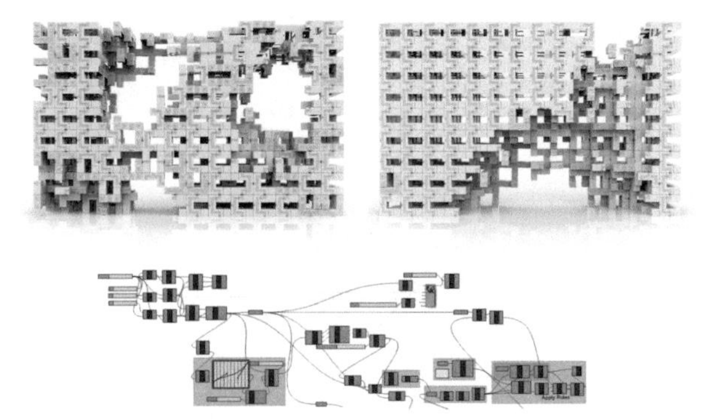

Fig. 7. Infinite Porosities generated by the script

Fig. 8. Geometric topology and final module

Neil holds the opinion in his book Digital Tectonics: Random component ought to permit the existence of weaker designs. Then new rules, such as structural efficiency, should be taken into consideration. From the above point of view, structure feasibility is a chance to guide the randomness to be ordered.

First of all, we learn from the traditional brick masonry system by optimizing the grid from 90° to 45°. The FE force simulation proves that 45° performs much better concerning force distribution (Fig. 10). After that, we add a new algorithm which defines the distribution of the force as a small truss in this system. The script will overlap all the center lines of each rotation and creates an overall truss system automatically, then calculate the force value of each position. After that, the algorithm will put each rotation in the specific position to calculate the force value of it and then select the one whose load bearing condition is closest to the previous value. In this way, the script will select the most efficient rotation for each position (Fig. 9).

According to the force simulation (Fig. 11), red and blue in the diagram indicates compress and tension respectively. It's evident that the structure behaves much better after the optimization of the script. We played the rules in-between the real structure efficiency and aesthetic requirements of the language. Which means by deciding the height, thickness of the wall and the position of holes on the wall, the adaptive system will generate new porosity pattern but always remains the best structural performance. Thus, the design of the wall is not random anymore but implies how the force goes through in each wall (Fig. 12).

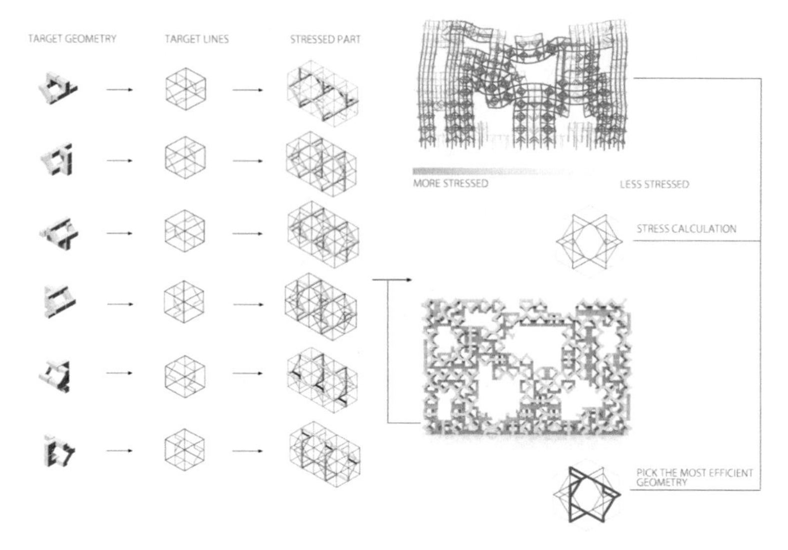

Fig. 9. Algorithm optimization based on structural stability

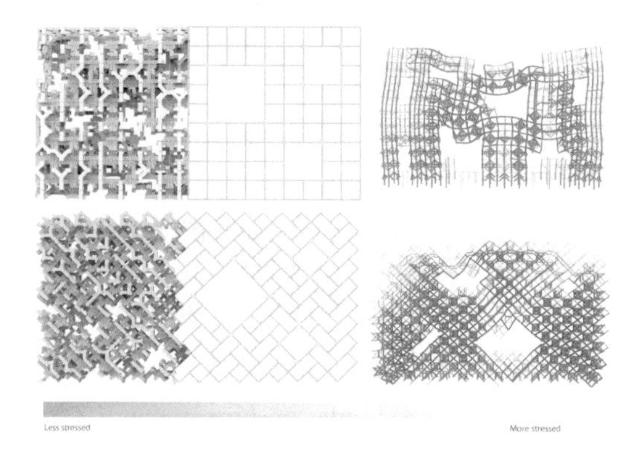

Fig. 10. Grid optimization

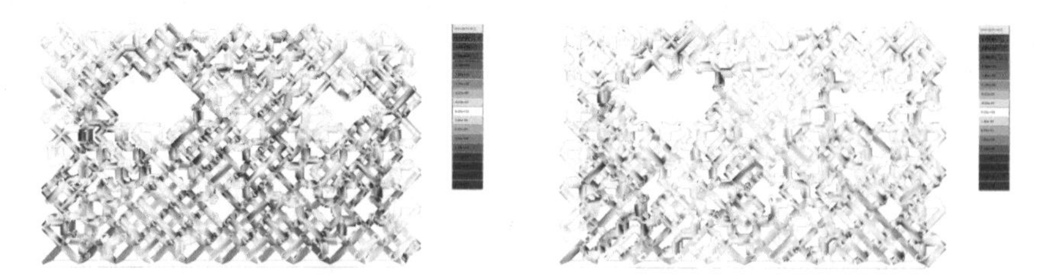

Fig. 11. Force simulation before (left) & after structure optimization (right)

Fig. 12. Different directed randomness based on structure optimization

4 Application in the Architecture Field

We use slip casting technique to fabricate the components with the help of special moulds. The outcome shows a quite good result-new components with the characterized texture and joints. After firing, the components become hard enough to bear the load. Meanwhile, we take advantage of the 45-degree grid to study the interlocking joints in-between the components. The joints include a fixed joint and a sliding joint to guarantee the assembly of each component. Those components will be oriented and locked by wooden pegs to form a smart interlocking system (Fig. 13).

Fig. 13. Fabrication of new component with characterized pattern and joints & The joint system

The single modular unit aligns with its different rotations to lead the way for continuously changing geometries and openings. The result shows aesthetic value in-between order and randomness rather than a single repetition. Like the cells following the specific rules to form tissues, those modular components follow the rules of generative language to obtain the best collection based on the structural efficiency (Fig. 14).

To do further explorations and simulate how the porous wall affects speed measurement of airflow, this set of comparative simulation test includes three variables, which are a wall with no openings, a wall with normal windows and the porous wall. By using the software called flow-designer, a wind tunnel is defined to test a pavilion section measuring width 2000 cm, length 1400 cm and height 4280 cm.

From simulation 1, it's evident that the porous wall gives more wind circulation to the interior space. The graph in simulation 2 demonstrates how airflow blocked by different

Fig. 14. Final physical wall which is constructed at the Bartlett School of Architecture

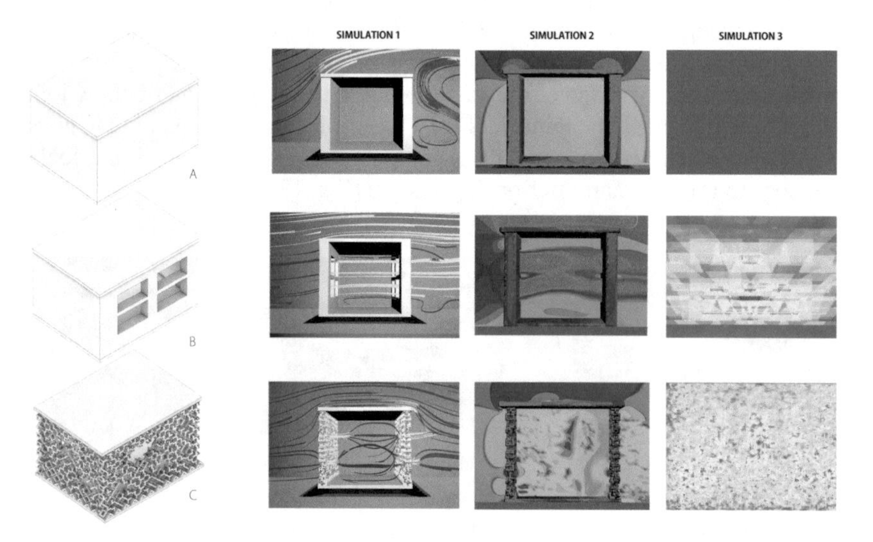

Fig. 15. Wind flow & radiation simulation based on different walls

walls velocity changes. The warmer air color represents a higher airflow pressure than cold ones. In comparison with the strong wind ventilation caused by standard windows, the porous wall results in a more even and mild wind environment and brings the changed airflow to each corner of the room. The graph in simulation 3 indicates that the porous screen is proved to be a characterized shading system since it can proactively create more even light environment than the normal system (Fig. 15).

The above studies reveal the possibility of the porous screen to adjust the air movement and influence indoor temperature. From the perspective of energy saving, it can be used as the facade component to create a mild indoor micro-climate for special functions such as wine fermentation (Fig. 16).

Fig. 16. Design language applied to particular function like wine fermentation

5 Conclusion

Nature randomness and biomorphic form have always been the inspirations for architecture design. Thanks to the digital tectonics, the aesthetic value of the randomness will be developed into rich expressions rather than the standard and predetermined configuration.

The parametric algorithm is used as a media in our design research to direct the beauty of the nature porosity language to a graded porosity generative system where the structure stability, material property and the beauty from morphology in nature can co-exist in theory. The open-ended generative porosity language not only allows the application in different architecture scales but also achieves the mass production with the power of modular components.

References

1. Akçay, A.Ö., Alothman, H.: Fashion inspired by architecture: the interrelationship between Mashrabiya and Fashion World. J. Hist. Cult. Art Res. **7**(2), 328–351 (2018)
2. Hauer, E.: Erwin Hauer: Continua-Architectural Walls and Screens. Princeton Architectural Press, New York (2004)
3. Hofmeyer, H., Delgado, J.M.D.: Coevolutionary and genetic algorithm based building spatial and structural design. AI EDAM **29**(4), 351–370 (2015)
4. Knippers, J., Menges, A.: ICD/ITKE research pavilion 2011. In: Architectural material and Texture I, pp. 266–273 (2013)
5. Kotsopoulos, S.D.: From design concepts to design descriptions. Int. J. Archit. Comput. **6**(3), 335–360 (2008)

6. Leach, N., Turnbull, D., Williams, C.J.: Digital tectonics. In: The Randomness, pp. 43–57. Wiley (2004)
7. Needham, J.: Biochemical Aspects of Form and Growth (1925)
8. Perez, S.R.: The Synthetic Sublime (2004)
9. Roth, R., Pentak, S.: Design Basics: 3D. Cengage Learning (2012)

Adaptable Tool-Path Planning Method for 3D Concrete Printing Based on the Mapping Method

Qian Wan[1], Li Wang[1,2](✉), and Guowei Ma[1,2]

[1] School of Civil and Transportation Engineering, Hebei University of Technology, 5340 Xiping Road, Beichen District, Tianjin 300401, China
wangl1@hebut.edu.cn

[2] Tianjin Key Laboratory of Prefabricated Building and Intelligent Construction, Xiping Road, Beichen District, Tianjin 300401, China

Abstract. 3D concrete printing (3DCP) has been successfully and widely applied in the fields of civil structure, infrastructure, architectural decoration etc.,due to its unique advantages of automation and flexibility, and has shown great potential for development. One of the key steps in the execution of 3DCP is the tool-path planning process. However, 3DCP typically utilize the flowable composite materials which changes with time, and it is easy to induce interface joints or filling defects due to uneven path distribution when constructing the irregular-shaped construction. To solve the problem of printing path planning in 3DCP, this paper proposes tool-path planning based on mapping method, which integrates the rheology and necessary continuity of concrete materials into the printing process parameters, improves the mutual adaptability of printing process and material characteristics, improves the continuity of printing, the compactness of filling, and then ensures the stability and durability of printing structure.

Keywords: 3D concrete printing (3DCP) · Tool-path planning · Mapping · Adaptable

1 Introduction

The 3D concrete printing (3DCP) method has broken through the traditional construction method of casting components with formworks, so that the irregular shaped and non-standard components can be fabricated easily, the construction design has more freedom, and the structural design for 3DCP constructions can adopt more curvatures. In October 2019, an assembled 3D printing arch bridge with a span of 18.04 m was built in the campus of Hebei University of technology, Tianjin, China. Although 3DCP is in its preliminary exploration stages, in the construction industry, it shows a rapid development trend [1, 2]. Tool-path planning is one important step of 3D printing (3DP) execution which is based on the digital model of design calculating the trial of the print-head, usually nozzle mounted on the end-effector of the 3DP device. The current 3DP techniques include contour-parallel method and zigzag method. Although both these

P. F. Yuan et al. (Eds.): CDRF 2020, *Proceedings of the 2020 DigitalFUTURES*, pp. 255–264, 2021.
https://doi.org/10.1007/978-981-33-4400-6_24

methods satisfy the tool-path planning of regular cross-section model [3], some problems may occur when doing the irregular shaped construction models tool-path planning in the existent methods.

Tool-path planning is one of the important steps of 3DP technology including 3DCP, which generated the path of the print-head, i.e. the nozzle mounted on the 3D printer's end effector, based on the 3D-digital models. The current tool-path planning methods include the contour-parallel method and the zigzag method. Both of these are competent for the tool-path planning for the model with regular cross-section, e.g. the rectangle or circular sections, which plays an important role in experimental research of materials [3]. However, the current tool-path planning methods will inevitably encounter a lot of problems in respect of 3DCP which is dominated by the irregular shaped components.

The numerous tool-path turnings, tool-path interruptions, and the under fill are common problems of existent 3DCP tool-path planning methods for two main reasons: unreasonable tool-path shape and printing parameters. Contour-parallel tool-path is the offset curve of the model contour, when the size of model is not an integer multiple of the nozzle diameter, there will be a large amount of under fill caused by the self-intersection problem [4, 5]. Zigzag tool-path fills and scans the printing region with a batch of parallel lines, which may cause frequent tool-path turnings and interruptions. The printing parameters include the linear velocity of the print-head during the printing process (printing velocity), and the concrete materials extrusion velocity (extrusion velocity). Due to the fluidity of the concrete materials which change with age, the printing velocity and the extrusion velocity impact the actual width and thickness of materials extrusion. In the recent studies on 3DCP, few objects were studied for the printing parameters, usually using the constant value. However, If the actual extrusion width is inconsistent with the preset, the under fill (or over fill) may appear.

In order to improve the quality of 3DCP components, it is necessary to fully consider the shape of printing in the tool-path planning, to avoid the tool-path interruptions and the under fill, and to reduce the number of tool-path turnings. It's also necessary to evaluate the fluidity of concrete materials to determinate the best printing scheme. In this paper, a novel method of 3DCP tool-path planning method has been proposed, based on the transfinite mapping method and the full consideration of the fluidity. This method can generate tool-path without interruptions, under fill, and frequent tool-path turnings.

2 Mapping Tool-Path Planning Method

2.1 Transfinite Mapping Method

The transfinite mapping is a method of formulating a surface within the area surrounded by any four curves, the calculation plane is pushed to the physical plane by establishing the mapping function of the natural coordinates x-y and the calculation coordinates u-v [6, 7]. As shown in Fig. 1, in the physical panel, divides the region A into 4 boundary curves F1, F2, F3 and F4, and each boundary curves corresponds to the 4 edges of the unit square region F on the calculation panel u-v. Any point in the calculation panel u-v corresponds to one point in physical panel x-y.

Through the printing model is always established by the CAD software, and the printing panel always generated by clipper software, the boundary of printing region

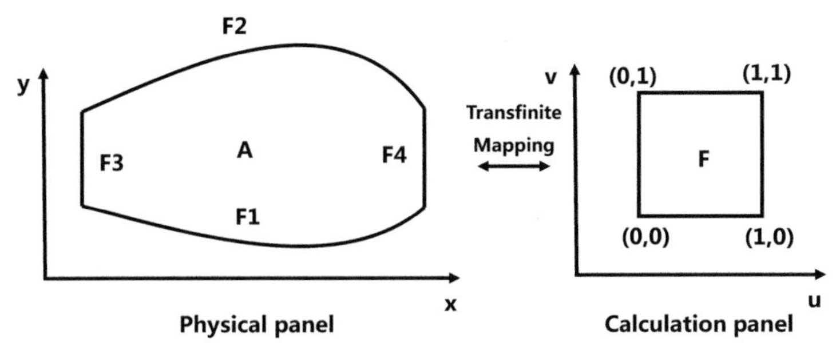

Fig. 1. Physical panel *x-y* and calculation panel *u-v*

always consists of finite points, shown as polygonal lines. So, in this method, polygonal lines are adopted to approximately describe the shape of the printing region. For a polygonal f, assuming the number of nodes is n, and the one node can be described with vector as $P_i = [x_i, y_i]^T$. Define the polygonal line length on P_i is the length between the 1st node to the ith node along the polygonal line. For range the polygonal line into [0, 1], the normalization polygonal line length l_i on P_i is shown in Eq. 1.

$$l_i = \begin{cases} 0 & i = 1 \\ \dfrac{\sum_{j=2}^{i} |P_j - P_{j-1}|}{\sum_{j=2}^{n} |P_j - P_{j-1}|} & i \neq 1 \end{cases} \tag{1}$$

Using the parametric representation Q, a parametric representation of the polygonal line f can be established in Eq. 2.

$$p_i = [x_i, y_i]^T = Q(l_i), l_i \in [0, 1] \tag{2}$$

For any point in f, there is a unique parameter l corresponding to it. For other positions on f except for nodes, it is determined according to the interpolation from the leading and succeeding nodes, as shown in Eq. 3.

$$Q(l) = \frac{l - a}{b - a} Q(b) + \frac{b - l}{b - a} Q(a), a < l < b \tag{3}$$

where: a (b) is the normalization polygonal line length of the closest front(back) node to $Q(l)$.

Using the parametric representation Q, the four boundaries F1, F2, F3 and F4 can be formulated as:

$$\begin{cases} f_{1i} = [x_i, y_i]^T = Q_1(l_i) \\ f_{2i} = [x_i, y_i]^T = Q_2(l_i) \\ f_{3i} = [x_i, y_i]^T = Q_3(l_i) \\ f_{4i} = [x_i, y_i]^T = Q_4(l_i) \end{cases} \tag{4}$$

where f_{il} is the ith node on F1.

$v = 0$ is one of four boundaries of F on calculation panel u-v corresponding to F1 on x-y panel, has only one variable parameter u and one constant parameter v. So, for F1 and F2, the parameter l can be exchanged with u, similarly, for F3 and F4, the parameter l can be exchanged for v, as shown in Eq. 5. That is a projector between region A in physical panel x-y and region F in calculation panel u-v.

$$\begin{cases} f_{1i} = [x_i, y_i]^T = Q_1(u) \\ f_{2i} = [x_i, y_i]^T = Q_2(u) \\ f_{3i} = [x_i, y_i]^T = Q_3(v) \\ f_{4i} = [x_i, y_i]^T = Q_4(v) \end{cases} \tag{5}$$

The mapping between physical panel x-y and calculation panel u-v is formulated as Eq. 6.

$$\begin{aligned} P(x, y) &= P_B(u, v) \\ &= (1 - v)Q_1(u) + vQ_2(u) + (1 - u)Q_1(v) + uQ_2(v) \\ &-(1 - u)(1 - v)F(0, 0) - (1 - u)vF(0, 1) \\ &-uvF(1, 1) - u(1 - v)F(1, 0) \end{aligned} \tag{6}$$

where $P_B(u, v)$ is the mapping function. $F(u, v)$ is the corresponding point of (u, v) in physical panel x-y, specially, $F(0, 0)$ is the 1st point of F1, $F(0, 1)$ is the 1st point of F2, $F(1,0)$ is the last point of F1, $F(1,1)$ is the last point of F2.

2.2 Mapping Tool-Path Planning Method

Mapping tool-path planning method (MTPPM) is a novel tool-path planning method based on transfinite mapping method. Firstly, dividing the boundary of compensated digital construction model cross-section into 4 curves as F1, F2, F3, and F4. Secondly, establishing the physical panel x-y and calculation panel u-v and the zigzag liked "module" tool-path. Finally, using mapping function P_B to calculate the actual tool-path. The specific steps including: model and parameters input, generation of base-line, establishing of boundary formulations, generation of module tool-path, calculation of actual tool-path, calculation of printing parameters, and NC programs output.

(1) Model and Parameters Input

In MTPPM, the model is the 2D CAD model of construction, and the smooth curves in model exchanged by polygonal lines, denoted as Ω. The parameters include: the diameter of nozzle (D_n); the number of tool-paths (n_p). The number of tool-paths can be determined by the geometry of model and the diameter of nozzle, specially, can be preliminarily calculated by the ratio of the narrowest dimension of the model (D_{min}) to D_n.

$$n_p = \lfloor D_{\min}/D_n \rfloor \tag{7}$$

(2) Generation of Base-line

In 3DP technology, the tool-path is not in contact with the printing region boundary to avoid the over fill. The base-line is the offset line of model boundary, and the distance between model boundary and base-line is $D_n/2$. The base-line denoted as Ω'.

(3) Establishing of Boundaries Formulations

The base-line Ω' was divided into 4 parts, as F1, F2, F3 and F4, based on the interrupt nodes chosen on Ω'. Thus establishing the mapping function PB following Eq. 1–6. In this method, the interrupt nodes are chosen on UI manually.

(4) Generation of Module Tool-Path

A batch of zigzag tool-path with the number of n_p was generated in the unit-square region. The module tool-path can be calculated by Eq. 8, where $path_i$ is the ith module tool-path. The sequence of module tool-path is zigzag-like shape, i.e. if odd module tool-paths ($path_{odd}$) are positive sequence, the even module tool-paths ($path_{even}$) are inverted sequence. And vice versa.

$$path_i : v_i = \frac{i - 1}{n_p - 1}, u_i \in [0, 1] \tag{8}$$

(5) Calculation of Actual Too-Path

Choosing several nodes on the module tool-path, calculating the actual tool-path following Eq. 6, as shown in Eq. 9.

$$PATH_i = P_B(path_i) \tag{9}$$

(6) Calculation of Printing Parameters

Because the distance between each tool-path is variable, the printing parameters are also variable to ensure the complete filling, existing studies [8, 9] have shown that the relationship between printing parameters, printing velocity, extrusion velocity and extrusion width, can be regarded as linear, and can be expressed by Eq. 9.

$$v_f = \frac{\alpha D_n v_e}{h D_f} \tag{10}$$

where v_f is the printing velocity, v_e is the extrusion velocity, D_f if the extrusion width, h is the thickness of extrusion,α is an empirical parameter,determined by experiments.

Different color represents different extrusion width, the maximal width id $1.2D_n$, and the minimal width about is $0.8D_n$. Based on the extrusion graph and Eq. 10, the printing parameters at different position can be determined.

(7) NC Program Output

The NC program always includes the tool-path coordination, and the printing velocity. This study uses *Kawasaki* robot as a 3D printing device.

3 Case Study of MTPPM

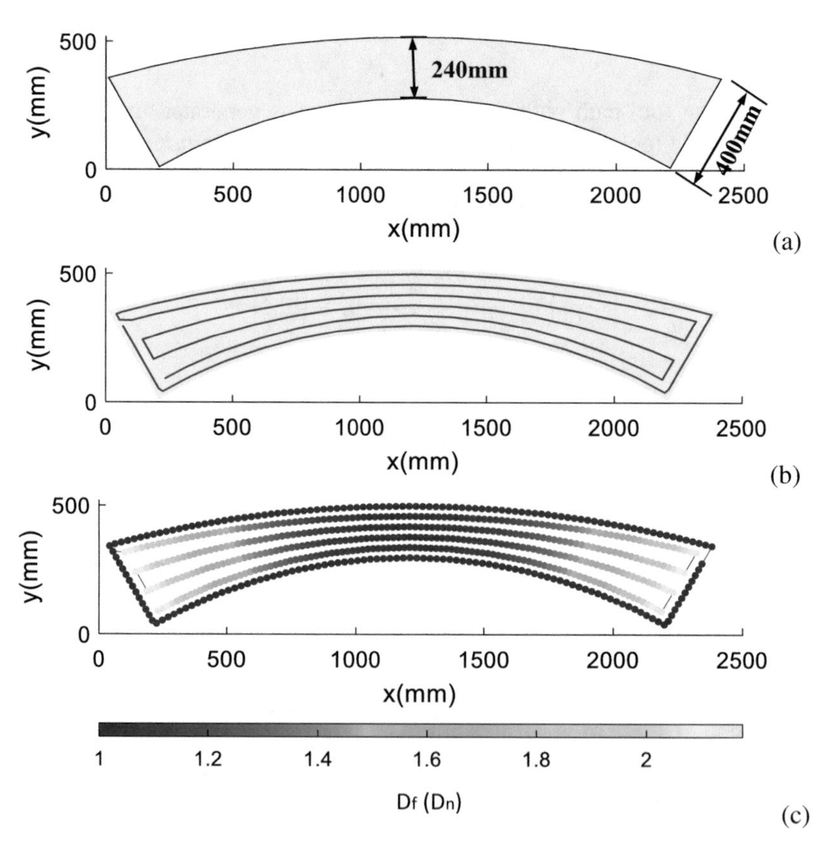

Fig. 2. Tool-path planning of e variable cross-section shaped arch, (a) the CAD model; (b) the tool-path; (c) extrusion width

In this section, taking the irregular shaped components of the arch bridge construction as an example, the tool-path planning of the three types of variable cross-section shaped arch, local irregular shaped arch and multi-component combined shaped arch are discussed respectively, and the feasibility and advantages of the MTPPM are demonstrated.

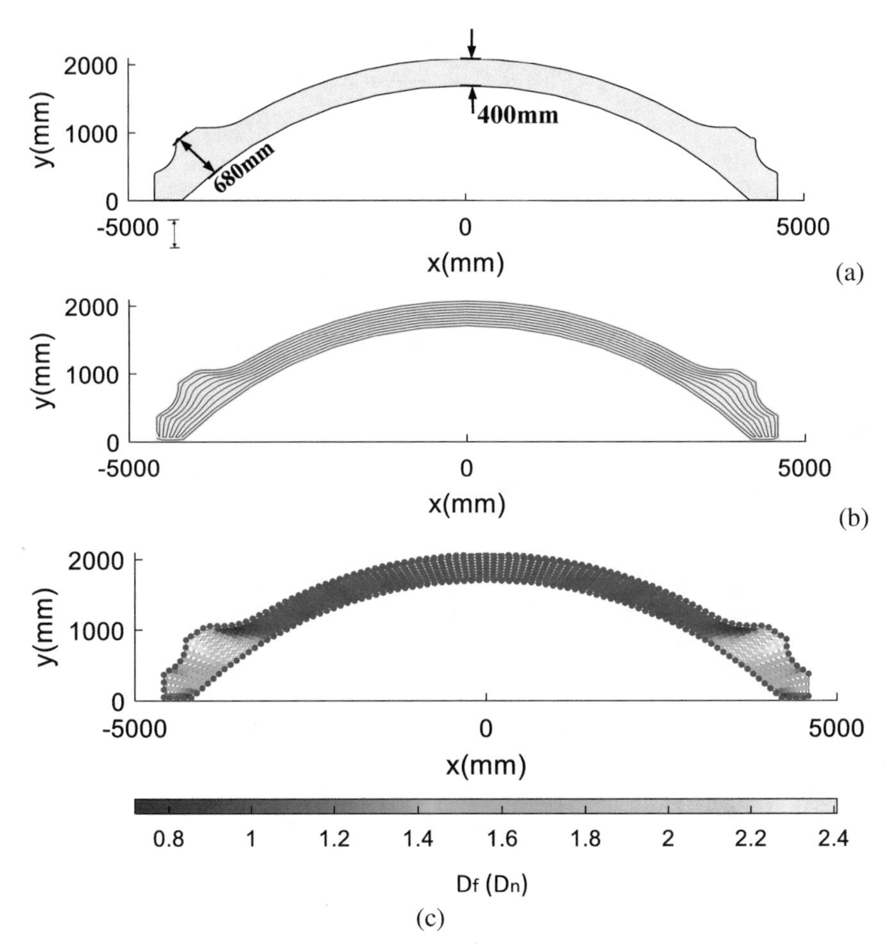

Fig. 3. Tool-path planning of local irregular shaped arch, (a) the CAD model; (b) the tool-path; (c) extrusion width

Case Study 1: The Variable Cross-section Shaped Arch

An arch whose cross-section changes gradually from the arch foot to the crown is called a variable-section shaped arch. Using this kind of special-shaped arch can effectively reduce the use of materials and make the structure stress more reasonable. The size of the variable cross-section shaped arch as shown Fig. 2(a), the narrowest length of the arch is 240 mm. To improve the precision of printing, a contour-parallel tool-path set with width of D_n. If the nozzle diameter is D_n, the number of expected contour-parallel tool-paths (number of mapping tool-paths) is $240/40-2 = 4$. The printing sequence is internal first, then external. The tool-path as shown in Fig. 2(b), and the extrusion width as shown in Fig. 2(c), the maximal extrusion is $2.2D_n$ and the minimal extrusion width is D_n.

Case Study 2: The Local Irregular Shaped Arch

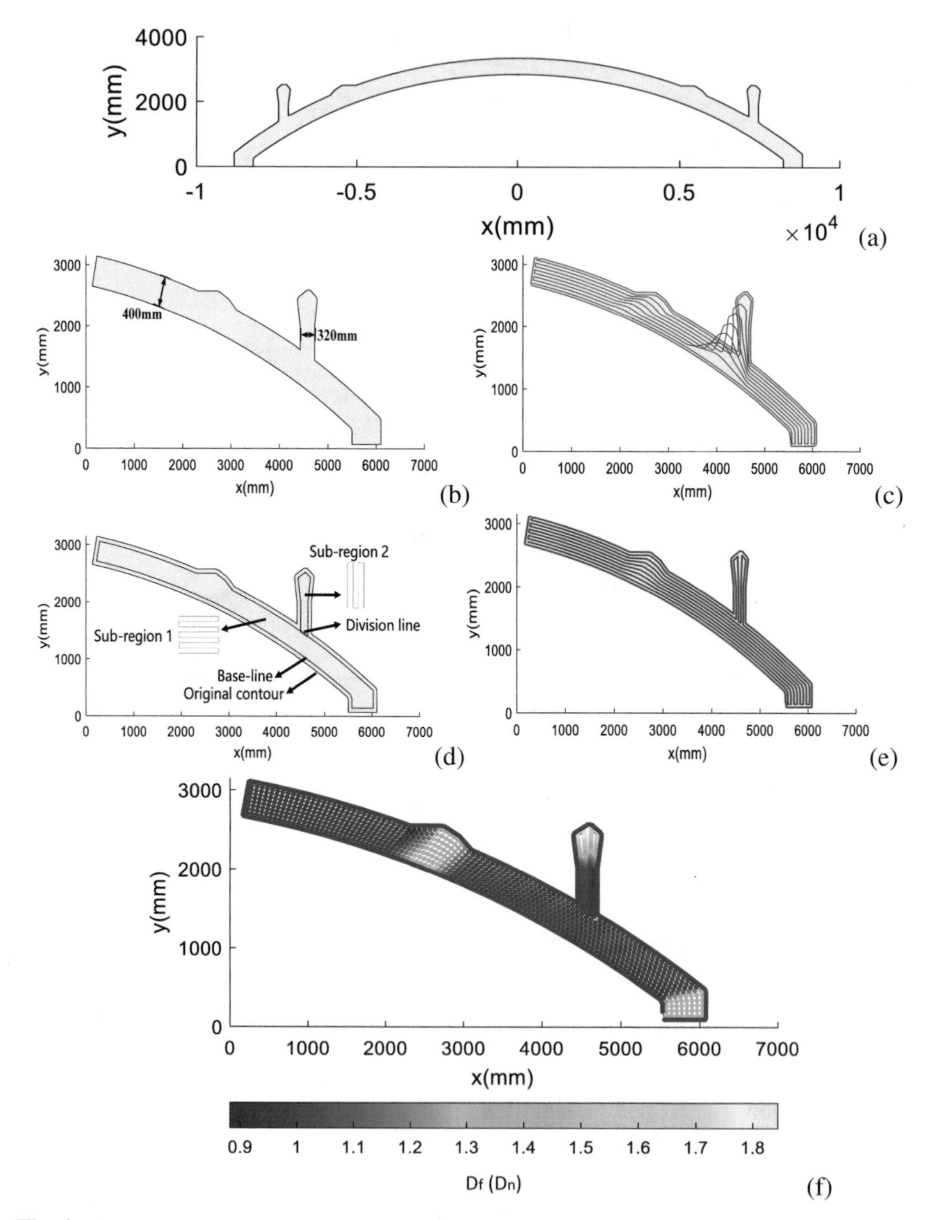

Fig. 4. Tool-path planning of the multi-component combined shaped arch, (a) the CAD model; (b) the right part of model;(c) the "overspill"; (d) division of filling area;(e) the too-path;(f) extrusion width

As shown in Fig. 3(a), the two ends of the local irregular shaped arch exist additional construction, as the base of other construction, e, g. the column supported the bridge deck. This type of construction using contour-parallel method will lead to large areas of under fill. The specifications of 3DCP using MTPPM is: nozzle diameter is 40 mm, setting one contour-parallel tool-path, the number of the mapping tool-paths is 8. As shown in Fig. 3(b, c), the most printing with extrusion width of D_n.

Case Study 3: The Multi-component-combined Shaped Arch

As shown in Fig. 4(a), the construction combining the arch with the columns is called multi-component combined shaped arch. The irregularity of this type of construction is greater than the local irregular shaped arch, and two protruding branches are formed at both ends of the arch. To avoid the "overspill" as shown in Fig. 4(c), the approach of division filling area into several sub-region have been introduced. Taking the right part of the arch shown in Fig. 4(b, d) as an example, the arch ring and the column are divided into two parts, and the tool-path of the arch-region and the column-region are respectively obtained by the MTPPM. The finial tool-path is the continuous combination of the two regions' tool-path.

4 Conclusion

The current 3DCP tool-path planning methods: zigzag method and contour-parallel methods inevitably cause tool-path turnings, interruptions and under fill, due the unsuitable tool-path shape and printing parameters, which limit the 3DCP in the actual engineering. This study proposed a novel 3DCP tool-path planning method, based on the transfinite mapping method. Based on the simple module tool-path calculates the extrusion variable tool-path, and adopts variable printing parameters to control the 3DCP process. MTPPM can generate continuous tool-path with no under fill and fewer tool-path turnings. To avoid the "overspill", the division filling region idea has been introduced. The three case studies prove that MTPPM is qualified for the 3DCP process.

References

1. Panda, B., Tay, Y.W.D., Paul, S.C., Tan, M.J.: Current challenges and future potential of 3D concrete printing. Materialwiss. Werkstofftech. **49**(5), 666–673 (2018)
2. Gosselin, C., Duballet, R., Roux, P., Gaudillière, N., Dirrenberger, J., Morel, P.: Large-scale 3D printing of ultra-high performance concrete – a new processing route for architects and builders. Mater. Des. **100**, 102–109 (2016)
3. Li, Z., Li, W., Ma, G.: Mechanical improvement of continuous steel microcable reinforced geopolymer composites for 3D printing subjected to different loading conditions. Compos. Part B: Eng. **187**(107796), 1–14 (2020)
4. Zhiwei, L., Jianzhong, F., Yong, H., Wenfeng, G.: A robust 2D point-sequence curve offset algorithm with multiple islands for contour-parallel tool path. Comput.-Aided Des. **45**, 657–670 (2013)
5. Held, M.: Voronoi diagrams and offset curves of curvilinear polygons. Comput.-Aided Des. **30**(4), 287–300 (1998)

6. Haber, R., Shephard, M.S., Abel, J.F., Gallagher, R.H., Greenberg, D.P.: A general two-dimensional, graphical finite element preprocessor utilizing discrete transfinite mappings. Intl. J. Numer. Methods Eng. **17**(7), 1015–1044 (1981)
7. de Oliveira Miranda, A.C., Martha, L.F.: Hierarchical template-based quadrilateral mesh generation. Eng. Comput. **33**(4), 701–715 (2015). https://doi.org/10.1007/s00366-014-0392-8
8. Tay, Y.W.D., Li, M.Y., Tan, M.J.: Effect of printing parameters in 3D concrete printing: printing region and support structures. J. Mater. Process. Technol. **271**, 261–270 (2019)
9. Xu, J., Ding, L., Cai, L., Zhang, L., Luo, H., Qin, W.: Volume-forming 3D concrete printing using a variable-size square nozzle. Autom. Construct. **104**, 95–106 (2019)

Working with Uncertainties: An Adaptive Fabrication Workflow for Bamboo Structures

Yue Qi[1(✉)], Ruqing Zhong[1(✉)], Benjamin Kaiser[2], Long Nguyen[1], Hans Jakob Wagner[1(✉)], Alexander Verl[2], and Achim Menges[1]

[1] Institute for Computational Design and Construction, Keplerstrasse 11, 70174 Stuttgart, Germany
dbddqy@gmail.com, zrq.zhongruqing@gmail.com, hans.jakob.wagner@icd.uni-stuttgart.de

[2] Institute for Control Engineering of Machine Tools and Manufacturing Units, Seidenstrasse 36, 70174 Stuttgart, Germany

Abstract. This paper presents and investigates a cyber-physical fabrication workflow, which can respond to the deviations between built- and designed form in real-time with vision augmentation. We apply this method for large scale structures built from natural bamboo poles. Raw bamboo poles obtain evolutionarily optimized fibrous layouts ideally suitable for lightweight and sustainable building construction. Nevertheless, their intrinsically imprecise geometries pose a challenge for reliable, automated construction processes. Despite recent digital advancements, building with bamboo poles is still a labor-intensive task and restricted to building typologies where accuracy is of minor importance. The integration of structural bamboo poles with other building layers is often limited by tolerance issues at the interfaces, especially for large scale structures where deviations accumulate incrementally. To address these challenges, an adaptive fabrication process is developed, in which existing deviations can be compensated by changing the geometry of subsequent joints to iteratively correct the pose of further elements. A vision-based sensing system is employed to three-dimensionally scan the bamboo elements before and during construction. Computer vision algorithms are used to process and interpret the sensory data. The updated conditions are streamed to the computational model which computes tailor-made bending stiff joint geometries that can then be directly fabricated on-the-fly. In this paper, we contextualize our research and investigate the performance domains of the proposed workflow through initial fabrication tests. Several application scenarios are further proposed for full scale vision-augmented bamboo construction systems.

Keywords: Adaptive fabrication · Computer vision · Construction tolerance · Natural material · Bamboo

P. F. Yuan et al. (Eds.): CDRF 2020, *Proceedings of the 2020 DigitalFUTURES*, pp. 265–279, 2021.
https://doi.org/10.1007/978-981-33-4400-6_25

1 Introduction and Research Context

1.1 Natural Material and Construction

Global average temperature has increased by more than one degree Celsius since pre-industrial times. Recent research (Churkina et al. 2020) has shown that using bio-based construction materials with good carbon storage capacity, such as bamboo and timber, can be part of a solution for the global warming issue.

Raw bamboo has remarkable mechanical properties and is a high-yield renewable resource (Atanda 2015). Bamboo poles as a construction material has not only been extensively used in vernacular buildings in its growth regions, but also attracted a lot of interest of contemporary architects around the world, such as Vo Trong Nghia, Kengo Kuma, Simon Velez and Markus Heinsdorff (Fig. 1). Though it has been explored in different construction typologies, such as small-scale residential buildings and pavilions, its performative potentials have not been fully exploited and it is still most commonly used as scaffolding in the east-Asian regions. Previous studies have demonstrated the potential of bamboo for pre-engineered structures (Bhalla et al. 2017), but challenges still exist for the fabrication process.

Fig. 1. Bamboo in architecture. (a) Bb Home, Vietnam, 2013 (Image Source: H&P architects); (b) Vinata Bamboo Pavilion, Vietnam, 2018 (Image Source: VTN Architects/Hiroyuki Oki); (c) Bamboo pavilion for Expo Shanghai, China, 2010 (Image Source: Markus Heinsdorff)

Both being plant materials, bamboo and wood have similar properties, and can be used in either natural or processed form. Since engineered wood technologies are far more developed than bamboo processing, analyzing the development of timber can be seen as an inspiration for bamboo application (Huang 2019). Natural timber has an intricate fibrous layout and can be used for evolutionary optimized structures (Self and Vercruysse 2017). However, it is hard to be applied to engineered structures because of its non-standardized material conditions and dimensional deviations during construction. Engineered timber has reliable product characteristics and is highly dimensionally stable. Nevertheless, its fibrous integrity is often lost and the standardization process requires effort and energy. For bamboo this is even more true: its natural hollow cylindrical geometry and high fibrous integrity constitute a highly optimized natural building material. Processing would compromise its structure, whereas its natural growth-dependent geometry deviations so far restrict many state-of-the-art construction processes.

1.2 Uncertainties of Natural Geometry and Deviation in Fabrication Process

Natural bamboo poles are historically highly dependent on manual operations in construction and difficult to combine with other standard building materials (Huang 2019). The main reasons are its material-related uncertainties, more specifically, its intrinsic geometrical variations. Bamboo rarely grows perfectly straight, the diameter of each pole is irregularly different.

The uncertainties of its natural geometry cannot be ignored, otherwise large deviations between desired- and fabricated structure will occur. Manual operations can also contribute to deviations. However, in most bamboo constructions, the deviations are manually corrected to reach relative, local accuracy.

It is of great importance to ensure global accuracy of fabricated structures, especially when they need to interface with other material groups. The success of integration requires tolerance of each material group to be within a specified range, e.g., the Chinese code GBT51233-2016 and GB50755-2012 set the maximum error to ±6.4 mm in length and ±3.2 mm in width for prefabricated timber plates and ±1 mm for steel components. Industrialized materials such as steel, concrete and engineered wood are used more frequently in construction because their geometrical and mechanical properties are more predictable (Lorenzo and Mimendi 2019). Although natural bamboo is a great candidate for building construction, more effort needs to be taken to control its tolerance.

1.3 Sensing System and Adaptive Fabrication Workflow

Digital fabrication processes firstly started from manufacturing industry for producing standardized products in assembly lines. Recently, digital fabrication has attracted great interest of the construction industry and opened up many possibilities. However, due to the fact that the design-to-fabrication workflow is usually unidirectional, such systems often can't react to unpredictable deviations.

In research, ideas are brought forward that deviate from static "digital chains". Behavioral and cyber-physical fabrication processes are defined by a set of adaptive rules and performative criteria (Brugnaro et al. 2016; Helm et al. 2017; Vercruysse et al. 2018) and afford the ability to deal with uncertainties from the real world (Bruyninckx et al. 2001; Jeffers 2016; Vasey et al. 2014). In order to respond to the uncertainties, the fabrication system needs to be able to sense the as-built structure. An open design system is also necessary in order to react to the updated information (Crolla 2017) and allow adaptive adjustment to the next fabrication tasks (Bruyninckx et al. 2001).

Employing sensory feedback can help to build a more reactive fabrication process. A few researchers have already shown that vision-based sensing systems bring many new possibilities to digital fabrication. The project "Bamboo^3" (Amtsberg and Raspall 2018) has employed vision sensing to get the individual bamboo section geometry in order to customize fittings for predefined joints. The project "Adaptive Part Variation" (Vasey et al. 2014) used vision feedback to make corrections on following elements in order to respond to the fabrication error of a cold bending steel rod structure. The project "Mesh Mould" from ETH (Dörfler 2018) used a vision-based sensing system to give a mobile robot more intelligence to adapt to the unpredictable material performance while welding steel rebars.

2 Research Aim

This research aims to develop an adaptive fabrication workflow for natural bamboo structures, which can compensate for cumulative deviations between the built- and the designed structure (Fig. 2), caused by material uncertainties. Such adaptation suggests an in-progress survey and the corresponding automated adjustment of the following fabrication tasks (Fig. 3), which are also responsible for varying levels of fabrication errors in other construction methods. With such an adaptive workflow, it is possible to efficiently build relatively precise bamboo structures. This opens up novel potentials for bamboo structures by enabling them to predictably interface with accuracy dependent building layers such as facades and roofs, and prefabricated construction elements made from materials such as glass and steel (Fig. 4).

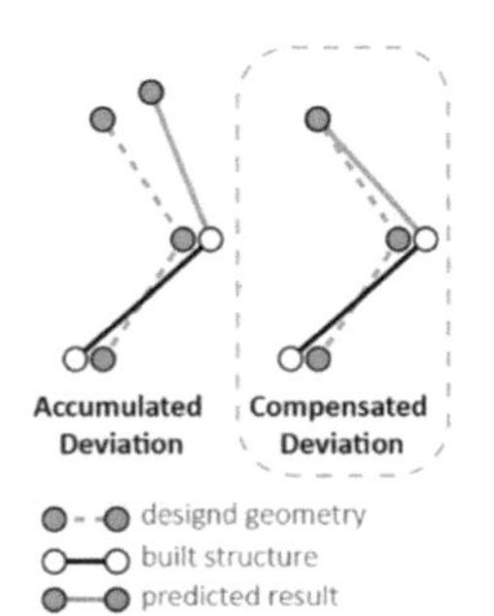

Fig. 2. Deviation can be compensated rather than accumulated

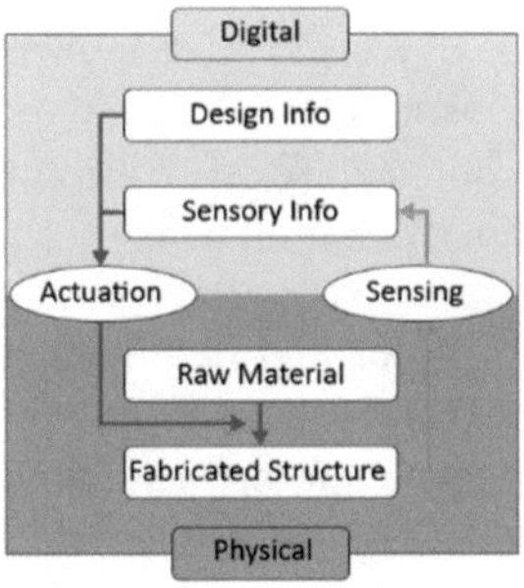

Fig. 3. Cyber-physical information flow of adaptive fabrication process

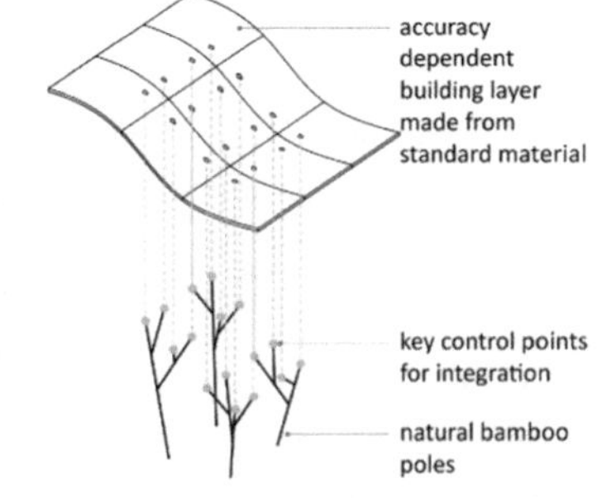

Fig. 4. The accuracy of key control points of the structure enables predictable interfacing between accuracy dependent building layers

3 Methods

The adaptive workflow is composed of multiple iterations of adding bamboo elements onto the existing structure (Fig. 5). Each iteration begins by checking the computational design model, retrieving information about the connection areas and the designed pose (position and orientation) of the next element to be added. The respective connection areas within the built structure and the new bamboo element are both scanned with a depth camera. After estimating geometrical parameters from the sensory data and comparing them with the computational model, the pose of the next element is optimized for compensating the error that occurred in previous iterations. The geometries of related connectors are generated as well. According to that, CNC instructions for fabricating custom connectors and marking connection areas on the bamboo are generated and directly streamed to the fabrication agent. Finally, the bamboo element is assembled onto the existing structure.

The following three sections explain in detail the sensing and adaptation strategies, as well as the connection system used in this project.

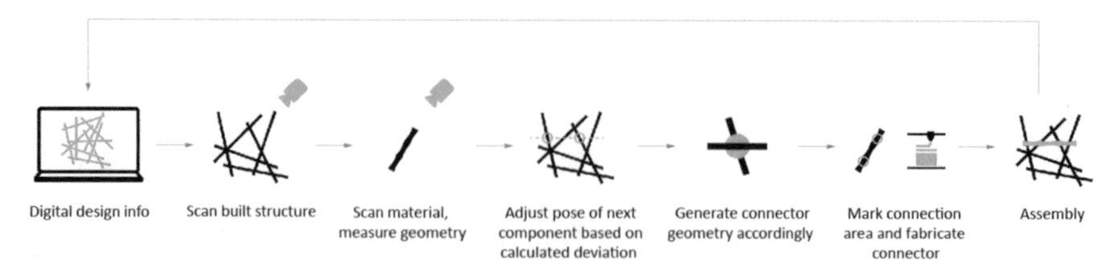

Fig. 5. Fabrication workflow with in-process survey

3.1 Computer Vision

Setup. We use a comparably affordable Intel RealSense D415 as our sensor for initial tests. As an RGB-D camera, it provides color- and depth frames. The color image can be used for detecting objects of interest, and the depth information can be used for constructing 3D point clouds of these objects so that their poses and geometries can be measured. The performance of the sensor is tested with the depth quality tool provided by the RealSense SDK. The RMSE (root-mean-square error) of the depth value is around 0.36% in one meter, which is sufficient for initial tests.

In this project, the resolutions of color and depth image are set to 1920×1080 and 1280×720 respectively. For the 3D reconstruction, the factory-calibrated camera intrinsic parameters are used. During the scanning process (Fig. 6, Fig. 7), the camera is mounted on a tripod and an average of 20 frames of depth images are used to increase the precision of the depth data. A board with an asymmetric circle grid pattern is used for the alignment of the sensory data and the digital model.

Given the images, the task of the computer vision algorithm is to estimate cylinders representing certain parts of the bamboo poles. The cylinder is coded in seven parameters. They are given by a radius, a point on its axis, and a vector along the axis direction. Open source libraries OpenCV (Bradski 2000), Point Cloud Library (Rusu and Cousins 2011) and Ceres-Solver (Agarwal et al. 2018) are used in this project.

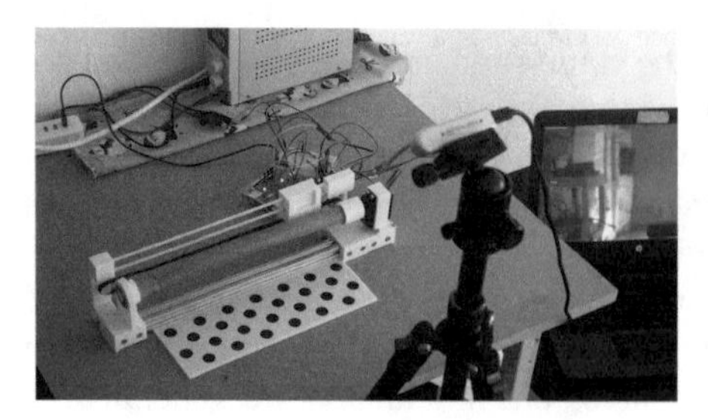

Fig. 6. Scanning bamboo pole

Fig. 7. Scanning built structure

Detection and Segmentation. One important step to extract useful information from the image is to segment it out of other irrelevant data. In our application, pixels of bamboo connection areas need to be segmented. We avoid using data of the whole pole since it leads to local inaccuracy when the pole is globally not straight.

The vision algorithm begins with detecting the circle grid pattern, which provides a reference coordinate system. The centers of those circles are detected by implementing the OpenCV function *findCirclesGrid*. With a known size of the pattern, the camera-to-pattern transformation is solved with the Perspective-n-Point algorithm using the OpenCV function *solvePNP*, with the mode *CV_ITERATIVE*.

Then all the pixels are iterated through and the corresponding 3D points in the reference coordinate system are calculated, so that it can be aligned with the digital model. The 3D point **p** is solved from the equation of the pinhole camera model:

$$d\begin{bmatrix} u \\ v \\ 1 \end{bmatrix} = \mathbf{K}\begin{bmatrix} 1\,0\,0\,0 \\ 0\,1\,0\,0 \\ 0\,0\,1\,0 \end{bmatrix}\mathbf{T}_{co}\mathbf{p}, \tag{1}$$

where d is the depth value, u and v are the pixel coordinates, **K** signifies the camera calibration matrix, and $\mathbf{T}_{\mathrm{co}}$ signifies the object-to-camera transformation matrix.

As the center line and radius of the cylinder segment is roughly known, the distance from the point to the cylinder is calculated and if it is below a specified threshold, it is classified as a point on the cylinder.

The following figures give examples of the results. The reference objects and the bamboo segments (marked purple) are correctly segmented (Fig. 8, Fig. 9) and aligned with the digital model in the 3D-modelling software Rhinoceros (Fig. 11, Fig. 12).

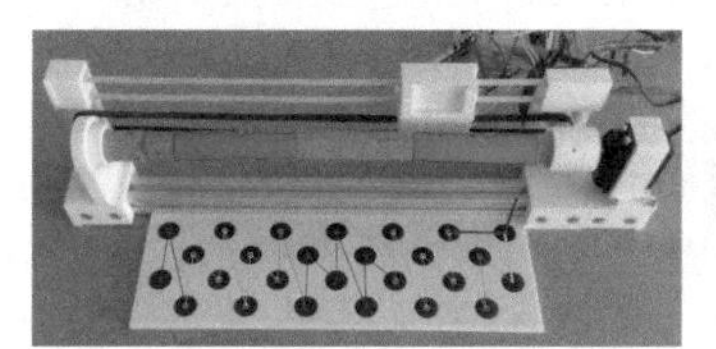

Fig. 8. Detection of bamboo pole

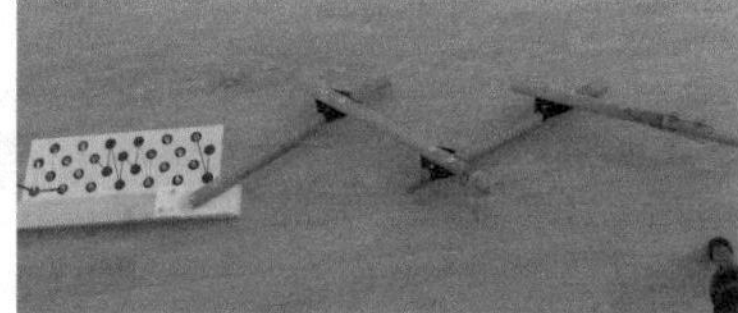

Fig. 9. Detection of built structure

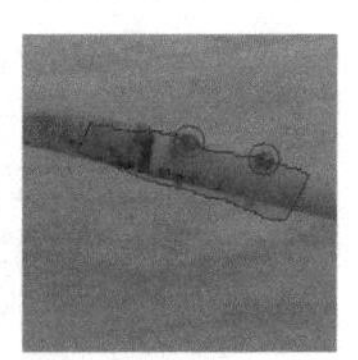

Fig. 10. Detection of drilled holes

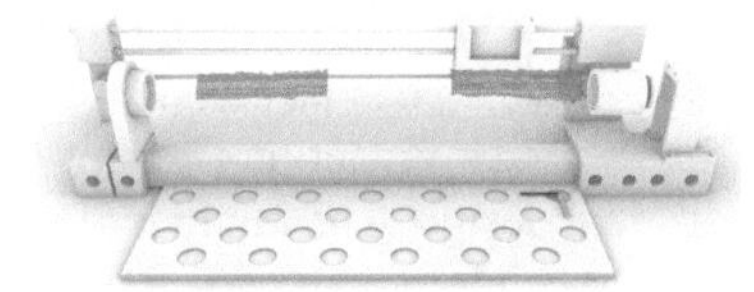

Fig. 11. Aligned point cloud of bamboo pole

Fig. 12. Aligned point cloud of built structure

Apart from that, since the connection system that we use requires drilling on the pole, the positions of holes also need to be sensed and compared with their design positions (Fig. 10). As the holes are small and therefore difficult to detect, we insert a bolt into each hole with its head painted black. A rectangular region of interest is set and those two black heads are detected using circle Hough Transform, which is also an implemented function in the OpenCV library. The two centers are used in the following adaptation process.

Least Squares Cylinder Fitting. After having the point clouds, we use least square method to fit a cylinder representing that bamboo segment into our data. The seven cylinder parameters are given by minimizing the sum of the squared distance from all the points to the cylindrical surface:

$$\mathrm{r}, \mathbf{p}, \mathbf{v} = \mathrm{argmin}_{r,\mathbf{p},\mathbf{v}} \sum_{i=1}^{n} (r - D(\mathbf{p}_i, \mathbf{p}, \mathbf{v}))^2, \quad (2)$$

where r is the radius, $\mathbf{p}$ is the point on the cylinder axis, $\mathbf{v}$ is the vector along its axis, $\mathbf{p}_{\mathrm{i}}$ is the i-th point from the point cloud, $D(\mathbf{p}_{\mathrm{i}}, \mathbf{p}, \mathbf{v})$ signifies the distance function from $\mathbf{p}_{\mathrm{i}}$ to the axis, which is formulated as:

$$D(\mathbf{p}_i, \mathbf{p}, \mathbf{v}) = \frac{|(\mathbf{p}_i - \mathbf{p}) \times \mathbf{v}|}{|\mathbf{v}|}. \quad (3)$$

The nonlinear least squares problem is solved using Ceres-Solver. Before that, random sample consensus (RANSAC) implemented by Point Cloud Library (PCL) is used for removing the outliers as well as providing an initial guess for the seven cylinder parameters.

3.2 Adaptation

After comparing the scan result with the design information, the deviation can be noticed. The pose of the next bamboo element should adapt to the current situation and compensate for the existing error. The concept to achieve that is to move the next bamboo pole to such a location, where it is able to connect with the existing deviated structure while keeping future connection areas as originally designed.

Figure 13 shows one scan result from our experiment, where $\boldsymbol{l}_1$ and $\boldsymbol{l}_2$ signify the cylinder axes (aligned with the design model) given by scan result of the next bamboo element, $\boldsymbol{l}_{\mathrm{b}}$ signifies the cylinder axis given by scan result of the built structure, d_{c} signifies the desired distance between the center lines defined by the connection system, $\boldsymbol{h}_1$ and $\boldsymbol{h}_2$ signify the axes of drilled holes on the built structure. $\boldsymbol{P}_1$ and $\boldsymbol{P}_2$ are two target points that the center-line of the next bamboo pole should ideally pass through. $\boldsymbol{P}_1$ indicates the connection position with the built structure, which can be calculated once $\boldsymbol{h}_1$, $\boldsymbol{h}_2$, $\boldsymbol{l}_{\mathrm{b}}$ and d_{c} are known, while $\boldsymbol{P}_2$ indicates the desired connection position with the future bamboo element, which is given by the original design. The number of target points doesn't need to be limited to two since there might be more than two connections on one pole and the end points of the pole might also be required to reach some specific position in space. When those objectives are conflicting with each other, they can be set to different weights to indicate the priorities.

The adaptation process is formulated again as an optimization problem:

$$T = \mathrm{argmin}_T \sum_{i=1}^{n} (w_i \cdot D(\mathbf{p}_i, T(l_i)))^2, \quad (4)$$

where $\boldsymbol{T}$ signifies the transformation of every line $\mathbf{l}_{\mathrm{i}}$ to its ideal state $\mathbf{l}_{\mathrm{i}}'$, n signifies the number of target positions, w_{i} signifies the weight of the i-th target position and $D(\mathbf{p}, \mathbf{l})$ signifies the distance function from point to line, which is similar to (3).

The problem can be again solved with a least squares solver. However, as all the data are loaded in the Grasshopper3D plugin for Rhinoceros, we used its evolutionary

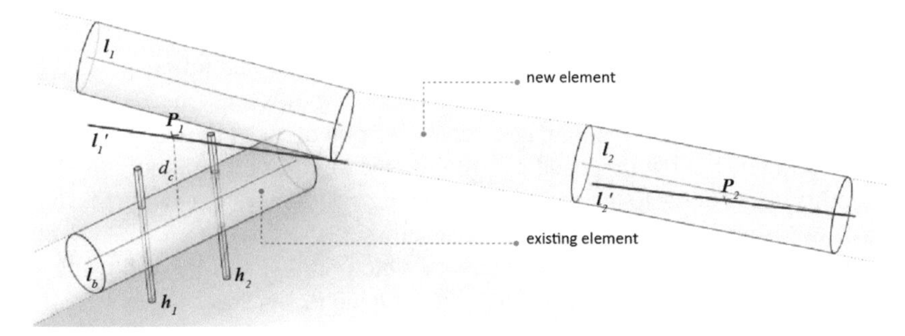

Fig. 13. Geometrical parameters of sensing results used in adaptation process

optimization tool Galapagos for convenience. The rotational part of the transformation is parameterized by axis-angle representation. Together with the translational part, six parameters are optimized with the evolutionary solver.

The result of adaptation provides all the information needed for generating the geometry of a tailor-made connector as well as the locations for drilling the pole.

3.3 Connection

There are many types of connections for natural bamboo structures (Fig. 14). They can be divided into two main groups: traditional and modern. Mortise-tenon joints and lashing joints belong to traditional connection group in conventional bamboo buildings and they are usually used in combination. Modern bamboo buildings have higher requirements on joints, and therefore, metal connectors such as bolting joints and steel member joints are more frequently used. Meanwhile, many other connection possibilities are investigated in recent studies. To name a few, these are CFRP (carbon fiber reinforced plastics/polymer) reinforced joints, wooden-clamp joints and 3D-printed joints (Hong et al. 2019).

Fig. 14. Bamboo connection types. (a) Lashing joint. Erber Research Center, Thailand, 2014 (Image Source: Chiangmai Life Construction/Create Up Co., Ltd. and Markus Roselieb); (b) Bolting joint. Bamboo Bridge, Indonesia, 2016 (Image Source: ASF-ID/Andrea Fitrianto); (c) Steel member joint. Bamboo Pavilion for Expo Shanghai, 2010 (Image Source: Markus Heinsdorff/Tong Ling Feng); (d) 3D-printed joint. Sombra Verde's 3D Printed Bamboo Structure, Singapore, 2018 (Image Source: Airlab @SUTD/Carlos Bañón)

As for this project, the design of connection is also constrained by the adaptive fabrication approach. First, the geometry of the connector should adapt to different angles and distances between two bamboo poles and their different diameters. Second, to enhance the performance of the joint and deploy the potentials of the system's precision, the poles should be held with a bending-stiff connection. For fabrication, this shifts the responsibility of accuracy to the advanced digital fabrication system, while the dexterity of human still is best suited to fasten the connectors.

Figure 15 shows the design of connectors which are used in the following experiments. All of them are 3D-printed with PLA filament, but have the potential to be fabricated out of wood by robotic milling to increase their performance and minimize fabrication time. The geometry is generated automatically by a custom grasshopper script with the parameters of two cylinders as inputs.

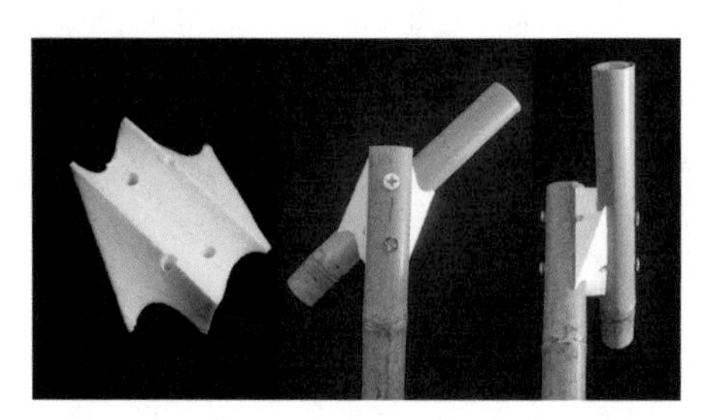

Fig. 15. 3D printed connector prototype

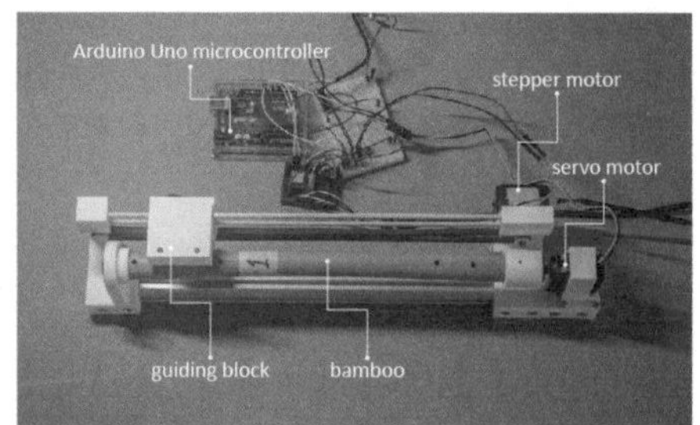

Fig. 16. Low-cost task-specific machine

Two holes need to be drilled on each bamboo pole to be assembled, which can also give indications to human during assembly process. The positions of the holes on the cylindrical surface can be calculated from the cylinder parameters and sent to the CNC machine or industrial robot for the drilling. For this project, we made a two-axis machine (Fig. 16), which is low-cost and specific to this task. A servo motor is used for rotating the bamboo pole into the desired orientation, while a stepper motor together with belt and pulley is used for moving a guiding block to the desired position along the axis of the pole. The precise turning of the bamboo pole and positioning of the guiding block makes it possible to manually drill at required locations. Both motors are controlled by an Arduino Uno microcontroller.

Other connection design options which can avoid drilling on the bamboo pole to keep its intrinsic fiber layout intact are under development, for example, a tailor-made clamping joint. However, without drilling, the connecting areas on the pole need to be properly indicated by other methods.

4 Experiments

4.1 Zigzag Structure

Setup. The first experiment is intended to validate the effectiveness of the above mentioned in-progress survey and adaptation methods. A simple zigzag geometry is chosen as the building goal. It consists of five bamboo components with 30 cm length. The diameter of the pole is ranging from 18 to 22 mm. Two groups of raw bamboo poles are prepared. Both of them are used for building the same designed structure in the same sequence with three different methods: Method A (Fig. 5) is exactly the proposed workflow of this project as described in the method chapter, while Method B and C (Fig. 17) are used to provide references for the performance validation. Unlike Method A, in both Method B and C, in-progress survey is not performed and the measurement of the bamboo poles happens only once before all connectors are fabricated. While Method B employs vision-based techniques to scan the raw material once before assembly, in Method C, the poles are assumed to be straight and their diameters are measured with a caliper manually. The latter is what typically would happen in bamboo construction.

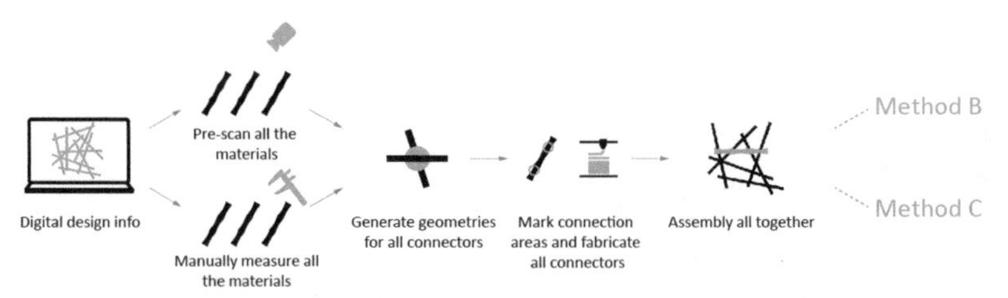

Fig. 17. Method B (pre-scan): material measured with vision-based sensing. Method C (manual measurement): material measured with a caliper. As comparison to proposed method of this project (Method A), both of them don't employ in-progress survey

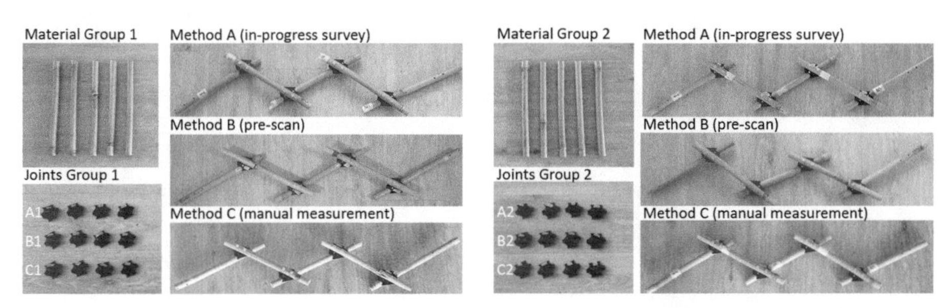

Fig. 18. Raw material, joint nodes and assembled prototypes of group one

Fig. 19. Raw material, joint nodes and assembled prototypes of group two

Result. With two groups of material and three different methods, six zigzag structures are made and documented (Fig. 18, Fig. 19). All of them are again scanned and compared to their designed states. With all the methods, the results (Fig. 20, Fig. 21) showed noticeable deviation from their designed final state. The primary cause of the deviation is likely to be the limited accuracy of the used sensor and computer vision algorithm. However, the results of Method B and C show significantly larger deviation of the fourth and fifth bamboo pole. The center line of every pole is estimated again with the cylinder fitting method, and its position error (distance between midpoints) is plotted (Fig. 22, Fig. 23). In comparison to an overall accumulating trend of error with Method B and C, the errors of Method A fluctuate at a relatively low level as the building process continues, which matches our expectation. A comparison between the results of Method B and C indicates that "pre-scan" helps to reduce the error in magnitude, but cannot prevent its accumulation.

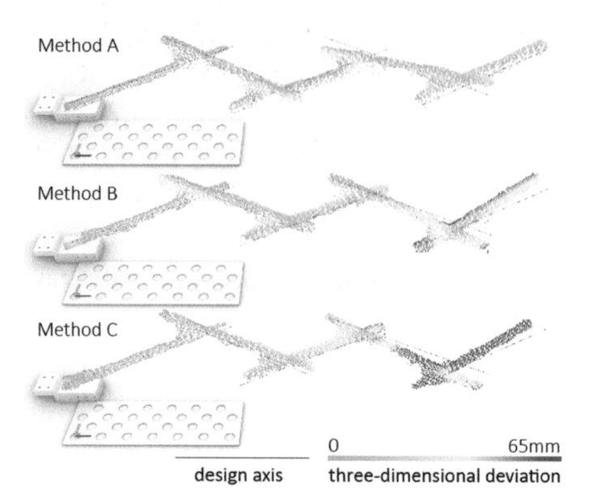

Fig. 20. Result point clouds of group one. Three-dimensional deviation up to 65 mm

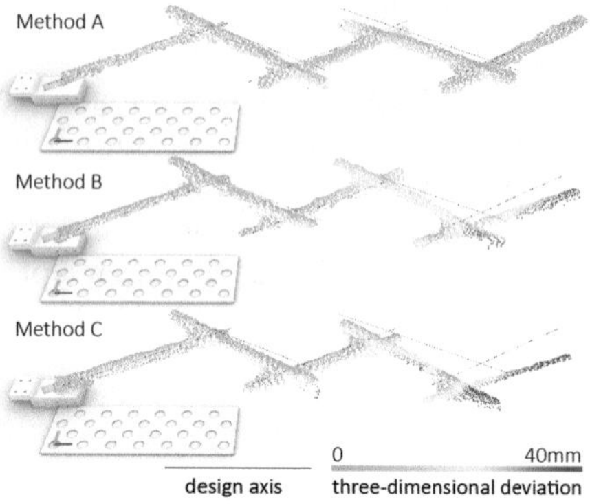

Fig. 21. Result point clouds of group two. Three-dimensional deviation up to 40 mm

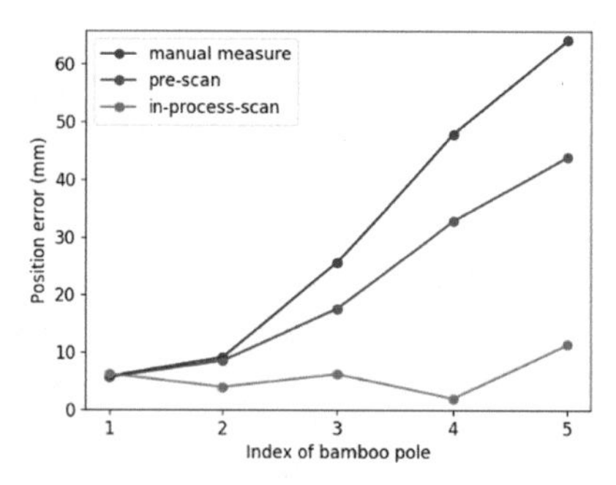

Fig. 22. Position error of group one

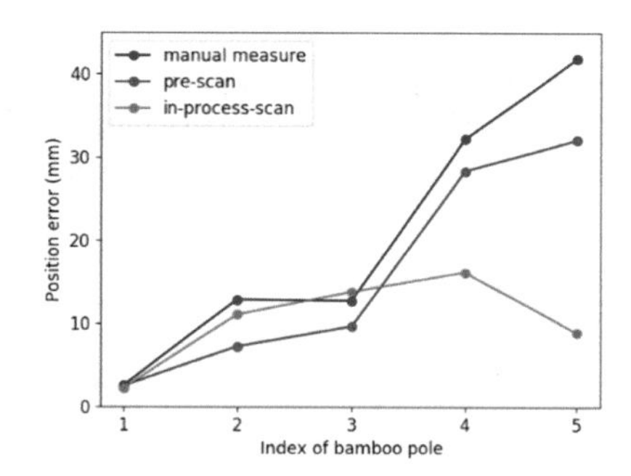

Fig. 23. Position error of group two

4.2 Tree-like Structure

Setup. The second experiment is intended to apply the proposed method to achieve a more complex three-dimensional structure. As the top bamboo poles need to interface with prefabricated elements, high accuracy needs to be achieved. A tree-like structure (Fig. 24) is designed to reach the 4 target points in space. Since prefabricated elements typically are cut in size and drilled in advance, their positions impose a hard constraint on the natural bamboo assembly. After the bamboo structure is fabricated, customized pole-to-plate connectors are 3d-printed according to the final step of sensing. They can correct deviations up to 10.0 mm at the end of the poles.

Result. Figure 24 and 25 show the result of second demonstrator. The structure is 45 cm tall, cover a space of 26.0 cm × 26.5 cm. The average deviation of 4 final points is 8.26 mm according to the final survey. Through adapting the geometries of pole-to-plate connectors to those deviations, the interfacing is successful.

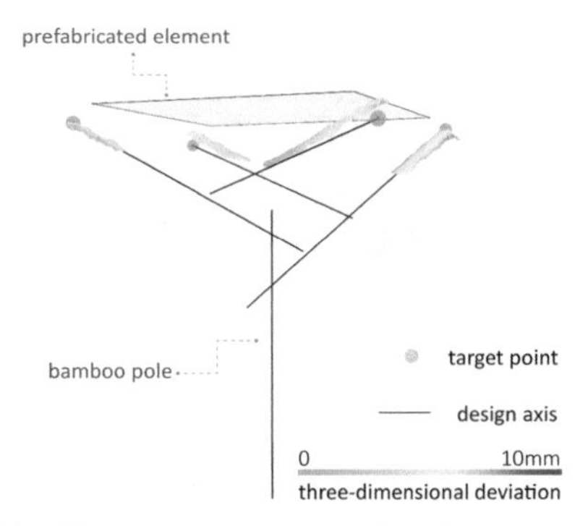

Fig. 24. Designed target points for interfacing with the prefabricated plate. Average deviation of 4 final points is 8.26 mm

Fig. 25. Finished structure. Successfully interfacing with plate through adapting the geometries of pole-to-plate connectors

5 Discussion

In conclusion, we have presented a feedback driven adaptive workflow for efficiently surveying and correcting deviations during the construction of natural bamboo structures. This is an important step towards reliable and predictable digital fabrication methods for large-scale bamboo architecture. The current study is limited by the accuracy of the sensor, detection algorithm, manual drilling and node valence. Still we show that tolerances can be significantly reduced even with low-cost sensors and equipment. Such approach can lead to predictable and semi-automated bamboo constructions, which is hardly achieved in other state-of-the-art bamboo fabrication workflows (cf. Crolla 2017; Amtsberg and Raspall 2018). The application of bamboo in construction can thus be expanded (Fig. 26). For instance, it may enable bamboo structures to interface with some prefabricated components made of wood, metal or glass, so that they can be used as building substructures for roofs and facades, or even be used on top of existing structures and fit them well. It may also enable the construction of bamboo structures in complex shapes (e.g., space frames) efficiently and reliably. To demonstrate the approach in full architectural scale, further development is necessary in areas of sensing hardware, digital feedback automation, and intuitive human-machine interfaces for on-site information visualization.

Fig. 26. Application scenarios of bamboo pole based structures integrated with conventional building materials. (a) Substructure of bridge; (b) Structure in multi-story construction; (c) Facade substructure; (d) Space frame

Acknowledgements. The presented work was conducted within the ITECH master thesis program at the University of Stuttgart, Germany. We would like to thank the following for their support and feedback during the course of research: the Integrative Technologies and Architectural Design Research (ITECH) program, researchers at the Institute for Computational Design and Construction (ICD) and Institute for Building Structures and Structural Design (ITKE), researchers at Institute for Control Engineering of Machine Tools and Manufacturing Units (ISW). This work was partially supported by the State of Baden-Wuerttemberg, the European Regional Development Fund and by the German Research Foundation under Germany's Excellence Strategy – EXC 2120/1 – 390831618.

References

Agarwal, S., Mierle, K., et al.: Ceres Solver (Version 1.14) [Open Source Code] (2018). http://ceres-solver.org

Amtsberg, F., Raspall, F.: 'Bamboo3'. In: Fukuda, T., Huang, W., Janssen, P., Crolla, K., Alhadidi S. (eds.) Learning, Adapting and Prototyping, Proceedings of the 23rd International Conference of the Association for Computer-Aided Architectural Design Research in Asia (CAADRIA), Tsinghua University, Beijing, China, 17–19 May 2018, pp. 245–254 (2018)

Atanda, J.: Environmental impacts of bamboo as a substitute constructional material in Nigeria. Case Stud. Constr. Mater. **3**, 33–39 (2015)

Bhalla, S., West, R.P., Bhagat, D., Gupta, M., Nagpal, A.: Pre-engineered bamboo structures: a step towards sustainable construction. In: Sivakumar Babu, G.L., Saride, S., Basha, B.M. (eds.) Sustainability Issues in Civil Engineering. Springer Transactions in Civil and Environmental Engineering, pp. 339–344. Springer, Singapore (2017)

Bradski, G.: The OpenCV library. Dr. Dobb's J. Softw. Tools **25**, 120–125 (2000)

Brugnaro, G., Baharlou, E., Vasey, L., Menges, A.: Robotic softness: an adaptive robotic fabrication process for woven structures. In: ACADIA. 2016: Posthuman Frontiers: Data, Designers, and Cognitive Machines, Proceedings of the 36th Annual Conference of the Association for Computer Aided Design in Architecture (ACADIA), Michigan, 27–29 October 2016, pp. 154–163 (2016)

Bruyninckx, H., Lefebvre, T., Mihaylova, L., Staffetti, E., De Schutter, J., Xiao, J.: A roadmap for autonomous robotic assembly. In: Proceedings of the 2001 IEEE International Symposium on Assembly and Task Planning (ISATP2001). Assembly and Disassembly in the Twenty-first Century. (Cat. No. 01TH8560), Fukuoka, Japan, 28–29 May 2001, pp. 49–54 (2001)

Churkina, G., Organschi, A., Reyer, C.P.O., Ruff, A., Vinke, K., Liu, Z., Reck, B.K., Graedel, T.E., Schellnhuber, H.J.: Buildings as a global carbon sink. Nat. Sustain. **3**, 269–276 (2020)

Crolla, K.: Building indeterminacy modelling – the 'ZCB Bamboo Pavilion' as a case study on nonstandard construction from natural materials. Vis. Eng. **5**(1), 1–12 (2017). https://doi.org/10.1186/s40327-017-0051-4

Dörfler, K.: Strategies for Robotic in Situ Fabrication. ETH Zurich (2018)

Helm, V., Knauss, M., Kohlhammer, T., Gramazio, F., Kohler, M.: Additive robotic fabrication of complex timber structures. In: Menges, A., Schwinn, T., Krieg, O.D. (eds.) Advancing Wood Architecture: A Computational Approach, pp. 29–44. Routledge, Taylor & Francis Group, London (2017)

Hong, C., Li, H., Lorenzo, R., Wu, G., Corbi, I., Corbi, O., Xiong, Z., Yang, D., Zhang, H.: Review on connections for original bamboo structures. J. Renew. Mater. **7**(8), 713–730 (2019)

Huang, Z. (ed.): Application of Bamboo in Building Envelope. Green Energy and Technology. Cham, Springer (2019)

Jeffers, M.: Autonomous Robotic Assembly with Variable Material Properties. In: Reinhardt, D., Saunders, R., Burry, J. (eds.) Robotic Fabrication in Architecture, Art and Design 2016. STCEE, pp. 48–61. Springer, Cham (2016)

Lorenzo, R., Mimendi, L.: Digital workflow for the accurate computation of the geometric properties of bamboo culms for structural applications. In: MATEC Web of Conferences, vol. 275 (2019). https://doi.org/10.1051/matecconf/201927501024

Rusu, R.B., Cousins, S.: 3D is here: Point Cloud Library (PCL). In: IEEE International Conference on Robotics and Automation (ICRA), Shanghai, 9–13 May 2011, pp. 1–4 (2011)

Self, M., Vercruysse, E.: Infinite variations, radical strategies. In: Menges, A., Sheil, B., Glynn, R., Skavara, M. (eds.) Fabricate: Rethinking Design and Construction, pp. 30–35. UCL Press, London (2017)

Vasey, L., Maxwell, I., Pigram, D.: Adaptive part variation. In: McGee, W., de Ponce Leon, M. (eds.) Robotic Fabrication in Architecture, Art and Design 2014. Springer Transactions in Civil and Environmental Engineering, pp. 291–304. Springer, Cham (2014)

Vercruysse, E., Mollica, Z., Devadass, P.: Altered behaviour: the performative nature of manufacture Chainsaw Choreographies + Bandsaw Manoeuvres. In: Willmann, J., Block, P., Hutter, M., Byrne, K., Schork, T. (eds.) Robotic Fabrication in Architecture, Art and Design 2018, pp. 309–319. Zürich, Springer (2018)

Freeform Volumetric Fabrication Using Actuated Robotic Hot Wire Cutter

Paul Loh[1(✉)], Yuhan Hou[1], Chun Tung Tse[1], Jiaqi Mo[1], and David Leggett[2]

[1] Melbourne School of Design, University of Melbourne, Melbourne, Australia
paul.loh@unimelb.edu.au
[2] Architectural Research Laboratory, Melbourne, Australia

Abstract. This paper discusses the design, fabrication and operational workflow of a novel hot-wire cutter used as an end effector for a robotic arm. Typically, hot wire cutters used a linear cutting element which results in ruled surfaces geometry. While several researchers have examined the use of hot wire cutter with cooperative robotic arms to create non-ruled surface geometry, this research explores the use of an actuated hot wire cutter manoeuver by a single robotic arm to produce similar form. The paper outlines the machine making process and its workflow resulting in a 1:1 scale prototype. The paper concludes by examining how the novel tool can be applied to an urban stage design. The research set up a fabrication procedure that has the potential to be deployed as an on-site fabrication methodology.

Keywords: Robotic · Digital fabrication · Hot-wire cutting · Machine making

1 Introduction

The paper outlined the design and fabrication of a novel hot wire cutter used as an end effector for a robotic arm to create non-ruled surface geometry. Most hot wire cutters used a linear cutting element to achieve single ruled surfaces geometry (Bidgoli and Llach 2015). Rust et al. (2016) have explored using cooperative robotic arms to develop non-ruled surface geometry with linear hot wire cutting element. This research furthers the enquiry within the field by exploring non-linear cutting element for a hot wire cutter and thereby achieving non-ruled surface geometry with a single robotic arm. The novel tool utilises a curved hot wire cutting element with up to three controlled nodal points to shape Expanded Polystyrene (EPS). The actuated Interpolating Polynomial-Curve Hot Wire Cutter (IP-CHWC) used only a single robotic arm to achieve complex network surface geometry. Unlike conventional hot wire cutter, the IP-CHWC is numerically controlled and can perform variable cut sections with a single robotic movement to produce complex surface. In this research, the end-effector is used to cut EPS which is coated with resin-infused fibreglass to create a durable surface for architecture or sculpture application. The advantage of the novel end-effector provides a more significant degree of freedom for manufacturing complex form without the need for two collaborative robots.

P. F. Yuan et al. (Eds.): CDRF 2020, *Proceedings of the 2020 DigitalFUTURES*, pp. 280–289, 2021.
https://doi.org/10.1007/978-981-33-4400-6_26

The research is conducted through physical prototyping of the end-effector resulting in several full-scale furniture prototypes. We observed the potential of the fabrication techniques and tested the new tool through a design project: an urban performance stage design titled Motion Imprint. Critical to the design process is the integration of knowledge from the tooling to the design workflow. Later sections discuss the data extraction procedure and fabrication workflow as well as the difficulties encountered during the design of the end effector.

The design team utilised the geometric nature of an interpolating polynomial curve as a starting position. The IP-CHWC translates the parametric curve into actuated behaviour to articulate the cutting profiles. Here, the geometric description is aligned with material behaviour to develop a productive mechanical system. The paper concludes by examining future research and applications associated with the novel tool.

2 Background

Numerous researchers have explored hot wire cutting technique using a robotic arm (Pigram and McGee 2011; Martins et al. 2019; Kaftan and Stavric 2013). The technique has proven to be an effective fabrication methodology that offers a low cost and flexible production of non-standard volumetric components that could be used for prototyping mound for casting panels (Martins et al. 2019). Typically, nichrome wire is used as a cutting element which is stretched taut over a supporting frame. A spring mechanism is used to ensure there is no slack in the wire when it is heated. An alternative technique to cut EPS is to use a thicker gauge nichrome wire and bend it into shape as a sculpting knife. This technique is further developed by using two cooperative robotic arms to create a non-ruled surface (Søndergaard et al. 2016). In this research project, the two cutting techniques are combined into a single end-effector for robotic cutting.

2.1 Research Objectives

The IP-CHWC utilised the basic principle of interpolating polynomial curves to create a UV network surface that is not based on ruled surface geometry (Pottman 2007, Burry 2011). The IP-CHWC determines the horizontal curves (U) while a set of design parameters determines the vertical curves (V), see Fig. 1. The technique relies on the hot wire to be shaped into a desired interpolating curve. As the wire passes through the EPS, IP-CHWC continues to transform its curvature. The resulting cut is a smooth and non-developable surface.

3 Interpolating Polynomial Curve

The hot wire behaves like an interpolating polynomial curve. Mathematically, such curve used Lagrange polynomials as a basic polynomial; Lagrange polynomials force the curve to pass through the nodal points, unlike a Bezier curve. Lyche and Morken (2008) highlight the difference between the property of the curve and its parametric representation. They define a curve as the collection of all the different parameter representations of

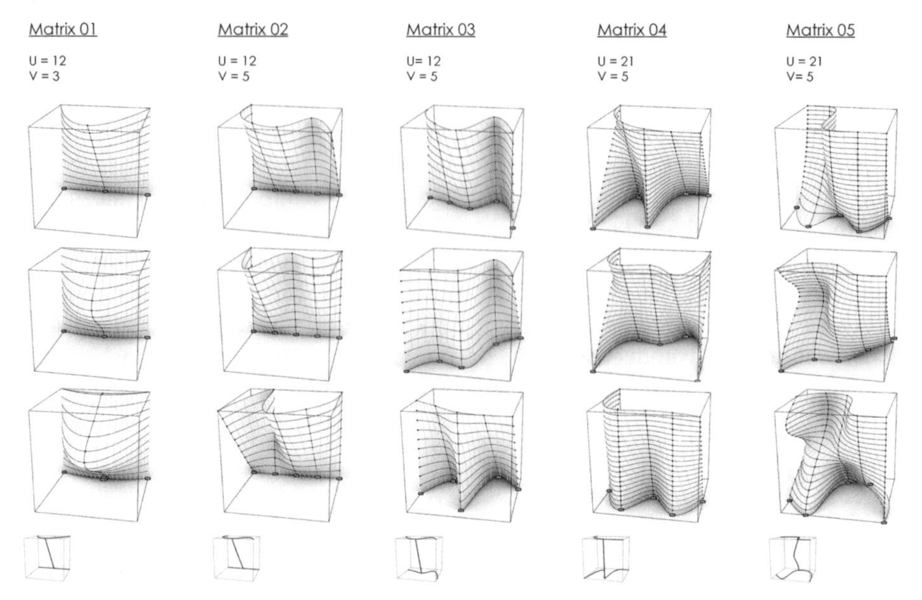

Fig. 1. Matrix showing variable surface based on five-point interpolating polynomial curves to form network surfaces

the curve. Here, it is essential to distinguish the two as in this project; both are present simultaneously to provide the geometric description of the resulting form. The parametric curve is the underlying principle of the machine as the vector along the line of the hot wire connected at the nodal points. The resulting curve exists in the digital simulation of the form as a series of UV network surface.

4 Design of IP-C Hot Wire Cutter (IP-CHWC)

Six hot wire cutter prototypes are developed during the research, see Fig. 2. Of the six prototypes, V3 and V4 utilised tension only to shape the nichrome wire while the remaining used a combination of tension and rigid armatures to control the curvature. The connection points between the nichrome wires act as the nodal points which control the curvature of the cutting shape. It became apparent that a series of hot wire knife acting in compression is needed to allow the cutter to slice through the dense EPS. The hot wire knife enables both pushing and pulling of the nodal point.

Figure 3 illustrates the current design of the IP-CHWC. The end effector is connected to the robotic arm via a connector [A]. The aluminium frame [B] holds three stepper motors [C] which each drives a corresponding hot wire knife [D]. The outer frame holds the 0.4 mm thick nichrome wire [E] retracted by two further stepper motor [F]. The nichrome wire passes through a pin-head size hole at the end of the hot wire knife [G].

The difficulty in this research lies in the design of the bespoke hot wire knife, see Fig. 3 (Right). The knife-edge is lined with a 1 m long (1.5 mm thick) nichrome wire, adhered to the edge with a high melting point silicone mastic. With a 12 V power supply, the cutting knife reaches a temperature of 150 °C. The current design of the knife is

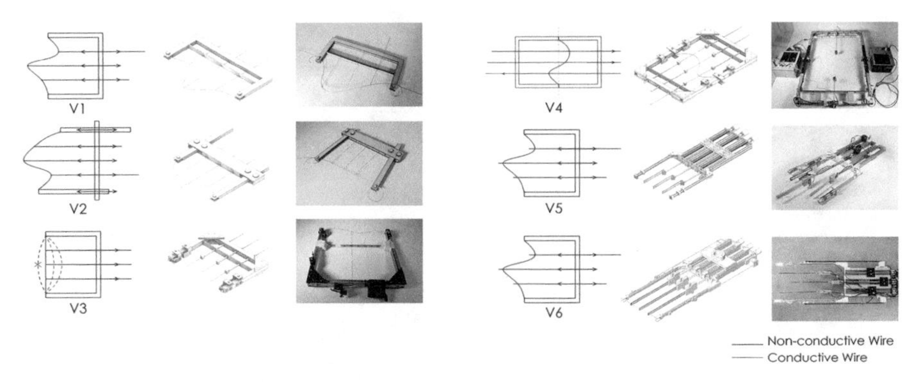

Fig. 2. Prototypes developed from the research, V6 is the most developed working prototype

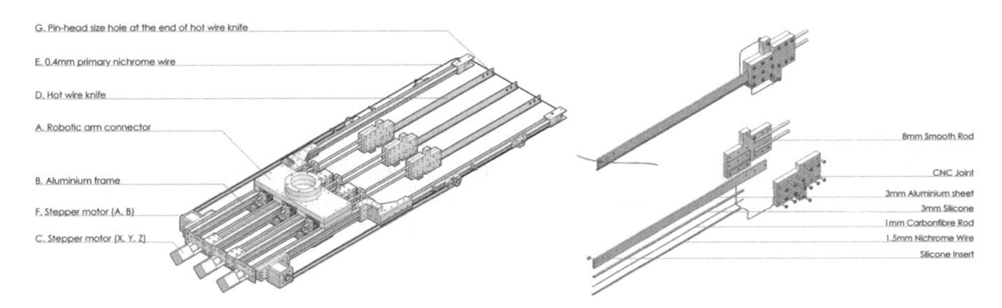

Fig. 3. Left, design of V6 prototype. Right, detail of hot wire cutting knife

temperamental as when the wire is heated, and it expands faster than the silicone mastic. If the nichrome wire [E] accidentally meets the knife [D], it causes a short circuit which would foul the machine.

5 Parametric System and Mechanisms

The fabrication procedure translates the UV network surface into G-code for the stepper motors and the robotic movement, see Fig. 4. Using KUKA PRC, the movement of the robotic arm with variable velocities can be precisely scripted and simulated before cutting the EPS. Firefly plugin for Grasshopper 3D is used to transform the nodal point of the UV curves which forms the network surface geometry into G-codes. To operate the five stepper motors simultaneously, we use a Girbl-Panel controls Arduino Mega R3 with Ramps 1.4.

A given form is typically sectioned to produce a set of contour curves on the X-Y plane, for example from wire 01(W01) to wire 104(W104), see Fig. 5. The profile of the U curve contains the same number of nodal points as the IP-C cutter, in this case, three only. The number of U lines decided the number of commands for steppers. A and B refer to the two-steppers which adjust the length of the primary nichrome wire, while X, Y and Z represent each of the three transforming nodal points driven by three cutting knives. The operation time frame is set as 12 s per command. The displacements

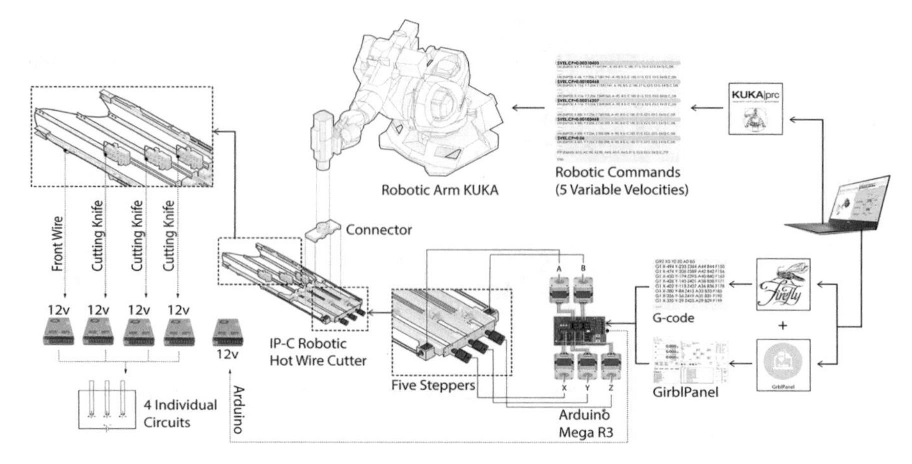

Fig. 4. Parametric system and mechanisms

required for X, Y, Z, A and B to move to the next position are different. However, they are based on identical operation time, as variable feed rate (FX, FY, FZ, FA and FB) across the surface. In other words, to translate the movement from W1 to W35, X, Y and Z will have variable feed rates as illustrated in Fig. 5.

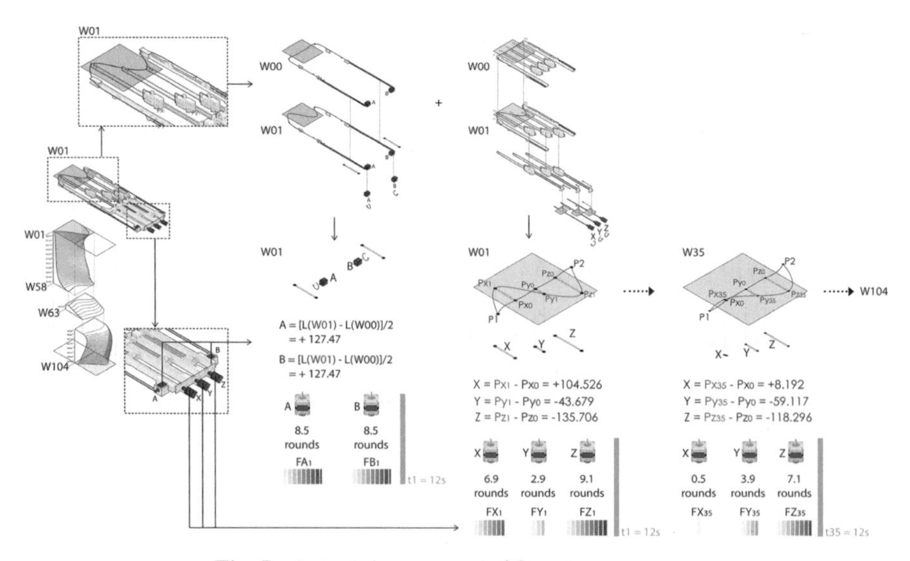

Fig. 5. Actuated movement of five stepper motors

6 Fabrication Workflow

Figure 6 illustrates the overall workflow of the technique. Through evaluation of the geometry input [A-B], the workflow synchronised the actuation of the IP-CHWC [D-F] with the robotic workflow [E-G] to produce a fabricated outcome [H-I]. The following sections will unfold these procedures in detail.

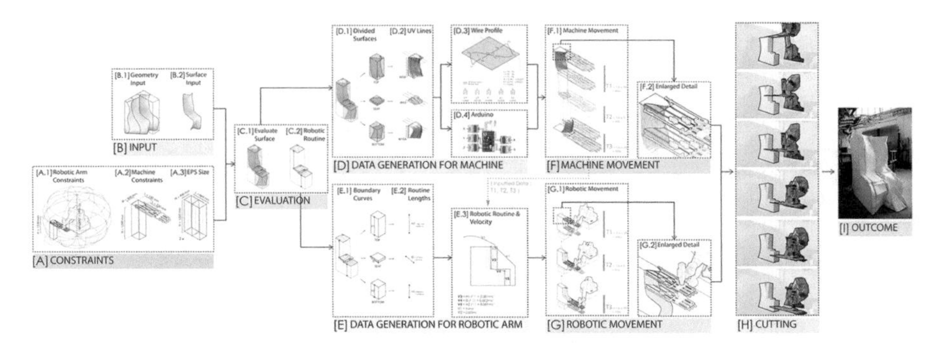

Fig. 6. The overall workflow of IP-CHWC

6.1 Geometry Input and Evaluation

The IP-CHWC and the robotic arm both provide constraints to the size of the EPS stock, see Fig. 7[A]. The height of the stock is limited by the maximal reach of Kuka KR120 R2500 [A.1]. The frame width of IP-CHWC determines its width (<300 mm), while the stock length (<330 mm) is constraint by the length of the cutting knives [A.2]. Here, we test the IP-CHWC to fabricate an outdoor seat design on an EPS stock: 1200 mm deep x 400 mm wide x 1200 mm tall [A.3]. The digital freeform surface is divided to suit the stock size, Fig. 7[B]. Through evaluating the curvature of the surface, the robotic cutting procedure is divided into 3 continuous parts: top, seat and bottom, as T1, T2 and T3 respectively, see Fig. 7[C].

6.2 Actuated Movement of the IP-CHWC

The intersection of the UV curves forms the nodal point of the cutting profile, Fig. 8 [D]. Driven by the movement of the cutting knives' displacement, each U curve is translated into a line of command for the IP-CHWC. The total cutting time of T1, T2 and T3 are calculated on the number of command lines over the operation period (set as 12 s for this research), see [F.1]. The cutting time of T1, T2 and T3 are used as inputs to calculate the robotic velocity.

6.3 Synchronisation of IP-C with Robotic Arm

The boundary curve of the surface (H1, D and H2) is used to generate the robotic movement in Kuka PRC, see Fig. 9[E.1 and E.2]. The travelling speed is critical as

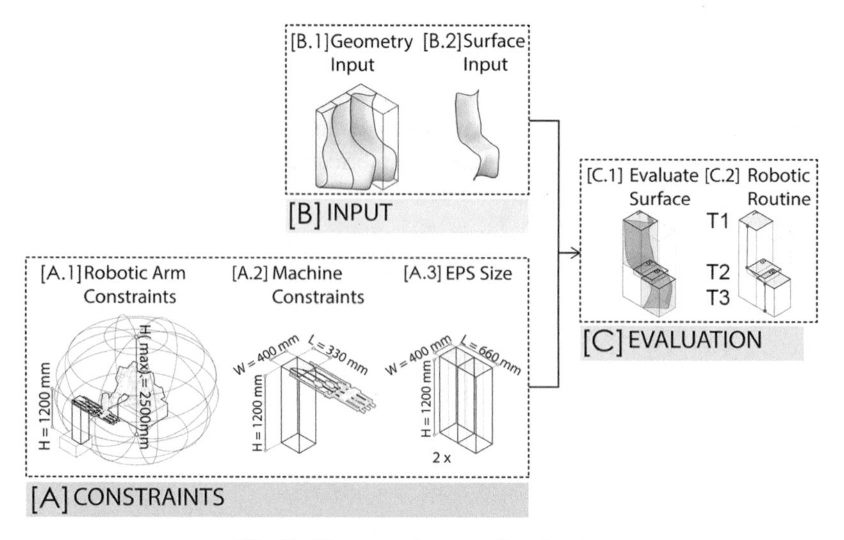

Fig. 7. Geometry input and evaluation

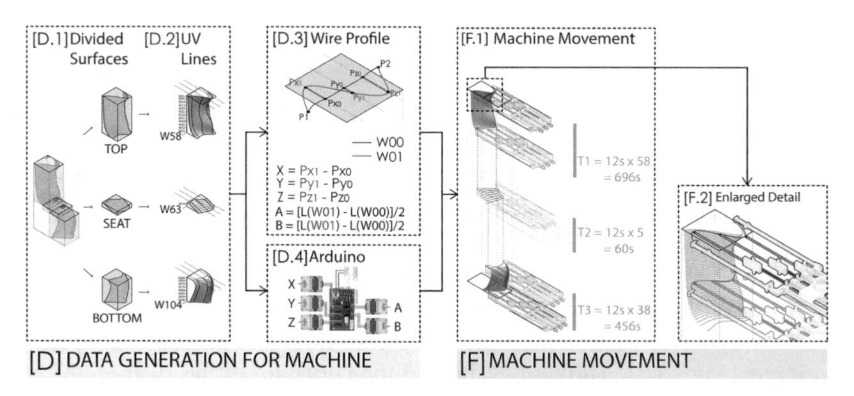

Fig. 8. Data generation for the IP-CHWC

it needs to synchronise with steppers motion described in Sect. 5.2. Thus, the robotic velocity V3, V4 and V5 for cutting the continuous parts are set by dividing the lengths of routine (H1, D and H2) with same travelling time corresponding to T1, T2 and T3 [E.3]. The synchronised movement of IP-CHWC and the robotic arm is further simulated in SprutCAM before proceeding to actual cutting [G].

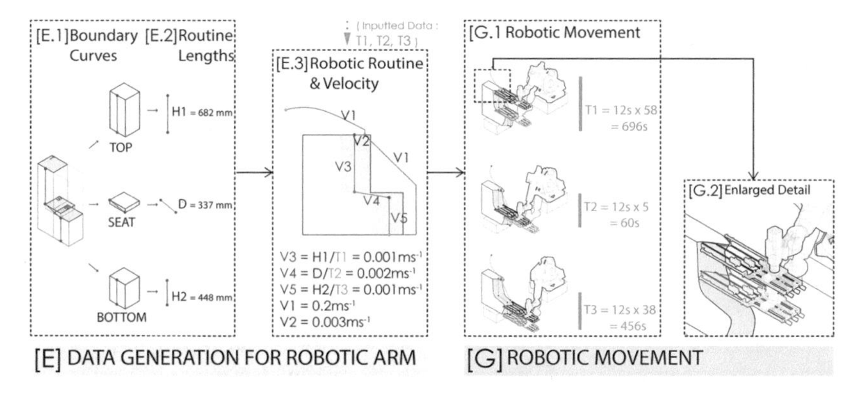

Fig. 9. Data generation for the robotic movement

7 Future Enquiries and Potential

Current research has identified several limitations of the system. Future research will explore an increase cut edge to the boundary to avoid a framing of the cut profile; this enables a more continuous curvature between cuts. The joint between the knife cutter and the frame needs to stiffen significantly to avoid the cutting knife from twisting and bending under the force of the robotic movement. Another significant improvement will be to re-design the circuit to minimised uneven heat distribution and potential for short-circuiting of the system.

7.1 Design Implication: Radical on-Site Fabrication

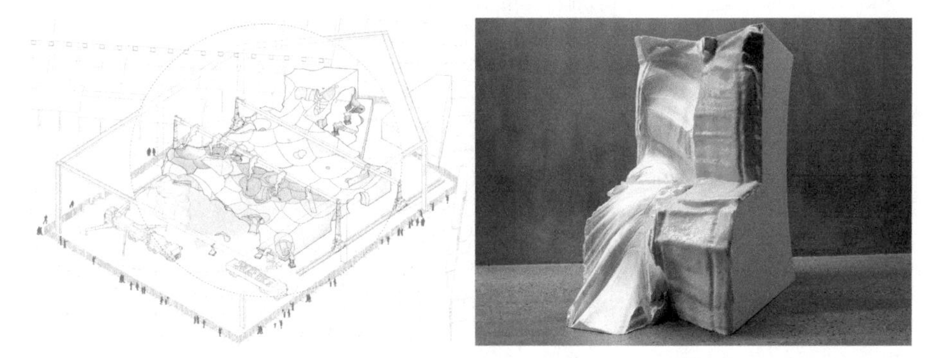

Fig. 10. Left, construction sequence for on-site fabrication. Right, 1:1 scale prototype with the resin-infused coating to cut EPS

The design team tested the tools in a speculative design, titled Motion Imprint. The proposal is an urban performance stage located in the city of Melbourne. Using Kuka KR120 working envelope as a design constraint, we speculate on how the design can be fabricated based on the limitations and opportunities of the tool through an on-site fabrication scenario, see Fig. 10. The cut EPS is coated with a 2-part polyurea which is then covered with fibreglass and infused with resin. A 1:1 scale prototype following the geometric description is developed as a proof-of-concept, Fig. 10 (Right).

8 Conclusion

The research presents a novel hot wire cutter that has the potential to produce non-ruled free form curved surface using a single robotic arm. The design team aligns both the material system necessary to construct the device with the fabrication workflow to extract data from the parametric curve. In this research, the parametric curve is used to drive the time-based transformation of cutting profile to create an intricate doubly curved surface. Data is processed through the electronic system, scripted and transferred into mechanical movement. The synchronised movement of the robot arm and IP-C hot wire cutter enable the production of a complex form with better time efficiency than standard CNC milling procedure currently used in the construction industry. The research speculates on how IP-CHWC could be adopted as on-site construction methodology, allowing design, fabrication and building construction to be fully integrated into a unified system.

Acknowledgement. The team likes to acknowledge the technical contribution of Ryan Yuanye Huang and the structural advice from Sascha Bohnenberger of Bollinger-Grohmann.

References

Bidgoli, A., Cardoso-Llach, D.: Towards a motion grammar for robotic stereotomy (2015)

Burry, M.: Scripting Cultures: Architectural Design and Programming. Wiley, New York (2011)

Kaftan, M., Stavric, M.: Robotic fabrication of modular formwork. In: International Conference of the Association for Computer-Aided Architectural Design Research in Asia. Research Publishing Services (2013)

Lyche, T., Morken, K.: Spline methods draft. Department of Informatics, Center of Mathematics for Applications, University of Oslo, Oslo (2008)

Martins, P.F., Nunes, S., Fonseca de Campos, P., Sousa, J.P.: Rethinking the Philips Pavilion through robotic hot wire cutting. -An experimental prototype (2019)

Pigram, D., McGee, W.: Formation Embedded Design: a methodology for the integration of fabrication constraints into architectural design (2011)

Pottmann, H.: Architectural geometry. Bentley Institute Press (2007)

Rust, R., Jenny, D., Gramazio, F., Kohler, M.: Spatial Wire Cutting: Cooperative robotic cutting of non-ruled surface geometries for bespoke building components (2016)

Søndergaard, A., Feringa, J., Nørbjerg, T., Steenstrup, K., Brander, D., Graversen, J., Markvorsen, S., Bærentzen, A., Petkov, K., Hattel, J., Clausen, K., Jensen, K., Knudsen, L., Kortbek, J.: Robotic hot-blade cutting. In: Reinhardt, D., Saunders, R., Burry, J. (eds.) Robotic Fabrication in Architecture, Art and Design 2016, pp. 150–164. Springer, Cham (2016)

Material Disruption

Mary Spyropoulos(✉) and Alisa Andrasek(✉)

Royal Melbourne Institute of Technology, Melbourne, Australia
s3488461@student.rmit.edu.au, alisa.andrasek@rmit.edu.au

Abstract. This paper examines the role of computational simulation of material processes with robotics fabrication, with the intent of examining its implications for architectural design and construction. Simulation techniques have been adopted in the automotive industry amongst others, advancing their design and manufacturing outputs. At present, architecture is yet to explore the full potential of this technology and their techniques. The need for simulation is evident in exploring the behaviours of materials and their relative properties. Currently, there is a distinct disconnect between the virtual model and its fabricated counterpart. Through research in simulation, we can begin to understand and clearly visualize the relationship between material behaviours and properties that can lead to a closer correlation between the digital design and its fabricated outcome. As the first phase of investigation, the material of clay is used due to its volatile qualities embedded within the material behaviour. The input geometry is constrained to rudimentary extruded forms in order to not obscure the behaviour of the material, but rather allow for it to drive the machine-making process.

Keywords: Material behaviour · Material systems · Simulation · Complex geometries · Computation · Additive manufacturing · Robotic fabrication

1 Context/Introduction

Currently, architecture is constructed through static modes of inhabitation and engagement originating from the use of the same fundamental materials and construction techniques that were being practiced in the 19th and 20th centuries. In the present-day landscape of the construction sector, there is great potential for innovation through the greater exploration of materiality and embedding big data and new information networks into material-based organized systems of architecture. The expansion into technology enhanced approach aims to challenge the current construction industry, which has become one of the least efficient industries in regard to material usage, material waste and durability. Buildings that are constructed today are being built faster and cheaper as a result of economic competition. The byproduct of this process results in buildings that are more static, material-dense, less durable and less material intelligent then the current technology and knowledge would allow for.

Through collaboration with multidisciplinary research such as material science and advanced manufacturing, this research looks at the specificity of embedding various data systems into material compositions. In conjunction, the notions of integrated fabrication

P. F. Yuan et al. (Eds.): CDRF 2020, *Proceedings of the 2020 DigitalFUTURES*, pp. 290–296, 2021.
https://doi.org/10.1007/978-981-33-4400-6_27

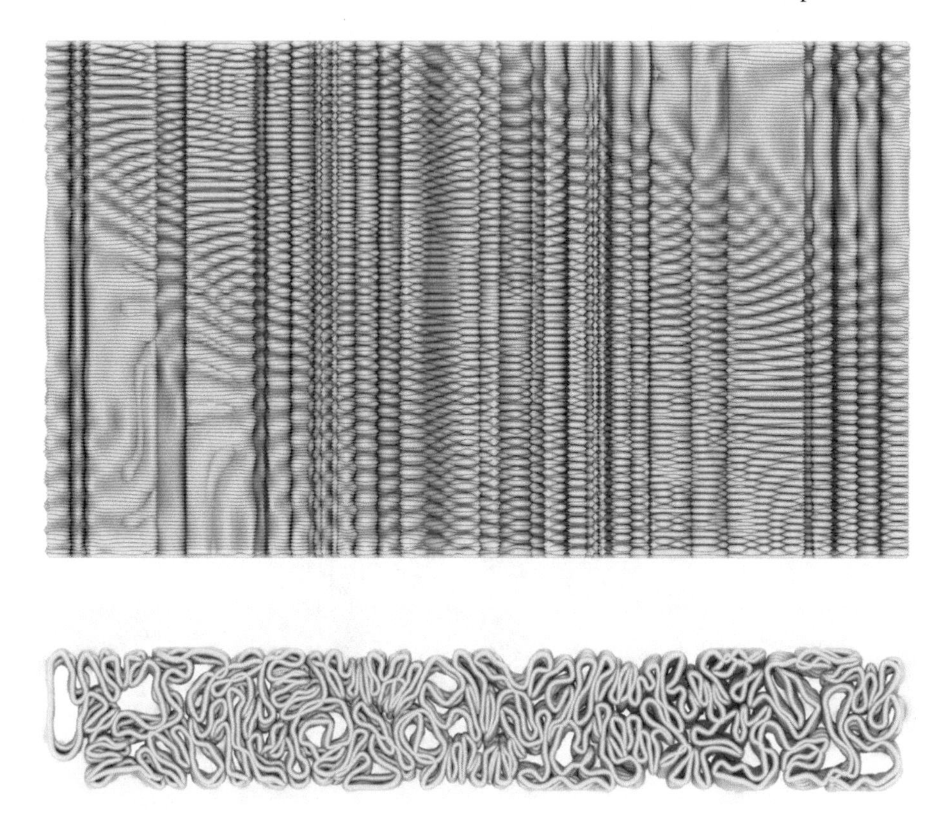

Fig. 1. Wall prototype

techniques, such as robotics and machine vision, will begin to operate in negotiation with materials to evolve fabrication logics and capabilities. This will allow for enriched material palette, with new materials designed by material science in vitro, and existing materials redeployed with novel capacities, for expanded possibilities for architectural applications.

2 Raw Material Behaviour

The initial testing for this research sets out to use only constrained rudimentary geometry. This technique is used to privilege the behaviour of the material over a predetermined geometrical form. Minimal cylindrical and rectangular shapes provide an unobscured base for robotic toolpaths. This allows for tracking of qualitative changes in material behavior and uncovering its capacity to be a driver of dominant geometric features.

Programmability of tool paths allows for variations in speed, rate of extrusion, and extruder nozzle diameter during the robotic printing process. In conjunction to this process tool paths are further manipulated through algorithms that add noise and vibration amplifying the x and y axis of the path. Due to the non-linear nature of clay material, each factor contributes to the varying results produced, from initial testing of material behaviour alone.

Fig. 2. The collapse

Analysing material behavior resulting from programmable fabrication is providing feedback data for the simulation process. There are many and varying qualitative outcomes embedded within each individual printed outcome. These initial tests provided an insight into the various effects of directionality and slumping which occur through manipulating the toolpaths layer height in conjunction with how fast the robot is printing. Toolpaths were also manipulated through noise algorithms to inform where the deposition of clay would occur. Introducing more algorithmic noise into the toolpath would result in greater material deposition by the robot. This produces emergent qualities of slumping, bulging, weaving and pleating. Tolerance to cantilevering is particularly relevant factor for structural stability. This is particularly evident in Fig. 2 which formed a complete collapse due to pushing the limits of how far the material could support itself when cantilevered.

In less noise manipulated areas directionality is embedded in the printing process through the rate at which the robot would distribute the material. This is due to the consistency in the deposition size which would juxtapose the irregularity in deposition when slumping qualities would occur. This produced differences in scales and density of the patterns due to the varying degree of deposited materials (Fig. 3)

Newer simulation as per Fig. 1 begins to look at the relationship between density and amplitude to inform how porosity could occur in the typology of a wall. Algorithms were produced to control the growth and variances of pattern in the wall to allow for a distribution and directionality of porous qualities.

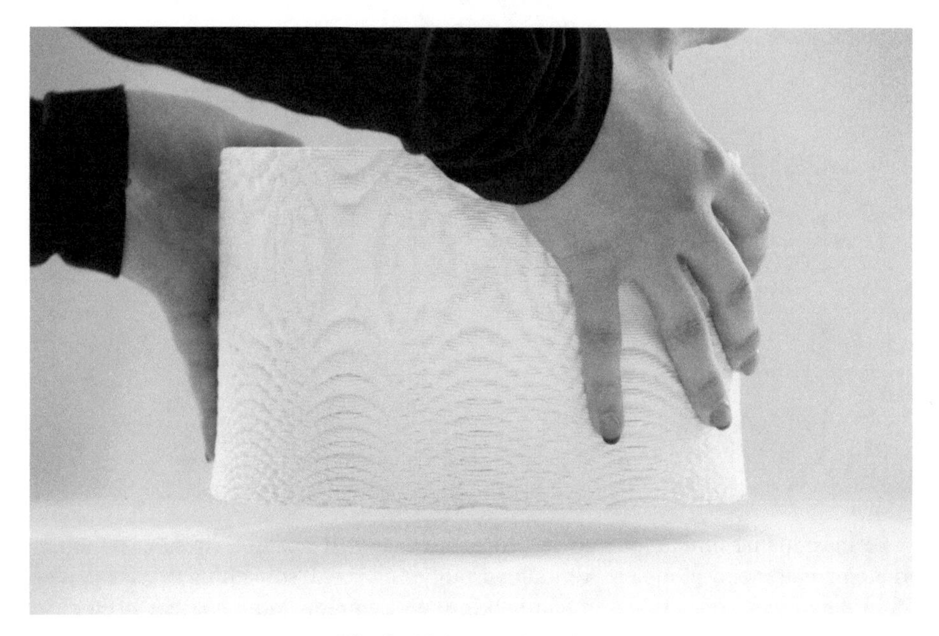

Fig. 3. Noise manipulation

3 Digital Simulation

These series of tests form the basis of a feedback loop between robotic fabrication and computational simulation. A series of prototypes allow for investigation of the evolution of the physical material behaviours and its properties. This process provides data and feedback for computational simulation.

Current computational simulation still lacks physics of gravity and friction, and therefore capacity to simulate interaction between the layers. Such abstract and idealized interpretation of material is closer to simulation of less volatile materials such as PLA plastic prints. This in part is due to the volatility and continually varying water levels within the material composition. Water levels are constant contributors to variability, even when mixed with the same ratio, with the time between pugging the mixture, and the external air temperature being the factors with most influence.

Fig. 4. Digital simulation prototype

By introducing simulation into the process of design, it creates a virtual experience of producing material prototypes. Although prototyping is still something that is inevitable within the process of fabrication and making, we can reduce the number of iterations which are required to be created as physical prototypes. It creates a more streamlined process, although one with more rigorous potential and a greater opportunity for design search in order to truly understand the potential of material behaviors to contribute to the aesthetic, structural, acoustic, thermal and other properties (Figs. 4 and 5).

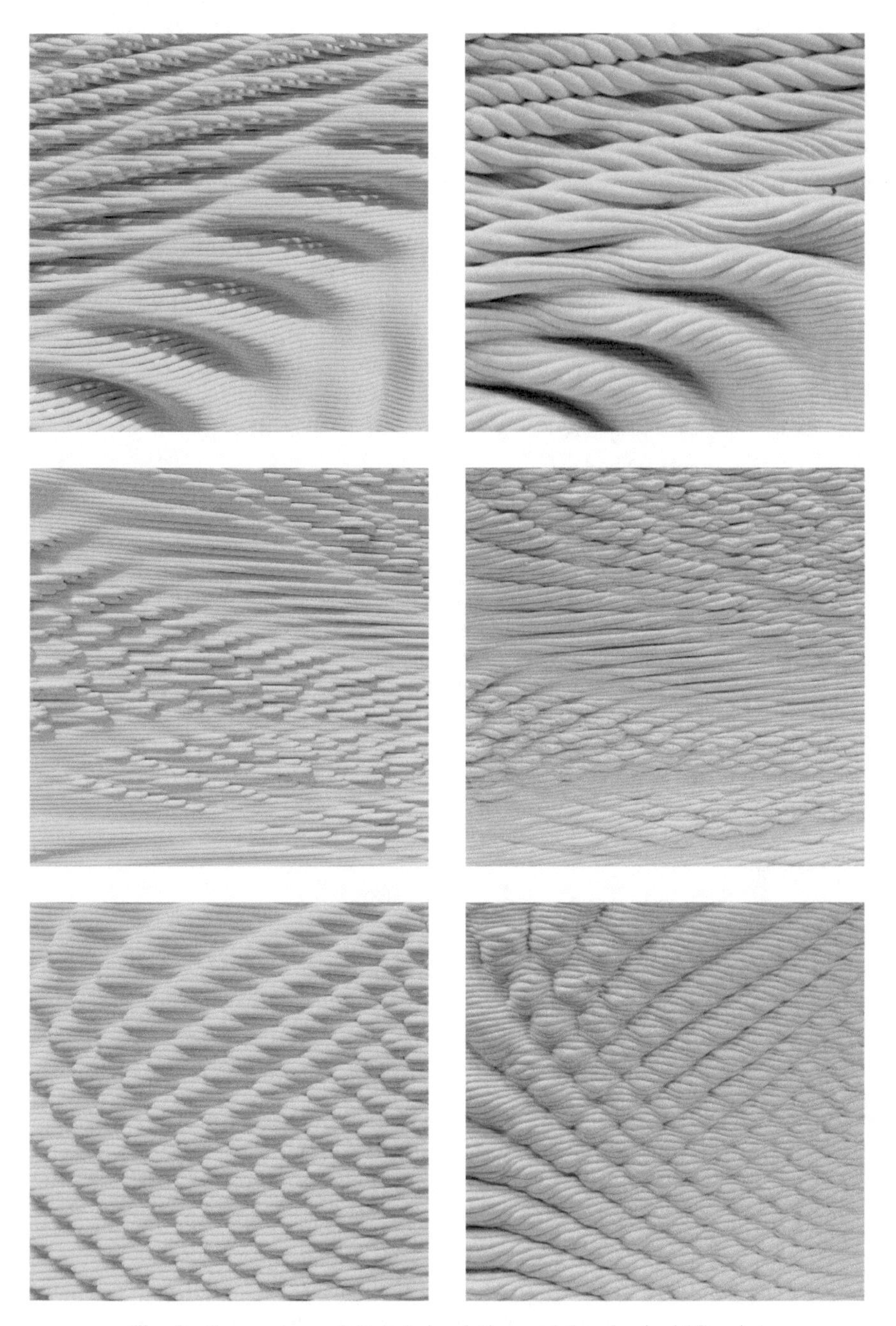

Fig. 5. Comparison of digital simulation and the physical 3D print.

4 Conclusion

Fundamentally, this research proposes the synthesis of material science and advanced fabrication techniques through understanding both physical material properties and its adaptation to computational processes. This allows for a new convergent design and fabrication platform, that can deploy new material compositions within fabric of architecture, while enhancing myriad of performance aspects. Like in the recent automotive industry developments enabled by simulation, big data and machine learning, it can offer radical acceleration in design alternatives, and shortcut physical prototyping time and number of iterations, thus reducing material waste and CO2 emissions in construction.

Such convergence of material and information ecologies, can enable design of architecture with greatly enhanced performance and novel aesthetics.

References

1. Manninger, S., del Campo, M.: Plato's columns: platonic geometies vs. vague gestures in robotic construction. In: ACADIA 2017: Disciplines & Disruption, pp. 374–381 (2017)
2. Carpo, M.: Breaking the curve: big data and design. Artforum, pp. 169–173 (2014)
3. Retsin, G., Garcia, M.J.: Discrete computational methods for robotic additive manufacturing. In: ACADIA 2016: Posthuman Frontiers (2016)
4. Leach, N., Farahi, B., Antonelli, P.: 3D-Printed Body Architecture. Wiley, New York (2017)
5. Menges, A.: Material Synthesis, Fusing the Physical and the Computationa. Wiley, New York (2015)
6. Andrasek, A.: High resolution fabric of architecture, Doctor of Philosophy (PhD), Architecture and Urban Design, RMIT University (2018)

Super Composite: Carbon Fibre Infused 3D Printed Tectonics

H. Mohamed[1], D. W. Bao[2], and R. Snooks[1](✉)

[1] Royal Melbourne Institute of Technology (RMIT), Building 100, Melbourne 3000, Australia
{hesam.mohamed,roland.snooks}@rmit.edu.au

[2] Centre for Innovative Structures and Materials, School of Engineering, RMT University, Melbourne 3001, Australia
nic.bao@rmit.edu.au

Abstract. This research posits an innovative process of embedding carbon fibre as the primary structure within large-scale polymer 3D printed intricate architectural forms. The design and technical implications of this research are explored and demonstrated through two proto-architectural projects, Cloud Affects and Unclear Cloud, developed by the RMIT Architecture Snooks Research Lab. These projects are designed through a tectonic approach that we describe as a *super composite* – an approach that creates a compression of tectonics through algorithmic self-organisation and advanced manufacturing. Framed within a critical view of the lineage of polymer 3D printing and high tech fibres in the field of architectural design, the research outlines the limitations of existing robotic processes employed in contemporary carbon fibre fabrication. In response, the paper proposes an approach we describe as *Infused Fibre Reinforced Plastic* (IFRP) as a novel fabrication method for intricate geometries. This method involves 3D printing of sacrificial formwork conduits within the skin of complex architectural forms that are infused with continuous carbon fibre structural elements. Through detailed observation and critical review of Cloud Affects and Unclear Cloud (Fig. 2), the paper assesses innovations and challenges of this research in areas including printing, detailing, structural analysis and FEA modelling. The paper notes how these techniques have been refined through the iterative design of the two projects, including the development of fibre distribution mapping to optimise the structural performance.

Keywords: 3D printing · Fibre composite · Additive manufacturing

1 Introduction

The focus of this paper is the application and development of a series of technical innovations that enable the design and realisation of complex, intricate architectural forms through a carbon fibre reinforced 3D printed polymer fabrication approach. This *Infused Fibre Reinforced Plastic* (IFRP) method offers a viable approach to constructing super composites that have intrinsically complex geometries and compressed tectonics. This strategy is part of a larger trajectory of design research undertaken by the Snooks Research Lab that explores the compression of the tectonic relationship between skin, structure, services, and ornament, through the composite fabrication of intricate algorithmic architecture (Snooks 2020) (Fig. 1).

P. F. Yuan et al. (Eds.): CDRF 2020, *Proceedings of the 2020 DigitalFUTURES*, pp. 297–308, 2021.
https://doi.org/10.1007/978-981-33-4400-6_28

Fig. 1. Cloud Affects, 2019, detail photographs, Snooks Research Lab.

This research contributes to a growing international body of work and community of architects who are exploring the application of 3D printing in architecture. The research presented here has evolved from an exploration of the design implications of 3D printed translucent plastics and how these can be reinforced to enable their robust application to architecture. While Fused Deposition Modeling (FDM) is a relatively mature 3D printing technology, its application to large-scale architectural fabrication requires numerous innovations in terms of hardware, printing techniques, software development, architectural detailing, structural design, and the design of tectonics. This paper outlines some of these developments including printing topologically complex forms, the subdivision of form into printable parts, non-parallel printing techniques, the infusion of carbon fibre within printed conduits, structural analysis and design of carbon-infused plastic parts and jointing systems.

2 Context

2.1 Fibre Composites in Architecture

Advancements in compounding and fabrication techniques have paved the way for synthetic polymer composites to be deployed as a new building material. While the use of fibre composites in architecture dates to at least 1956 with the Monsanto House, recently there has been a re-emergence of polymer composites in architectural construction and discourse. This is evident in the work of architects such as Greg Lynn; with a long track record of innovative applications of composites through which he has realised complex curvature architectural forms. He describes "intricacy" as a tectonic relationship between elements with the continuous and gradient change to produce a dynamic spatial effect

(Lynn and Friedman 2003). Lynn envisages a shift in architectural fabrication in which fusion by heat and chemical attachment replaces assemblage through mechanical connections (Lynn 2011a, b). Use of fibre reinforced polymer composites in Lynn's work allows for the gradual transformation of geometry to expose structure and alignment of load paths with the flow of surface.

2.2 Robotic Fabrication and High-Tech Fibre Structures

The repetition in production of fibre composite parts laminated onto traditional moulds create considerable efficiencies in sectors such as the marine industry to invest in reusable moulds. However, the architectural application of this approach is problematic when there is little repetition as the cost of the mould becomes prohibitive. Over the past decade architectural research groups including the Institute of Computational Design and Construction (ICD) in Stuttgart, and the Digital Building Technologies (DBT) lab at the ETH in Zurich have developed new robotic techniques for the fabrication of fibre reinforced polymers that avoid the use of moulds. These techniques, and the projects they are applied to, are important precursors to the research posited here. However, they embed specific limitations on architectural form. The ICD Stuttgart fibre winding projects, including the Research Pavilion (2010) and Buga Fibre Pavilion (2019) use customised jigs to create ruled surface geometry. An alternative approach explored by the ICD is robotic taping/gluing of a fibre structure to a temporary inflatable formwork (Doerstelmann et al. 2014). The DBT lab at ETH has experimented with a two-step fibre reinforcement of 3D printed form, where a composite of polymer and fibre is extruded directly on to a 3D printed base geometry (Kwon et al. 2019). Although this CFRP technique advances design research into fibre reinforced polymer structures, this method limits the topological scale and intricacy of the architectural forms that can be fabricated.

2.3 Big Area Additive Manufacturing

This paper is focused on Fused Deposition Modelling (FDM) of polymers within the field of additive manufacturing, also known as 3D printing. Big Area Additive Manufacturing or (BAAM) refers to the fabrication of large-scale components through FDM techniques. In the past few years, BAAM has become a desirable technique for relatively fast and efficient fabrication of components in different industries such as aerospace, automotive, product design and more recently within the fields of architectural design and building construction.

Fig. 2. Cloud Affects, 2019 (Left), and Unclear Cloud, 2020 (Right), Snooks Research Lab.

3 Methodology: Intricate Tectonics and Composite Fabrication

3.1 Super Composite

This research extends the concept of composites from material (fibre composite) to tectonics by compressing surface, structure, services and ornament into a single irreducible assemblage – a super composite (Fig. 3). This strategy is leveraged to develop intricate tectonics for expressive architectural forms. The super composite approach is designed through an agentBody self-organising generative algorithmic process that draws on the logic of swarm intelligence. It operates through multi-agent systems (agentBody algorithms form part of the Behavioral Formation design methodology developed by Roland Snooks since 2002). The agentBody generative design process simultaneously creates both the structure and form, as the agentBody consists of a skin and embedded structural skeleton (Snooks et al. 2020). The fabrication logic of the super composite leverages the geometric capacity of 3D polymer printing and the structural capacity of carbon fibre.

3.2 A Lineage of 3D Printed Sacrificial Formwork Projects

A super composite is a 3D printed hybrid of skin and embedded structural conduits that are infused or cast with structural material. The precursor to this approach was developed through a series of projects beginning with the NGV Pavilion (2016) Hawthorn Public Art (2016) by Studio Roland Snooks. These projects developed a tectonic, based on the logic of 3D printed sacrificial moulds, or formwork, into which high strength concrete structural elements were cast (Fig. 4) (Snooks 2018). This research has more recently been expanded to explore the implications of infusing carbon fibre into 3D printed sacrificial formwork conduits through the Cloud Affects and Unclear Cloud projects.

The infusion of carbon fibre within 3D printed conduits represents a novel approach to fibre composite fabrication that departs from the contemporary carbon fibre fabrication approaches, such as fibre winding and taping described above. Instead, this research employs a fibre infused sacrificial formwork technique to create a negotiated relationship between the inner and outer skins where form and structure negotiate one another to enable intricate topologically complex forms.

Fig. 3. Super composite: compressed structure, services, and skin

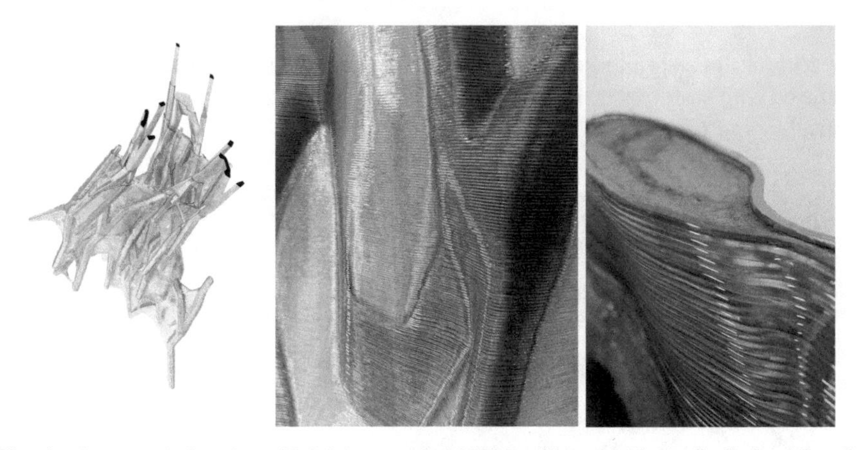

Fig. 4. Concrete infused sacrificial formwork, NGV Pavilion, 2016, Studio Roland Snooks

3.3 Sacrificial Formwork for Carbon Fibre Infusion

In this approach, a multilayered sacrificial formwork, or mould, is used to accommodate both structure and services within embedded conduits. This creates a complex lattice network which is made from a, primarily tensile, structural skeleton at the core surrounded by a secondary layer to allow for building services and compression foam insulation. This lattice network is embedded in a translucent outer skin. These two inner and outer skins laminate to one another periodically and are connected through flange geometries to create sufficient structural integrity in the skin at the macro scale (Fig. 5 right).

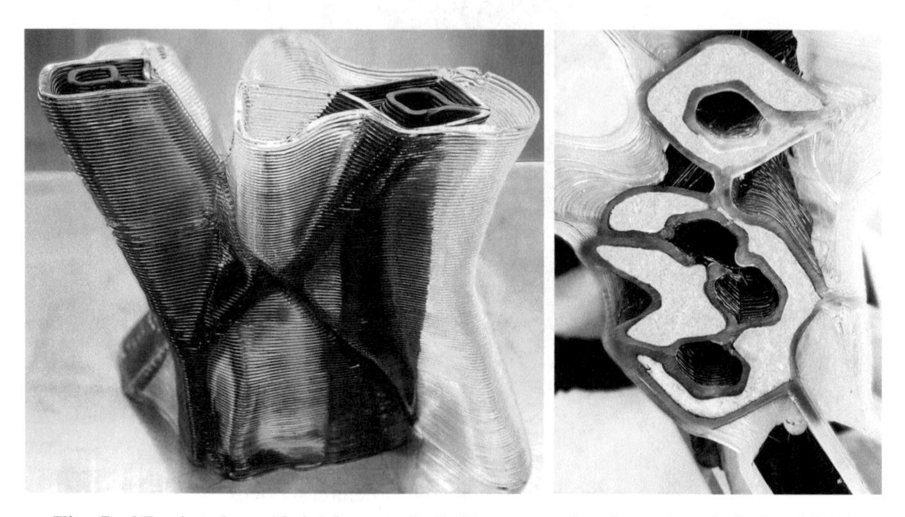

Fig. 5. 3F printed sacrificial formwork (left), preparation for carbon infusion (right)

Within the inner printed conduit, continuous carbon fibre elements are positioned and infused with an epoxy resin bonding the carbon fibre, steel connections and polymer sacrificial mould (Fig. 6).

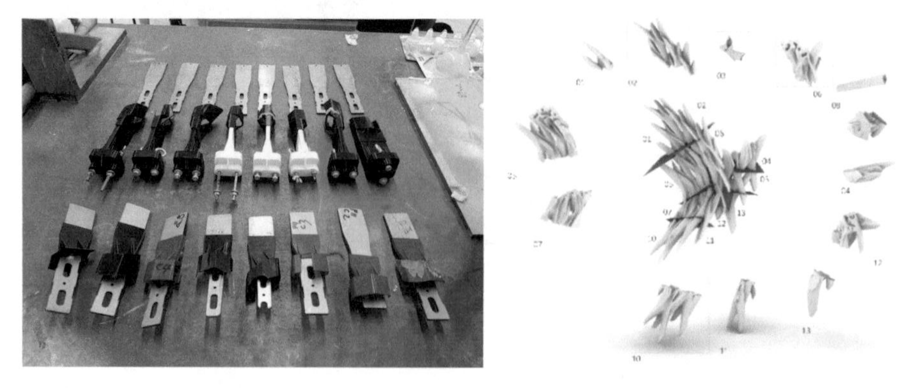

Fig. 6. Customised jigs for steel connections (left), assembly division & joints (right)

3.4 BAAM Techniques and Geometric Implications

Advancement in BAAM processes at the RMIT robotics lab in recent years has enabled the realisation of large scale algorithmically designed architectural prototypes such as Cloud Affects. Some of these technical enhancements, like improved start-stop printing and non-parallel/non-horizontal layer printing, are necessary to achieve substantial topological freedom of geometry and extreme cantilevering capacity. These improvements have also posed new challenges to the 3D printing process.

3.4.1 3D Printing Non-continuous Surfaces

Robotic polymer pellet extruding enables relatively fast 3D printing through a high material flow rate. However, compared to filament printing, there is no material retraction in this process when a tool path stops. This results in the dragging of excess material between the endpoint of a tool path and the subsequent start point, creating polymer strings within the printed part. These strings cause accumulated defects between layers, affect the surface quality and results in possible print failure in large prints. Despite challenges created by this basic issue, start-stops are necessary for the fabrication of non-continuous surfaces, in particular surfaces with topological complexity. Through a design and fabrication feedback loop, this research introduced a new technique to resolve this by using flange connections between the independent layers of structure, services and skin conduits. This technique minimises the strings within the geometry by minimising start-stops and maximising continuous toolpath length, which results in enhanced print quality. As well as surface quality improvement, these flange connections increase structural connections between the conduits and transfer the loads more evenly to the fibre structure located in the core conduit.

3.4.2 Directionality, Geometric Freedom and Extreme Cantilevering

Conventional FDM printing where each layer is deposited horizontally, and perpendicular to the vertical axis of the tool, poses significant limitations to geometrical directionality. Small layer width to height ratios in large-scale polymer printing is essential to keep the weight of the 3D printed components down. However, this reduces the capacity to print cantilevering geometries with large draft angles. Non-parallel printing leverages the 6 axes of the robot to deposit non-horizontal layers in order to gradually change the layer heights and print plane orientation within the printing process. This enables the fabrication of geometries with more complex surface directionality and higher draft angles. This technique accompanied by temporary printed support material enabled significant cantilevers in the Cloud Affects project (Fig. 7).

These advancements require high accuracy and calibration for both hardware and software tools. Over the past 6 years of research and development, the RMIT robotics lab, in collaboration with industry partners, has developed several generations of polymer extruders with high accuracy and large deposition capacity. The latest generation of these extruders implements new functions and technical capacities such as a new shut-off valve system to reduce polymer strings as well as multiple heating and cooling zones in the extrusion barrel for controlled melting of the polymer. These hardware improvements have been in parallel with the development of software to better control and diagnose all

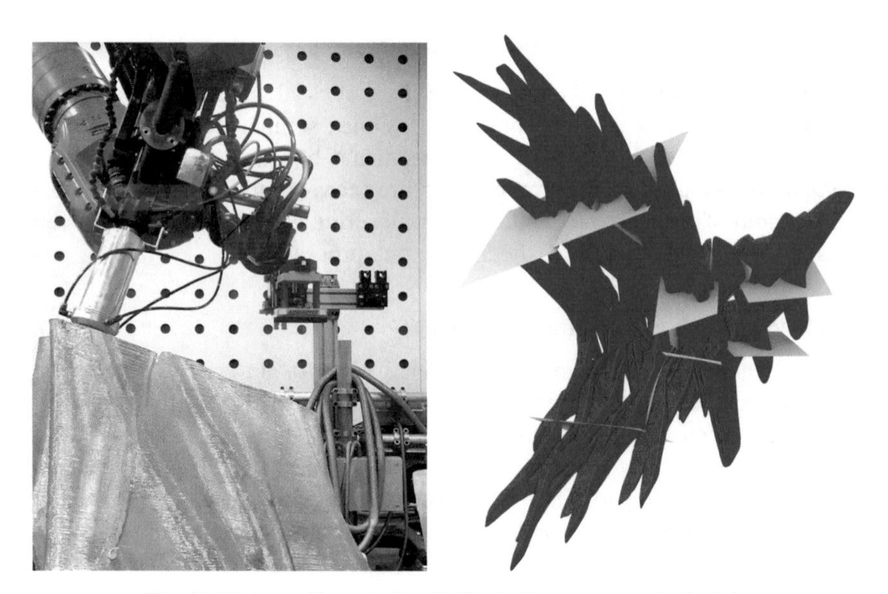

Fig. 7. Non-parallel printing (left), draft angle analysis (right)

aspects of the printing process, across multiple robotic platforms, materials and extruder types.

3.5 Structural Design Implications

The posited super composite of 3D printed polymer with infused carbon fibre structural members creates an integral link between the form, articulation and structure of the project. This requires both the geometry of the agentBody and its computational behavior to be designed through structural heuristics and their interaction with FEA analysis (Snooks 2020). This process establishes an inherently structural lattice which is subsequently fine-tuned and sized through an FEA analysis approach. This process evolved through the two projects to increase the accuracy of the analysis and efficiency of the structure.

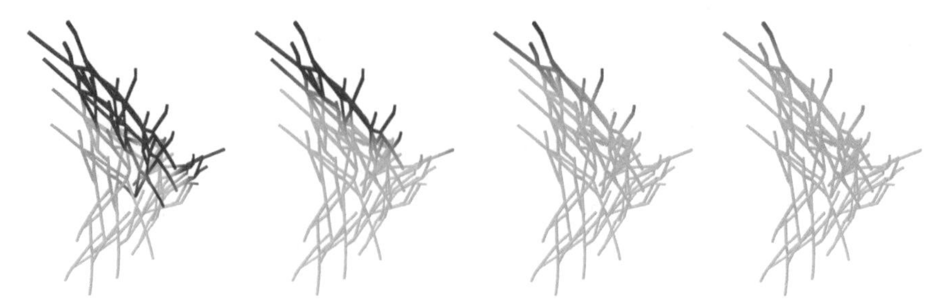

Fig. 8. Real-time structural feedback for fine-tuning of the amount and position of carbon fibre (front view)

The structural setup of the two projects consists of an irregular cantilevering carbon fibre structure with prefabricated steel connections. Our approach to Cloud Affects was to simulate the stresses in the structural members. The overall shape of the structure is analysed as a truss successively using the FEA tools Millipede and Abaqus. We examined the structural strength of the assembly at the early form-finding stages using the Millipede in Rhino/Grasshopper platform. Global deformations and bending moments of the structure have been calculated and simulated for real-time fine-tuning of the carbon fibre sizing during the schematic design phase. This analysis took the overall weight of the structure, material properties of the resin and fibre and connection locations into account to calculate values of deformation at the extremes of the structure, where there is significant cantilevering. These simulations also determined the variable bending moments throughout the structure, the higher zones of stress indicated in red in Fig. 8, which enabled the distribution of variable section sizes of the carbon fibre elements. Later, we collaborated with the RMIT Center for Innovative Structures and Materials to calculate the structure in the phase of design development using the engineering FEA tool Abaqus. As opposed to the bending simulation in Grasshopper Millipede, the explicit-dynamic analysis in FEA engineering software, Abaqus provided more accurate results for the design development. It offered the simulation and analysis of the bending process and values of maximum magnitude, mean stress and displacement by including different parameters such as overall structure weight, proportions of carbon fibre to resin, steel properties and other related boundary conditions. The gradient colour legend indicates the range of values for structural analysis result (Fig. 9). The accurate data helped us refine the design to ensure the appropriate safety and redundancy.

Fig. 9. Accurate simulation & in-depth analysis of carbon fibre structure (rear view)

Although this analysis enabled us to design the overall structure by ballparking the deformation amount and weak moments of the structure, this method considers the carbon fibre as the only structural element, deliberately ignoring the strength of the 3D printed geometry to simplify the FEA approach. It, therefore, overestimated the required structural strength of the carbon fibre composite, resulting in oversized structural conduits, carbon fibre cross-section, resin volume and steel connections. Consequently, this process has been refined in the Unclear Cloud project for which we have developed a hybrid analysis that includes both the carbon fibre network and the compressive capacity of the polymer skin. This improvement in the simulation of the structure is helping us to create an accurate estimation of the required structural strength to reduce the amount of polymer, resin and fibre needed for the project.

The carbon fibre elements infused within the 3D printed conduits are continuous while the network analysis is undertaken as discrete segments of the structural lattice. Consequently, the discrete segments cannot be optimally sized without consideration of the continuous extent of the carbon element. This has led to the development of an approach, which we describe as *fibre mapping*, that optimises the layout of the continuous carbon fibre elements. This approach maps, or navigates, the fibres through the network relative to the specific bending moments of the discrete segments. This was initially developed as a manual iterative process in the Cloud Affects, before being encoded into an algorithmic approach in the Unclear Cloud project (Fig. 10).

Fig. 10. Fibre mapping (right), improved bending moments (left)

3.6 Jointing

Logistics and polymer printing size limitations of the two projects require prefabricating parts which are assembled. Connecting these individual parts creates a substantial challenge and need for robust joints that create a continuity of structure equal to the strength of the carbon fibre elements. Other polymer printing issues such as shrinkage and warpage increase these challenges by creating uneven joints between panels. One approach in dealing with this issue is to overlap and conceal the seam lines with a built-in tolerance to connect individual panels side by side, horizontally. This method was used in a number of previous projects, including SensiLab, and B515 Studios (Snooks et al. 2020). The issue with this approach is the lack of structural continuity between panels. Another previous project, Floe (2018), used a continuous structural network fabricated from folded steel plates and a shingle-style arrangement of 3D printed polymer panels overlapping each other and connecting only to the steel sub-structure – not to one another. This method overcomes the issue of warpage and shrinkage by overlapping the panels at the edges where maximum warpage and shrinkage occurs during the printing process. However, the steel structure and mechanical connections are exposed, creating a disruption to the fluidity of the form and establishing discrete structure and skin. Cloud Affects overcomes this by using concealed steel fixing plates that create structural continuity across panel joints. In this method the integrity of the structure is preserved by creating a physical bond between infused carbon fibre and steel, the steel connections are concealed within the structural conduits and are connected to each other mechanically with minimum disruption to the skin surface. A mechanical fixing method was required for this project as it needed to be assembled and disassembled at multiple sites (Fig. 11).

Fig. 11. Steel plate wall connection (left), implanted steel connections (right)

4 Conclusion

This research has outlined an approach to integrating carbon fibre structure within 3D printed polymer skins. This novel super composite strategy, of infusing carbon fibre within sacrificial formwork conduits, has evolved through a lineage of research projects at the Snooks Research Lab and contributes to a larger community of architects exploring robotic approaches to fabricating carbon fibre structures. To advance this research, we have developed innovative techniques and approaches in large-scale 3D printing, detailing necessary to infuse carbon fibre, and advances in digital tools to fabricate the necessary geometric complexity. The future direction of this work is intended to extend the research beyond proto-architectural demonstrator projects and explore the feasibility of this approach within the design and construction of larger-scale architectural projects.

References

Doerstelmann, M., Knippers, J., Koslowski, V., Menges, A., Prado, M., Schieber, G., Vasey, L.: ICD/ITKE research pavilion 2014–15: fibre placement on a pneumatic body based on a water spider web. In: Architectural Design, 5th edn. vol. 83, pp. 60–65. Wiley, New York (2015)

Kwon, H., Eichenhofer, M., Kyttas, T., Dillenburger, B.: Digital composites: robotic 3D printing of continuous carbon fibre-reinforced plastics for functionally-graded building components. In: Willmann, J., Block, P., Hutter, M., Byrne, K., Schork, T. (eds.) Robotic Fabrication in Architecture, Art and Design 2018, ROBARCH 2018. Springer, Cham (2019)
Lynn, G.: Animate Form, pp. 40–41. Princeton Architectural Press, New York (1999)
Lynn, G.: Composites, Surfaces, and Software, Yale School of Architecture, New Haven, pp. 21–22 (2011a)
Lynn, G.: Chemical Architecture. DIALOG, pp. 27–29 (2011b). www.jstor.org/stable/41765685. Accessed 27 May 2020
Lynn, G., Friedman, T.: Intricacy. Intricacy: A Project by Greg Lynn FORM, pp. 27–29. Institute of Contemporary Art, Philadelphia (2003)
Miller, S.: Critical Mass: A Studio Tour with Greg Lynn. Composites and Architecture, DIALOG (2016). http://compositesandarchitecture.com/?p=4296. Accessed 25 May 2020
Snooks, R., Harper, L. (eds.): Printed assemblages. In: Burry, J., Sabin, J., Sheil, B., Skavara, M. Making Resilient Architecture, Fabricate 2020, pp. 202–209. UCL Press, London (2020)
Snooks, R.: Sacrificial formation. In: Wit, A., Daas, M. (eds.) Towards a Robotic Architecture, 5th edn. Novato, California (2018)
Snooks, R.: Behavioral Formation: Volatile Design Processes and the Emergence of a Strange Specificity. ACTAR, New York (2020)

Machine Learning for Fabrication of Graded Knitted Membranes

Yuliya Sinke(✉), Sebastian Gatz, Martin Tamke, and Mette Ramsgaard Thomsen

The Royal Danish Academy of Fine Arts, Schools of Architecture, Design and Conservation, CITA - Center for IT and Architecture, Philip de Langes Alle 10, 1435 Copenhagen, Denmark
ybar@kadk.dk

Abstract. This paper examines the use of machine learning in creating digitally integrated design-to-fabrication workflows. As computational design allows for new methods of material specification and fabrication, it enables direct functional grading of material at high detail thereby tuning the design performance in response to performance criteria. However, the generation of fabrication data is often cumbersome and relies on in-depth knowledge of the fabrication processes. Parametric models that set up for automatic detailing of incremental changes, unfortunately, do not accommodate the larger topological changes to the material set up. The paper presents the speculative case study *KnitVault*. Based on earlier research projects *Isoropia* and *Ombre*, the study examines the use of machine learning to train models for fabrication data generation in response to desired performance criteria. *KnitVault* demonstrates and validates methods for shortcutting parametric interfacing and explores how the trained model can be employed in design cases that exceed the topology of the training examples.

Keywords: Machine learning · Fabrication · Graded membranes · Knitting

1 Introduction

New interfaces between design and fabrication have fundamentally changed and extended the practice of architects, expanding our design agency from the designers of artefacts to the designers of material systems [17]. The direct design interface to digital fabrication at the material scale allows a new degree of material address and enables the steering of material behaviour in respect to performance criteria. In contrast to composites, functionally graded materials are singular multi-materials that vary their consistency gradually over their volume [8]. Due to their optimized material composition, they better respond to given conditions and a particular set of design and environmental constraints.

Although the integration of graded material properties into design is advantageous, these systems require a deep understanding of fabrication processes and depend on the detailed parametric and simulation models to steer the material composition. This in-depth knowledge of dedicated fabrication systems often requires specialist training and manufacturing setups can be unintuitive and repetitive. The current way to solve this

P. F. Yuan et al. (Eds.): CDRF 2020, *Proceedings of the 2020 DigitalFUTURES*, pp. 309–319, 2021.
https://doi.org/10.1007/978-981-33-4400-6_29

is setting up parametrically defined interfaces that link the specification of geometry and material with fabrication steering data [5, 6, 9, 14]. These workflows employ a simulation-model-validation loop, where the performance of the graded materials is validated through simulation and physical testing of fabricated probes. However, the predefined fixed topology modelling logics are often linear and inflexible, thus can not readily be changed on the fly [4].

2 Machine Learning for Complex Graded Material Systems

The integration of machine learning into design systems enables new measures for feedback in the architectural design chain. These methods allow to rethink the organisation of the parametric model and break its inherent reductionism [17, 18]. Where machine learning is finding application in design generation, design optimisation and analysis [20], special interest lies with the application of machine learning for fabrication. Here, the training of models to detect and automate simpler parts of the steering process could lead to more intuitive specification processes. The process of design specification for fabrication lends itself well to machine learning. By being inherently data-heavy training data can be provided on an ever-expanding data set. In this way the quality of the model's outcome increases with each fabrication cycle, evolving its mapping, complexity and scope while remaining intuitive and design integrated. The second potential of employing machine learning for manufacturing is its adaptability. The promise here is a higher degree of versatility in the design process where trained models can predict fabrication data with a higher degree of the topological variance than parametric models. The following case study examines the training of a fabrication model, its validation through physical prototyping and its application on topologically different situations that would otherwise be impossible through traditional parametric interfaces.

3 Knit as a Model for Material Specification

KnitVault takes a point of departure in the research project *Isoropia* [16]. Developed for the 2018 Venice Biennale, it represents a CNC knitted membrane architecture based on a system of arches pulled into tension by a cable-net system. Being a site-specific installation, it required the bespoke variation of each membrane through functional material grading [12]. In order to increase the structural performance of the arches, the knit variation took place through the methodical evaluation of fibre length and its deduced extendability to increase the depth of tensioned cones. The design model incorporated a parametric interpretation of a series of stitch pattern boundaries and translation of these into encoded pixel images as a means of material automatic specification.

The difficulty in evaluating the performative effect of the patterns and the inflexibility of the parametric set up is the starting point for this study. The process of specifying for CNC knitting is highly cumbersome relying on an in-depth understanding of the fabrication parameters and their limitations. Design protocols remain non-intuitive and require special expertise for programming the fabrication files, complicated by many parameters such as yarn friction and tensioning, knit speed, roll down settings and the performance of the resulting fabric. Additionally, the material specification encoding is limited to typical garment templates, such as sweaters and trousers etc. The lack of integrated simulation tools means that the prediction of material outcome: the shape, size and performance of knitted materials are still mainly evaluated and tuned through physical prototyping [17]. The use of functionally graded knit structures increases design complexity which inevitably leads to extensive prototyping. Only recently related research in computer graphics has investigated more user-friendly design to manufacturing interfaces for users who have no or only limited knowledge in knitting [2, 11, 13]. As a consequence, state of art in design-to-manufacture workflows for knit production are not meeting the current architectural design and fabrication paradigms, where predictability and the simulation of a design's performance on all levels before making and construction are key [19]. The challenge of knit simulation lies in loops' intricate structure and their interaction at multiple scales. If quite precise yarn-to-yarn modelling methods exist in textile engineering, these are only applicable on a sample scale, due to computational capacity limits. Besides, methods for abstracting knit behaviour through digital models are not yet fully understood.

Machine Learning offers an alternative approach for specifying graded knitted membranes. Here, the learning of relations between fabricated surfaces and corresponding design data replacing the simulation of highly detailed knit surfaces, their specification accompanied by the challenging encoding of knit parameters. As knit is interfacing with the machine through the image-based file, machine learning image-based algorithms are applicable here to generate CNC-fabrication data. KnitVault builds on and develops methods first examined in the probe Ombre [15] where the machine learning approach was first tested. In the following sections, it will be described how the methodology first established in *Ombre* developed further to *KnitVault*.

4 Ombre - Model for Light Filtering Textile Specification

Ombre investigates the specification of a light filtering textile. Constructed from two yarn types; the first a reflective monofilament and the second absorbent elastomer, *Ombre* is knitted in a single knit *jacquard*[1]. Here the technique is used to create a complex non-binary pattern as the floats affect and change the translucency of the knitted textile. The shifting between two yarns creates complex non-linear correlations between light and shadow that are not readily intuited by the designer. To explore the correlation between the shadow and the material specification, Ombre uses image-based sensing through photographs under an artificial sun to record the shadow effects of the textiles. This data is then used to determine the textiles' light filtering performances and map these to the corresponding fabrication bitmap patterns (Fig. 1). In *Ombre*, GANs [7] are employed to train the model and build the link between the shadow effect, fabric density and the corresponding fabrication bitmaps. Machine learning allows here to construct relations between sets of data when these cannot be described in a linear straightforward manner. The study proved the viability of predicting fabrication data for complex graded materials

[1] knitting technique often used in garments and fashion, in which two or more types of yarns are knitted together and allows the intermittent foregrounding of one yarn while the other yarns "float" behind the knitted surface [21]. The technique allows the designer to create highly complex patterns, while "hiding" the unused yarns on the backside of the fabric.

and demonstrated the relevance of image-based machine learning methods to train direct design to fabrication models. However, where the study focussed on simple performance criteria of light filtering, the further question of transferring the method onto *another performance criteria* provoked the development of the project into *KnitVault*.

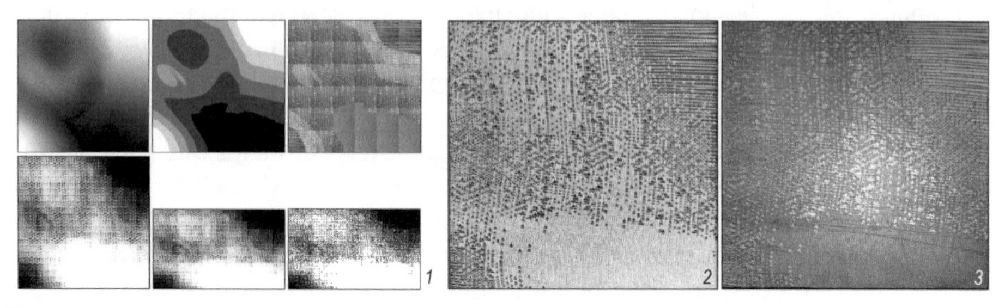

Fig. 1. *Ombre* 2019 - speculative probe demonstrating the link between the desired shadow and the fabrication files for knit production. 1 - design intent. 2 - resultant knitted fabric. 3 - shadow cast from the knitted fabric

5 KnitVault - Geometrical Performance Through Machine Learning

5.1 Experiment Description and Setup

In *KnitVault* the machine learning for fabrication approach is further extended to work on three-dimensional configurations. Rooted in *Isoropia* project, the aim is to study the functional grading of knitted textiles to control the deformation of the membrane under stress. The method is used to explore how the membrane can be better understood and how the trained model can be used on cases that exceed the topology of the training data.

The model is trained on a set of predesigned flat knitted textiles with varying patterns and fixed topology of a central node. All the textiles were knitted on an industrial knitting machine (STOLL 730T E7.2) with the same tension settings, in a single Dyneema® yarn[2], employing three knit stitch types (Fig. 2). This way the pattern geometry is the only changing parameter. Panels are then tensioned to a frame and loaded with 15 kg in the center, creating a 3d cone-like shape of different depths. The resulting shape is documented with a FARO 3d-laser scanner resulting in detailed point clouds to get the real-world geometrical performance. Pattern shapes were drawn manually, following the intention to create diversity in patterns. as vector files in turn transformed to bitmap files for fabrication (Fig. 3).

This shows that the *piquet* material is more stretchy (B expands by 13 cm) than *piquet lacoste* (A - by 12 cm). Stretch is also increased with *interlock* present in the stretched area (C - E), where E - the most stretchy of the investigated samples. The geometry of the patterns influences the stretch furthermore (C - I) (Fig. 4).

[2] High-tenacity and ultra-high molecular weight polyethylene (UHMwPE).

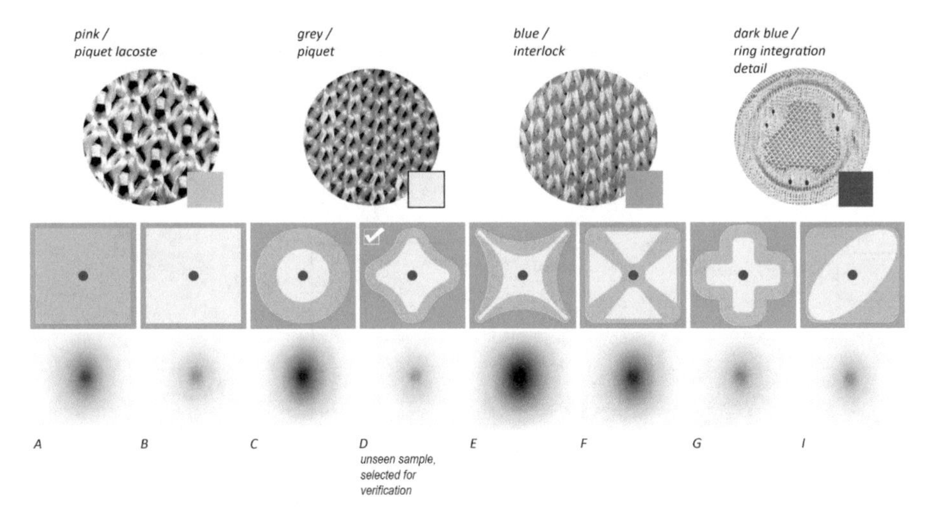

Fig. 2. Range of textile designs used in training with the close up of patterns compositions and their colour coding for the fabrication files

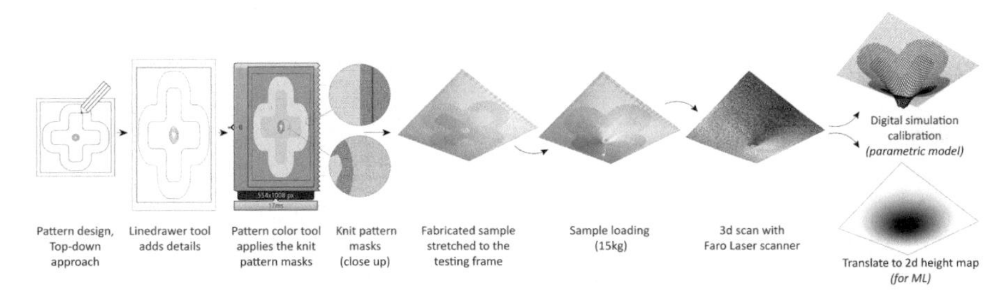

Fig. 3. The workflow of the sample design, fabrication and testing protocol (G as an example)

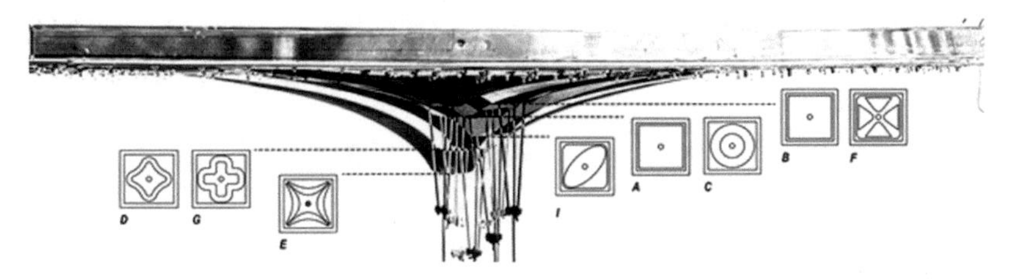

Fig. 4. Side view of 8 samples, performing the difference in depth of the expansion, under load

5.2 Description of Neural Network (NN) Workflow

The training of the neural network is based on the Pix2Pix architecture [10], implemented in Tensorflow [1]. Here the NN is learning a relationship between two sets of images: a training *input and an output,* where the 3D scans of the samples under stress are used for the input, and *pixel-based knit fabrication files* - for the output. Seven samples are used for training and one (D) - for validation. To interface the 3d data from point-clouds with the 2d image-based machine learning, clouds are translated into grayscale height maps (Fig. 2), where the darkness of the colour corresponds to the depth of the geometry (12–25 cm). For expanding the limiting teaching data, the image augmentation techniques [19] - flip, mirror, rotate, stretch and scale - are performed on the input and output files in parallel [3] (Fig. 5).

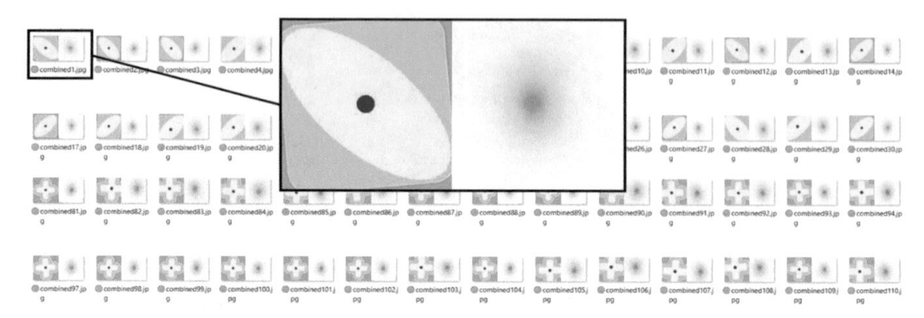

Fig. 5. Fragment of augmented input and output training pairs

The output images represent different pixel distributions - where each colour represents a particular knitting instruction for the machine. After training the NN for 150 epochs[3] on augmented fragments it is applied to the seven input full-size images and then verified on the "unseen" height map (Fig. 6).

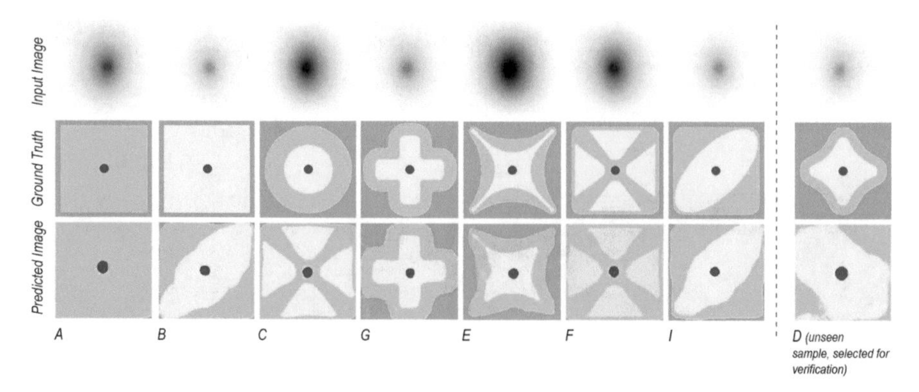

Fig. 6. Learned examples of the 7 seen samples and the 8th unseen knit sample

5.3 Results

The result of neural network application shows that it can correctly predict five of the samples (A, G, E, F, I) two of the predictions (B, C) are ambiguous (Fig. 6). This stems

[3] Cycles through the training data.

from a non-linear relationship between the pattern and the material stretching: alike samples have non-similar stretching behaviours and vice versa.

Although a 62% success rate is low compared to neural networks performances in other knowledge domains, it nonetheless seems reasonable, if assumed that ambiguity in the training data is the source of the error rather than the method itself.

The reassessment of the training data brings the understanding that one-dimensionality of the training data leads to logical errors where the high similarity of input files can be driven by very different output files, as visible through similar physical depths outcome for different knitting patterns (F/C "oval vs. cross"and I/B "oval vs. grey square" in Fig. 6).

5.4 Verification Through the Physical Prototyping

To verify the fidelity of the trained model the resulting pattern from the 8th "unseen" sample is physically knitted with previously established machine settings and tested on the loading rig (Fig. 7). Further, it will be referred to as ML sample.

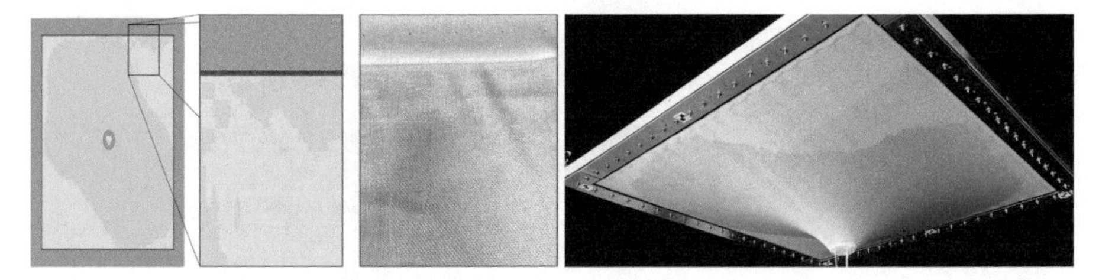

Fig. 7. ML generated fabrication file and the corresponding knitted textile

The loading of the 8th sample results in a depth of 25,1 cm (Fig. 8), which is relatively far from the predicted depth of 18.4 cm as in D. When looking at the per cent of stitch types between D and ML - they vary and we conclude that the ratio of stitches and their visual distribution in the surface affects the expandability of the textile under load. Also, the same conflation of inputs and outputs during training can be the reason for the depth difference.

Then, the visual examination was conducted to find patterns within the range of samples which are more similar in material composition and performance to the ML sample. I and E are the selected candidates for the discussion (Fig. 9). Although I sample, being visually similar to ML (asymmetric oval-like geometry) does not have an aligned depth in its physical sample (25 cm in ML vs. 16 cm in I). The closest in the depths of the physical sample is sample E, however, it remains challenging to explain the reasons, as it has different stitch ratio proportions and not-alike in its pattern geometry to ML. However, in general, the machine learning sample did achieve the predicted cone geometry, albeit with 6 cm tolerance.

We conclude that none of the identified parameters such as relations between the stitch ratios, pattern visual configuration (oval, cross, free shape etc.) alone is determining the stretch of the material. Yet it proves the complexity of the problem and confirms the

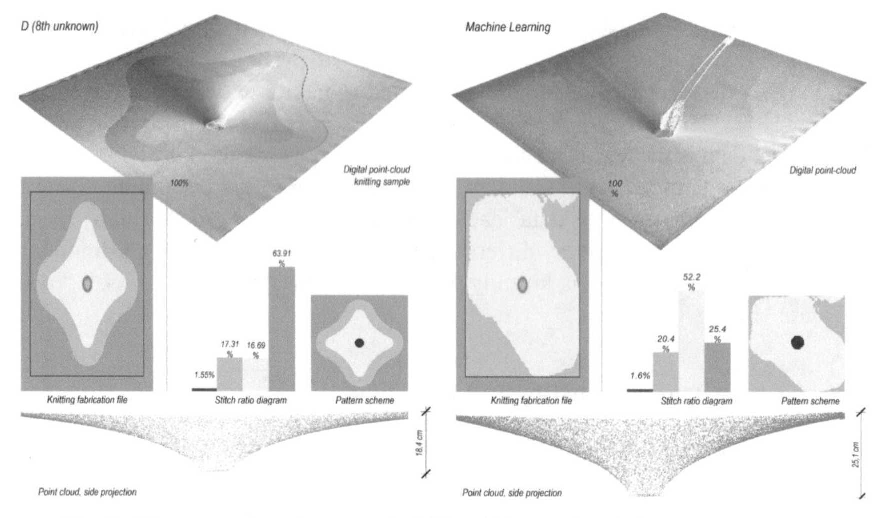

Fig. 8. The comparison between the ML and D sample knitting pattern samples

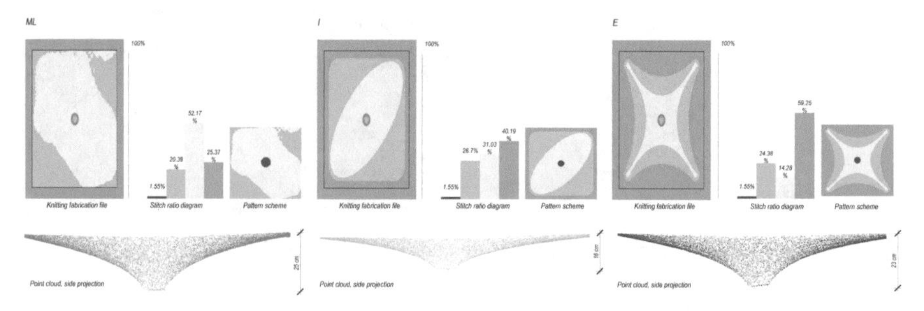

Fig. 9. Comparison between ML and the other two pre-defined designs (I and E)

expediency of exploring machine learning techniques for predicting fabrication files for challenging material behaviours.

5.5 Testing the Neural Network on the New Design Topology

The second part of the *KnitVault* experiment tests the adaptability of our trained neural network to predict outside of the learning scope. If previously the training data was based on the same single cone topology, the second test explores how double and triple cone configurations lead to novel fabrication files (Fig. 10).

Successfully, it shows the capacity of a neural network to translate from single to double and triple cones while keeping the logic of the colour encoding. This confirms the possibility to use NN to overcome the limitations of parametric workflows. While a parametric model needs to be restructured to deal with new inputs or new output typologies, a neural network is more flexible and can transfer from one problem domain to another without further specification. This approach is especially useful for modelling

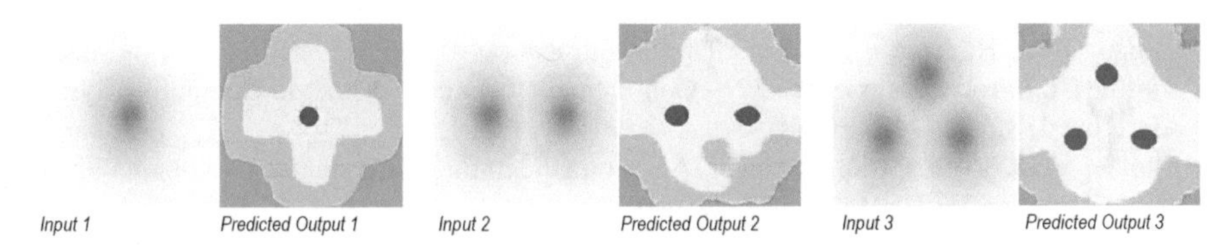

Fig. 10. 3Dot experiment to test the flexibility of the method

environments where input and output types are fixed but their degree varies over the time of design-to-production.

6 Conclusion

While the results of the first test of predicting the fabrication file from a desired stretching depth remains ambiguous, the success of the method's adaptability across different topologies is promising for future digital workflows. Generally, the project shows the possibility of bypassing domain-specific limitations of parametric models. Nevertheless, the used data set for the training was small and ambiguous and therefore the training results of the neural network are equally indistinct. Eventhough the heavy augmentation of a small data set is not advised in machine learning communities, it is symptomatic for the real-world problems in architectural applications, where big data sets are not always available. The question is how much data will be needed to train neural networks in architectural contexts to help to inform design decisions.

The second test of the adaptability of the method to different topologies, showed the possibility of transferring from one problem domain to another. Further probes have to be made to determine how far away from the training data a neural network can be used to predict the outputs.

The general approach of skipping the simulation-model-validation loop by training neural networks directly on the fabrication data can have useful implications for the other design-to-production workflows where a non-linear relation between design intent and fabrication data exists. For example, CNC milling usually includes cumbersome procedures of manual labour, when 3D models are translated to G-code. Nonetheless, there is a relationship between model and g-code which potentially can be learned by a neural network. Sharing fabrication data and design models for different fabrication machines (robots, CNC milling, 3d printing, knitting, etc.) can potentially help to overcome the limitations of small data sets and too much data augmentation for machine learning approaches in architectural digital fabrication.

Finally, the presented image-based approach can be extended to either parallel image-based approaches (optimizing for different problem domains simultaneously), voxel-based GANs or potentially even time-based voxel GANs in the future.

References

1. Abadi, M., et al.: TensorFlow: Large-Scale Machine Learning on Heterogeneous Distributed Systems (2015). https://www.tensorflow.org/

2. Albaugh, L., Hudson, S., Yao, L.: Digital fabrication of soft actuated objects by machine knitting. In: Proceedings of the 2019 Conference on Human Factors in Computing Systems. Glasgow, UK, pp. 1–13 (2019)
3. Bloice, M.D., Roth, P.M., Holzinger, A.: Biomedical image augmentation using augmentor. Bioinformatics **35**(21), 4522–4524 (2019)
4. Davis, D., Burry, J., Burry, M.: The flexibility of logic programming. In: Circuit Bending, Breaking and Mending: Proceedings of the 16th International Conference on Computer-Aided Architectural Design Research in Asia, Association for Computer-Aided Architectural Design Research in Asia (CAADRIA), pp. 29–38 (2011)
5. Deleuran, A.H., Schmeck, M., Quinn, G., Gengnagel, C., Tamke, M., Ramsgaard Thomsen, M.: The tower: modelling, analysis and construction of bending active tensile membrane hybrid structures. In: Proceedings of the International Association for Shell and Spatial Structures (IASS): Future Visions, Amsterdam, The Netherlands (2015)
6. Dörstelmann, M., Parascho, S., Prado, M., Menges, A., Knippers, J.: Integrative computational design methodologies for modular architectural fiber composite morphologies. In: Design Agency, Los Angeles, pp. 219–228 (2014)
7. Goodfellow, I., Pouget-Abadie, J., Mirza, M., Xu, B., Warde-Farley, D., Ozair, S., Courville, A., Bengio, Y.: Generative adversarial nets. In: Advances in Neural Information Processing Systems, pp. 2672–2680 (2014)
8. Grigoriadis, K.: Computational blends: the epistemology of designing with functionally graded materials. J. Archit **24**(2), 160–192 (2019)
9. Hensel, M.: Material and digital design synthesis: integrating material self-organisation, digital morphogenesis, associative parametric modelling and computer-aided manufacturing. AD Archit. Des. **76**(2), 88–97 (2006)
10. Isola, P., Zhu, J.-Y., Zhou, T., Efros, A.A.: Image-to-image translation with conditional adversarial networks. In: Proceedings of IEEE Conference on Computer Vision and Pattern Recognition (CVPR) (2017). https://doi.org/10.1109/cvpr.2017.632
11. Kaspar, A., Oh, T.-H., Makatura, L., Kellnhofer, P., Aslarus, J., Matusik, W.: Neural inverse knitting: from images to manufacturing instructions. In: Proceedings of International Conference on Machine Learning (ICML), Long Beach, California (2019)
12. La Magna, R., Längst, P., Lienhard, J., Fragkia, V., Noël, R., Baranovskaya, Y., Tamke, M., Ramsgaard Thomsen, M.: Isoropia : an encompassing approach for the design, analysis and form-finding of bending-active textile hybrids. In: Proceedings of the IASS Symposium 2018 Creativity in Structural Design (2018)
13. McCann, J., Albaugh, L., Narayanan, V., Grow, A., Matusik, W., Mankoff, J., Hodgins, J.: A compiler for 3D machine knitting. ACM Trans. Graph. **35**, 1–11 (2016)
14. Mendez Echenagucia, T., Pigram, D., Liew, A., Van Mele, T., Block, P.: Full-scale prototype of a cable-net and fabric formed concrete thin-shell roof. In: Proceedings of the IASS Symposium 2018, Boston (2018)
15. Ramsgaard Thomsen, M., Nicholas, P., Tamke, M., Gatz, S., Sinke, Y.: Predicting and steering performance in architectural materials. In: Proceedings of eCAADe 37 - Material Studies and Innovation. Bd 2, pp. 485–494 (2019)
16. Ramsgaard Thomsen, M., Sinke Baranovskaya, Y., Monteiro, F., Lienhard, J., La Magna, R., Tamke, M.: Systems for transformative textile structures in CNC knitted fabrics – Isoropia. In: Proceedings of the TensiNet Symposium 2019, pp. 95–110 (2019)
17. Ramsgaard Thomsen, M., et al.: Knit as bespoke material practice for architecture. In: Proceedings of ACADIA 2016: Posthuman Frontiers, Michigan, USA (2016)

18. Ramsgaard Thomsen, M., Tamke, M., Nicholas, P., Ayres, P.: Information rich design. CITA Complex Modelling, pp. 254–255. Denmark, Copenhagen (2020)
19. Sheil, B., Ramsgaard Thomsen, M., Tamke, M., Hanna, S.: Design Transactions: Rethinking Information Modelling for a New Material Age. UCL Press (2020)
20. Tamke, M., Nicholas, P., Zwierzycki, M.: Machine learning for architectural design: practices and infrastructure. IJAC **16**(2), 123–143 (2018)
21. Spencer, D.: Knitting Technology: A Comprehensive Handbook and Practical Guide. Woodhead Publishing, Lancaster (1996)

Author Index

P. F. Yuan et al. (Eds.): CDRF 2020, *Proceedings of the 2020 DigitalFUTURES*, pp. 321–322, 2021.
https://doi.org/10.1007/978-981-33-4400-6

Printed in the United States
By Bookmasters